T0338243

SHORT-RANGE WIRELESS COMMUNICATIONS

Wiley-WWRF Series

Series Editor: Prof. Klaus David, *ComTec, University of Kassel, Germany*

The *Wiley-WWRF Series* is a series of comprehensive and timely books based on the work of the WWRF (Wireless World Research Forum). This Forum is a global organization with over 130 members from five continents, representing all sectors of the mobile communications industry and the research community, with the mission to shape the wireless future. The authors are all active members of the WWRF. The series is focused on wireless communications, embracing all aspects from spectrum strategies, the physical layer and networking protocols, up to applications and services. Each volume of the series is a development of the white papers produced by the working groups of WWRF, based on contributions from members, and each describes the current research in the subject, together with an identification of future research requirements.

This book series is ideal for researchers from academia and industry, as well as engineers, managers, strategists, and regulators.

Other WWRF titles:

Rahim Tafazolli: *Technologies for the Wireless Future: Wireless World Research Forum, Volume 1*, 978-0-470-01235-2, October 2004

Rahim Tafazolli: *Technologies for the Wireless Future: Wireless World Research Forum, Volume 2*, 978-0-470-02905-3, April 2006

Klaus David: *Technologies for the Wireless Future: Wireless World Research Forum, Volume 3*, 978-0-470-99387-3, September 2008

Tierry Lestable and Moshe Ran: *Error Control Coding for B3G/4G*

Wireless Systems: Paving the Way to IMT-Advanced Standards, 978-0-470-77935-4, June 2009

SHORT-RANGE WIRELESS COMMUNICATIONS

EMERGING TECHNOLOGIES AND APPLICATIONS

Edited by

Rolf Kraemer

IHP GmbH, Germany

Marcos D. Katz

VTT, Finland

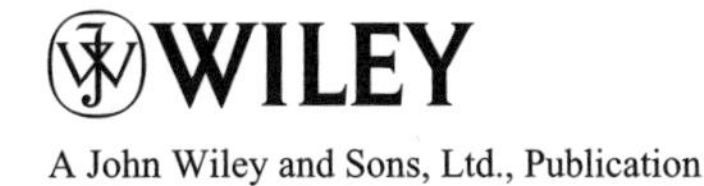

A John Wiley and Sons, Ltd., Publication

This edition first published 2009

Registered Office
John Wiley & Sons Ltd, The Atrium, Southern Gate, Chichester, West Sussex, PO19 8SQ, United Kingdom

For details of our global editorial offices, for customer services and for information about how to apply for permission to reuse the copyright material in this book please see our website at www.wiley.com.

Library of Congress Cataloging-in-Publication Data

Short-range wireless communications : emerging technologies and applications /
edited by Rolf Kraemer and Marcos Katz.
p. cm.
Includes bibliographical references and index.
ISBN 978-0-470-69995-9 (cloth)
1. Wireless communication systems. I. Kraemer, Rolf. II. Katz, Marcos D.
TK5103.2.S535 2008
621.384–dc22

2008034571

A catalogue record for this book is available from the British Library.

ISBN 978-0-470-69995-9 (H/B)

Set in 10/12pt Times by Aptara Inc., New Delhi, India.
Printed and bound in Great Britain by CPI Antony Rowe, Chippenham, England.

About the Editors

Rolf Kraemer received his Diploma and Ph.D. degrees in electrical engineering and computer science from the Rheinisch-Westfälische Technische Hochschule (RWTH) Aachen, Aachen, Germany, in 1979 and 1985, respectively.

He has worked for 15 years in R&D of communication and multimedia systems at Philips Research Laboratories in Hamburg and Aachen. Since 1998 he has been a Professor of Systems at the Brandenburg University of Technology, Cottbus, Germany. He also leads the Wireless Communication Systems Department of the Institute for High Performance Microelectronics (IHP), where his research focus is on wireless Internet systems spanning from application down to system-on-chip. He is co-founder of the start-up company lesswire AG, where he holds the position of Technical Advisor. He has published over 150 conference and journal papers, and holds 16 international patents.

Prof. Kraemer is a member of the IEEE Computer Society, the VDE-NTG, and the German Informatics Society.

Marcos D. Katz received his BSEE degree from Universidad Nacional de Tucumán, Argentina in 1987, and his MSEE and Ph.D. degrees from University of Oulu, Finland in 1995 and 2002, respectively. Dr Katz has worked in different positions at Nokia Networks, Finland from 1987 to 2001. He was at the Centre for Wireless Communications, University of Oulu, Finland in 2001–2002. From 2003 to 2005 he worked as a Principal Engineer at Samsung Electronics, in South Korea. Since 2006 he has been working as a Chief Research Scientist at VTT, the Technical Research Centre of Finland. Dr Katz has edited the books *Cooperation in Wireless Networks: Principles and Applications* (2006) and *Cognitive Wireless Networks* (2007). He has published over 80 conference and journal papers, has 30 granted international patents and more than 20 pending. In the years 2006–2007 Dr Katz was the vice-chair of Working Group 5 (short-range communications) of the Wireless World Research Forum and he has been the chair of the group since 2008.

Contents

Foreword

I'm very pleased to be able write the foreword and welcome you to the first in a new series of publications from Wiley of what we've called the WWRF Wireless Futures series. Each volume will address a key technology or concept that will play a major part in the wireless world of the future. WWRF (Wireless World Research Forum) was established in 2001 to bring together researchers from industry and universities to identify jointly the essential research questions and issues that the mobile and wireless research community needed to address. Now, with over 130 members from five continents and representing the most significant actors in the industry and the brightest stars in the academic firmament, we are reaching out to more research communities in developing markets and tackling key questions such as sustainability, efficient design and the future of networking. Membership of WWRF is available to all organizations that share our aims. Details can be found on our Web site http://www.wireless-world-research.org. To attain our vision of 7 billion people being served by 7 trillion devices by the year 2017, we need efficient and manageable short-range communication between the multitude of devices, both simple and sophisticated, that will fill our environment. The WWRF team on short-range communication (Working Group 5) have brought together all the key concepts, from those near to market to those still at the research stage, into a comprehensive and fascinating survey of the technologies that will be available to provide this. Their work continues, and if interested, you are strongly encouraged to get involved by contributing to, or attending, our meetings.

Dr. Nigel Jefferies
Vodafone Group R&D
WWRF Chairman

Foreword

The Wireless World Research Forum has coined the vision of developing technology for 'connecting 7 billion people with 7 trillion wireless devices by 2017'. The large amount of 1000 devices per human clearly is not going to be accounted for only by cellular technology, but a plurality of different wireless interfaces for a large set of different applications.

Cellular dominated the wireless industry over the past 20 years. However, in digital cellular's initial years cordless telephony was seen to be the volume driver, with cellular being more a niche application. Approximately 10 years ago the wireless local area network (WLAN) popped up as an important method for broadband cordless local connectivity. And as we can see from what is happening today, it has started to prove to be the door opener for broadband cellular to become widespread. So we see short-range cordless telephony and WLAN having paved the way for making people accustomed to using untethered communications. As the advantages of its use become apparent in a local environment, users want to have connectivity everywhere and start using the same service in a cellular wide area network (WAN).

However, today short-range communications cannot only be seen as a market developer for cellular WAN technologies. These days we want to use high-speed wireless connectivity for other use cases than continued connectivity, as for example, for very fast synchronization of data between devices, or for multimedia entertainment distribution in homes, cars, busses, trains and airplanes.

Other examples of application fields are RFID and wireless payment systems, which are becoming more and more important in various countries these days.

But the main volume driver for reaching the 1000 wireless devices per human will most likely be sensor networks. We will network our heating, lighting, security, windows, our vehicles, the roads, the rivers – mainly anything which it makes sense to monitor and control. This will drive the wireless industry into new volume markets of great future.

We are currently only at the advent of wireless technology, in particular in the case of short-range wireless, starting to blossom in nearly any field of our life. This book therefore is a great overview of some of the hottest new developments. It is a perfect companion for technical and marketing managers wanting to get an overview of what is coming, as well as for R&D engineers who get plenty of hints on where to dwell in much more depth. I hope you will enjoy the reading and become an expert in short-range wireless, a frontier which is just starting to become explored!

Professor Gerhard P. Fettweis
TU Dresden, Vodafone Chair

Preface

Short-range communications is one of the most relevant as well as diversified fields of endeavour in wireless communications. As such, it has been a subject of intense research and development worldwide, particularly in the last decade. There is no reason to believe that this trend will decline. On the contrary, the rapidly crystallizing vision of a hyper-connected world will certainly strengthen the role of short-range communications in the future. Concepts such as *wireless social networks*, *Internet of things*, *car communications*, *home and office networking*, *wireless grids* and *personal communications* heavily rely on short-range communications technology. No other communication branch has developed such a great variety of technical solutions as short-range communications. For several years the Wireless World Research Forum (WWRF) has explored concepts and technologies for short-range communications, mostly within Working Group 5 (WG5). The present book is the result of the latest coordinated activities in WG5 on several important topics of short-range communications. The editors of this book are the former and current chairs of WWRF's WG5.

The goals of this book are two-fold. In the first place, the book identifies and discusses in detail a number of emerging concepts and technologies for short-range communications. An objective overview of many technologies is provided, covering the main theoretical background and design approaches, without forgetting the equally important implementation sides. The book serves also as a motivating source for further research and development activities in short-range. The limitations of current approaches and challenges of emerging concepts are discussed. Furthermore, new directions of research and development are identified, hopefully providing fresh ideas and influential research topics to the interested readers.

Short-range communications is an overly broad subject to be covered in reasonable depth by this single book. The topics discussed here were jointly chosen by the participating academic and industrial partners and therefore they represent areas of current interest with high potential to be widely exploited in the future. Thus, rather than covering all possible technologies in short-range, this book reflects in its contents a number of emerging concepts and associated technologies, identified over the last years by a worldwide research community participating in WWRF. Therefore, the contents are well representative of the current and future trends in the field. The book is organized in four independent parts, as shown in Figure 0.1. **Part I** (Introduction) contains an introductory chapter defining in general the scope of short-range communications, followed by a chapter discussing basic design rules for modern short-range communication systems.

Part II (UWB Communications: State of the Art, Challenges and Visions) presents a comprehensive and well balanced view of ultra-wideband (UWB) communications, one of the key enabling technologies for short-range communications. The 15 chapters of Part II introduce

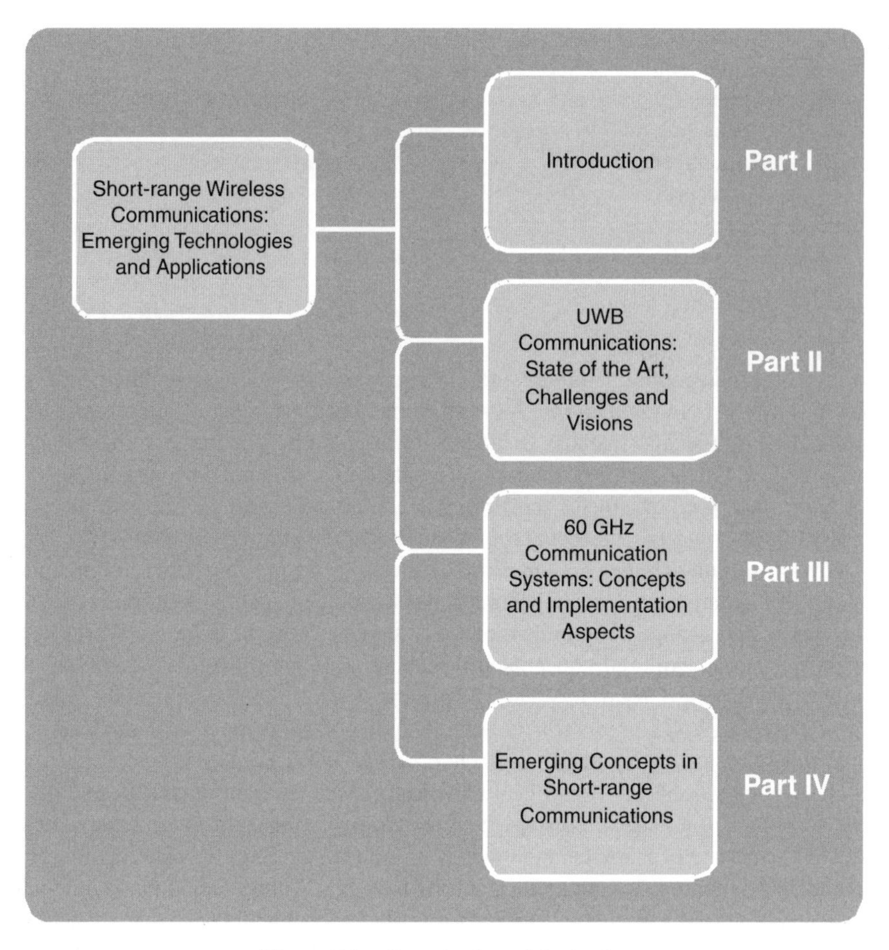

Figure 0.1 Organization of the book.

the most relevant aspects of advanced UWB transceiver design, including techniques such as pulse shaping, diversity, non-coherent detection, multiband modulation, synchronization, transmitter design, MIMO UWB and UWB channel modeling. In addition, UWB localization techniques are discussed. Moreover, implementation aspects are also taken into consideration, with subjects such as low-power UWB design and UWB A/D converters. UWB networking is also discussed, including UWB higher layers and UWB sensor networks for positioning and imaging, as well as coexistence aspects. Finally, regulation and standardization issues are also considered. The authors do not only discuss the state of the art in UWB, but also identify challenges and portray their visions on all these topics, aiming at motivating further research in these highly significant subjects.

Part III (60 GHz Communication Systems: Concepts and Implementation Aspects) is devoted to another rapidly-emerging short-range technology, millimeter-wave communications, particularly in the 60 GHz band. Seven chapters discuss fundamental theory, design techniques and implementation issues. 60 GHz systems are first approached from the regulation and services perspective, giving the reader a wider view of the possibilities of using these techniques for implementing very-high-date-rate wireless personal area networks. The key underlying techniques for modeling and designing 60 GHz systems are discussed in detail, including channel propagation, baseband algorithms, modulation techniques and system architecture. System concepts and circuitry design for transceivers are discussed. Fundamental implementation aspects are also presented, such as front-end-friendly air interface design, full CMOS integration and system-in-a-package. Adaptive arrays for 60 GHz are also discussed, including assembly techniques and compensation on nonlinearities. This part is concluded with techniques for enhancing power amplifier utilization in millimeter-wave multicarrier systems.

Part IV (Emerging Concepts in Short-range Communications) introduces novel and very promising concepts in short-range communications. Part IV consists of two chapters. The first deals with UWB-over-fibre, a recently proposed approach combining wireless and optical fibre media to transmit ultra-wideband signals. Such a concept does not aim only at increasing communication performance; it also increases the range and opens up a new world of applications. The second chapter also addresses a new concept which is lately gaining significant attention from the industry and research community, namely visible light communications. As high-efficiency white-light emitting diodes are increasingly being used for indoor illumination purposes, their additional use for communications has been contemplated, creating an optical downlink by modulating the visible light with information-carrying data. The basic techniques, their limitations as well as challenges and opportunities, are discussed in this chapter.

Finally, questions, comments or any feedback on this book can be addressed directly to the editors at kraemer@ihp-microelectronics.com and marcos.katz@vtt.fi.

Rolf Kraemer, *Frankfurt (Oder), Germany*
Marcos D. Katz, *Oulu, Finland*

Acknowledgements

This book is the result of coordinated efforts from leading researchers in Europe, America and Asia. The content of the book reflects the contributions and discussions in Working Group 5 (WG5) of WWRF (Short-range Communications) over the last years. We are grateful to the Steering Board and Vision Committee of WWRF for supporting this initiative.

The editors would like to thank the effort of each and every contributor to this book. Without the valuable contributions and enthusiastic participation of specialists around the globe this book would have never been possible. The authors of the chapters are (in alphabetical order): Antti Antonnen, Maria-Gabriella Di Benedetto, Yossef Ben Ezra, Troy Beukema, Olivier Bouchet, André Bourdoux, Steven Brebels, Isabelle Bucaille, Robert W. Brodersen, Geert Carchon, Chang-Soon Choi, Marcus Ehrig, Grahame Faulkner, Gerhard Fettweis, Gunter Fischer, Frank. H. P. Fitzek, Sinan Gezici, Eckhard Grass, Jelena Grubor, Valery Guillet, Frank Herzel, Ole Hirsch, Pertti Järvensivu, Thomas Kaiser, Timo Karttaavi, Markku Kiviranta, Antti Lamminen, Klaus-Dieter Langer, Kyungwoo Lee, Boris. I. Lembrikov, Hoa Le-Minh, Geert Leus, Kiattisak Maichalernnukul, Nadine Malhouroux-Gaffet, Srdjan Glisic, Aarne Mämmelä, Laurence B. Milstein, Andreas F. Molisch, Lorenzo Mucchi, Dominic O'Brien, Ian O'Donnell, Pascal Pagani, Maxim Piz, H. Vincent Poor, Jose A. Rabadan Borges, Walter De Raedt, Moshe Ran, Sebastien Randel, Juergen Sachs, Nuan Song, Jussi Säily, Christoph Scheytt, Klaus Schmalz, Bamrung Tau Sieskul, Isabelle Siaud, Claudio R. C. M. da Silva, Yaoming Sun, Reiner Thomä, Klaus Tittelbach-Helmrich, Anne-Marie Ulmer-Moll, Alberto Valdes-Garcia, A.-J. van der Veen, Antti Vimpari, Takao Waho, Joachim Walewski, Piet Wambacq, Mike Wolf, Eun Tae Won, Sven Zeisberg, Lubin Zeng, Rudolf Zetik, Feng Zheng and Peter Zillmann. We thank you all for sharing with us your technical expertise, and for the professionalism shown during the editing process!

We are particularly in debt with Yossef Ben Ezra, Daniel Dietterle, Frank Fitzek, Eckhard Grass (Editor of Part III), Tomas Kaiser (Editor of Part II), Dominic O'Brien, Moshe Ran and Isabelle Siaud for their support in preparing this book and for their active participation in the meetings of the short-range working group. We also thank the enthusiastic audience of WG5 for their support and enlightening discussions.

We are deeply thankful for the forewords by Dr Nigel Jefferies, Chairman of WWRF, and Prof. Gerhard Fettweis, former chair of WG5. We appreciate their support and strong prospective views.

We thank Wiley for their encouragement and support during the edition process. Special thanks to Tiina Ruonamaa, Anna Smart and Sarah Tilley for their kindness, patience and flexibility.

Finally, we are grateful to our colleagues at IHP and VTT for their motivating discussions and for creating a truly pleasing working atmosphere.

Part I
Introduction

1

Introduction

Rolf Kraemer[1] and Marcos Katz[2]
[1] *IHP, Germany*
[2] *VTT, Finland*

Short-range communications systems characterize a wide range of scenarios, technologies and requirements. There is no formal definition of such systems though one can always classify short-range systems according to their typical reach or coverage. We define short-range communications as the systems providing wireless connectivity within a local sphere of interaction. Such a space corresponds to the first three levels of the multisphere model as discussed in the Book of Visions of the Wireless World Research Forum (WWRF) [1]. Figure 1.1 depicts the multisphere concept, highlighting the levels associated with short-range communications, namely Personal Area Network, Immediate Environment and Instant Partners [2–4]. Short-range systems involve transfer of information from millimeters to a few hundreds of meters. However, short-range communication systems are not only systems providing wireless connectivity in the immediate proximity, but in a broader perspectivethey also define technologies used to build service access in local areas. The WWRF envisions that by year 2017 there would be seven trillion wireless devices serving seven billion people. Certainly, the overwhelming majority of these devices will be short-range communication systems providing wireless connectivity to humans and machines.

Together with wide/metropolitan area cellular systems, short-range systems represent the two main developing directions in today's wireless communications scene. In terms of design rules and target capabilities, short-range systems have certain commonalities as well as marked differences from their counterparts, cellular systems. Maximizing the supported data throughput is quite often one of the main design targets for both types of wireless networks though a detailed comparison between them is not straightforward. Figure 1.2 shows the evolution of data rate support in cellular, metropolitan, Wireless Local Area Networks (WLAN) and very short-range systems. We can see that a steady increase in the supported throughput at a rate of approximately one order of magnitude every five years [5].

Short-Range Wireless Communications Rolf Kraemer and Marcos D. Katz

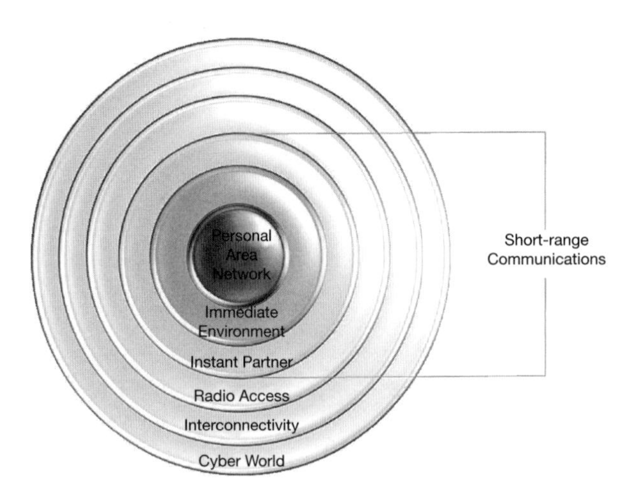

Figure 1.1 Short-range communications within the WWRF multisphere reference model.

There is a great deal of synergy between short-range and cellular networks, and in many cases exploiting their complementary characteristics results in very efficient solutions. The most important approaches to short-range are depicted in Figure 1.3, where a classification according to the operating range is shown. The short-range systems include Near Field Communications (NFC) for very close connectivity (range in the order of millimeters to centimeters), Radio Frequency Identification (RFID) ranging from centimeters up to a few hundred meters, Wireless Body Area Networks (WBAN) providing wireless access in the close vicinity of a person, a few meters typically, Wireless Personal Area Networks (WPAN) serving users in their surroundings of up to ten meters or so, Wireless Local Area Networks, the *de facto* local connectivity for indoor scenarios covering typically up to a hundred meters around the access point, Car-to-car communications (or Vehicle Area Networks) involving distances of up to several hundred meters and Wireless Sensor Networks, reaching even further.

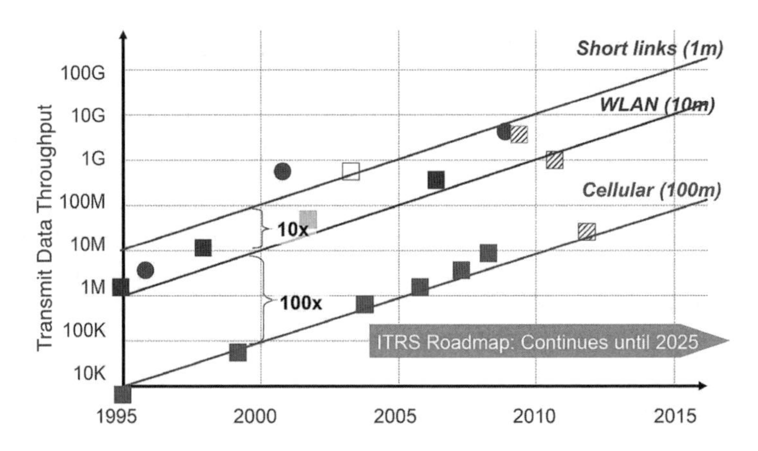

Figure 1.2 Data throughput evolution if mainstream communication systems [Courtesy of Wigwam project].

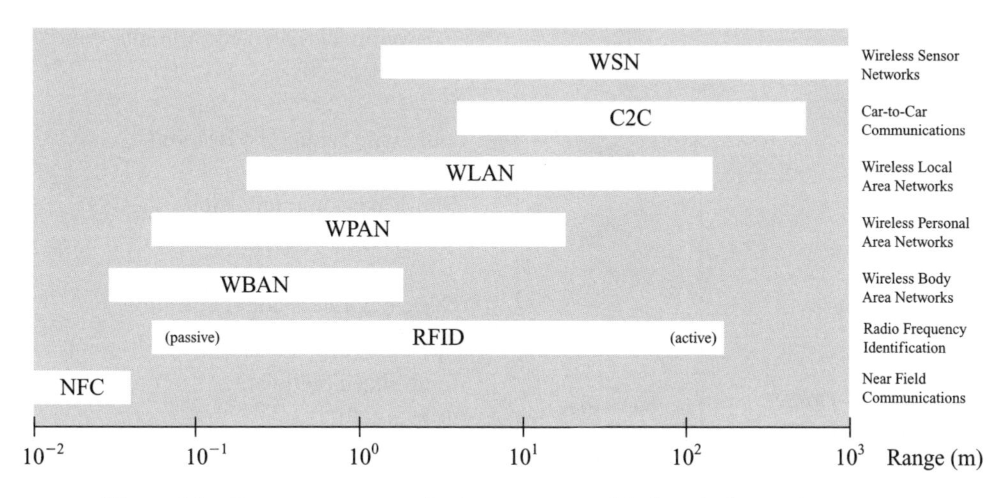

Figure 1.3 Short-range communications systems and their typical operating ranges.

Each of the aforementioned short-range approaches involves in general one or more specific air interface technologies (i.e. physical layer and medium access control layer) typically with the corresponding supporting standards defining most of the relevant technical details. A vast amount of literature exists for short-range technologies, in particular WLAN, WPAN, RFID and WSN are by far the most visible representative approaches. Ever since their inception, Wireless Local Area Networks (WLAN) and Wireless Personal Area Networks (WPAN) have had a leading role in the development and further diversification of short-range communications systems. Initial developments targeted simple point-to-point communication systems working as cable-replacements to connect wireless devices (e.g., computers, peripherals, appliances, etc.) Currently, there are several active and evolving standards particularly focused on short-range systems. The IEEE 802.11 standard series is the most popular example of coordinated and sustained development in short-range communications, particularly focused on WLAN. As for WPAN, the IEEE 802.15 series defines a family of wireless networking standards with different technologies, such as Bluetooth (802.15.1), High Speed (802.15.3) and Sensor Networks (802.15.4). There are several different physical layer descriptions in each of the 802.15 sections, for example 802.15.4a for UWB based wireless sensor networks or 802.15.3c Gigabit communication networks.

Figure 1.4 shows other aspects of short-range communication systems, highlighting the great diversity in possible air interface solutions, topologies as well as achievable data throughput and supported mobility. Physical layer technologies range from conventional radio (e.g., narrowband) to Ultra Wide Band (UWB) systems, exploiting single and multicarrier modulation schemes. Typical frequency bands span from some MHz to millimeter-wave systems (60 GHz and beyond). Moreover, optical communications are also attractive short-range technologies, including infrared and visible light communications. In general, short-range networks have *ad hoc* distributed architectures allowing direct and multi-hop connectivity among nodes. Centralized access is also possible, as in WLAN, which in fact supports both centralized and distributed topologies. Even though short-range systems are typically conceived for fixed or low-mobility environments, as in essentially all the indoor applications, new scenarios even foresee cases where high mobility is involved. Examples include car communications where

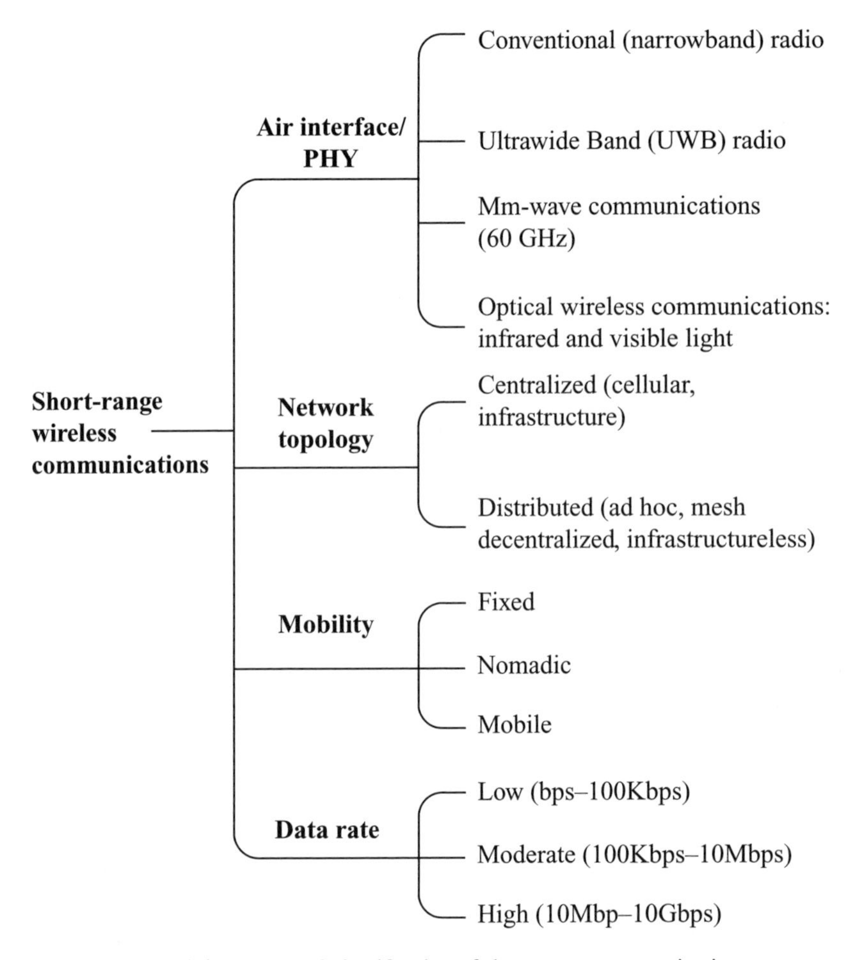

Figure 1.4 A general classification of short-range communications.

cars traveling in opposite directions attempt to exchange information, or information retrieval in the case of car-to-roadside communications. Data throughput requirements for short-range systems are also very broad, from some hundred bits per second in simple RFID systems up to 10 Gbps and beyond in WLAN systems, for instance.

In many cases low power consumption is one of the key requirements for short-range systems, particularly taking into account that transceivers are in many cases battery-driven. Particular attention is necessary when designing short-range devices while minimizing the power consumption at the three basic conditions of the transceiver, namely transmitting, receiving and idle states. Low cost is another important factor to take into consideration when designing short-range systems. Furthermore, minimizing size and weight are also quite often imperative engineering principles that need to be applied by the designer of such communication systems.

Another important aspect of short-range communications networks is their relationship to other existing wireless networks and their possible interaction. In very many scenarios it is indeed convenient to consider short-range networks as complementary to cellular networks, instead of the conventional approach considering the former in complete isolation from the rest of the networks. Countless efforts have been put into the research and development of *ad hoc* networks in the last twenty years. However, their presence today is effectively eclipsed by the omnipresent cellular network. It can be argued that one of the reasons for such a weak penetration is precisely the fact that these networks were designed largely ignoring the other mainstream network approach and their possible cooperation. As a matter of fact, these networks complement each other very well and a welldesigned cooperative strategy between both networks can bring significant benefits. Transmitting bits over short-range networks is much more energy efficient that doing it over cellular networks. In addition, short-range networks usually use spectrum-free bands (e.g., ISM bands).

WWRF has recognized the importance of short-range communications, creating a study group on this subject (WG5) back in 2004. Since then a number of highly relevant technologies have been actively explored jointly by academia and industry. The main focus of the short-range activities in WWRF is on air interface developments, and this is especially reflected in the contents of this book. Several white papers and briefings have been produced in recent years, dealing with the following subjects:

WWRF White Papers

- Multi-dimensional Radio Channel Measurements and Modelling for Future Mobile and Short-range Wireless Systems;
- UWB techniques and future perspectives;
- WBB over Optical Fiber;
- UWB Limits and Challenges;
- MIMO-OFDM for WLAN (TDD mode);
- Short-range Optical Wireless Communications;
- The Architecture of Mobile Internet;
- New Radio Interfaces for Short-range Communications;
- Gigabit LAN at 60 GHz;
- Cooperative Aspects of Short-range Systems.

WWRF Briefings

- Wireless Body Area Networks and Sensor Networks;
- High Throughput WLAN/WPAN;
- Gigabit WLAN Technologies;
- Visible Light Communications.

White papers and briefings on short-range communications at WG5 of WWRF

New research subjects are continuously identified, reflecting the interests of industry, academia and research institutes participating in WG5. The following subjects are currently being considered for further exploration in this group, leading to new briefing and white papers, thus becoming an integral part of the WWRF vision of future wireless communications.

New research subjects on short-range communications

- Car communications:
  - Car-to-Car,
  - Car-to-roadside,
  - Car-to-Infrastructure (jointly with WG4/cellular access group),
  - In-Car Communications.
- Ultra High Performance WLAN systems: Finding the limits of WLAN, targeting a bit rate of 100 GBps.
- Body Area Networks,
- Wireless Grids.

Identified new research items at WG5

The following chapters address aspects of the work on short-range communication of the last three years. They focus on several different aspects of UWB communications, ultra high speed communication at 60 GHz, visible light communication, UWB over fiber and also design rules for modern short-range communications systems. Our aim in publishing this book was to give a deeper insight in the important aspects and current research in short-range communication systems.

References

1. Wireless World Research Forum, http://www.wireless-world-research.org.
2. Tafazolli, R. (ed.) (2004) *Technologies for the Wireless Future: Wireless World Research Forum (WWRF)*, John Wiley & Sons, Ltd, Chichester, p. 580.
3. Tafazolli, R. (ed.) (2006) *Technologies for the Wireless Future: Wireless World Research Forum (WWRF)*, Vol. **2**, John Wiley & Sons, Ltd, Chichester, p. 520.
4. David, K. (ed.) (2008) *Technologies for the Wireless Future: Wireless World Research Forum (WWRF)*, Vol. **3**, John Wiley & Sons, Ltd, Chichester, p. 506.
5. WIGWAM Project: Wireless Gigabit With Advanced Multimedia Support, http://www.wigwam-project.com.

2

Design Rules for Future Short-range Communication Systems

Frank H.P. Fitzek[1] and Marcos D. Katz[2]

[1]*Aalborg University, Denmark*
[2]*VTT, Technical Research Centre of Finland*

2.1 Introduction and Motivation

With the introduction and rapid spread of the mobile phone, access architectures with centralized topologies (e.g., Cellular) have become dominant in mobile communications. The main task of a mobile phone was to access to a predefined connection point whenever it was turned on. Throughout the evolution of the mobile phone, short-range technologies were gradually integrated into the phones, resulting in multi-mode devices. The first solutions, based on infrared (IRDA) [1] and Bluetooth [2] technologies, were mainly intended to serve as a cable replacement, typically for carrying out rather simple jobs such as printing, exchanging data or attaching wirelessly peripheral devices (mainly for Bluetooth). Following the evolutionary steps of cellular air interfaces, short-range technologies also developed rapidly, supporting increasingly higher data throughputs, multiple access and advanced services. Unfortunately, in many cases short-range systems are still considered to be just a handy cable substitute. In addition to the mentioned short-range technologies, several others have been introduced, such as IEEE802.11 [3], ZigBee [4], ultra low power (ULP) Bluetooth, formerly known as Wibree, near field communications [5], and so on.

Short-range technologies are nowadays used for more than just connecting a mobile phone to another one or a PC. Short-range technologies are in fact already used to, or envisioned

Short-Range Wireless Communications Rolf Kraemer and Marcos D. Katz

to be used to:

1. form mobile social mobile networks,
2. to access wireless sensor networks and
3. to build cooperative wireless networks.

These application fields have different requirements for the short-range technology. In the following we briefly describe these application fields and derive the different requirements. It has to be noted that more application fields are emerging in these days, but the selected ones underline the need for specific technology features. Figure 2.1 summarizes the evolutive development of short-range communication systems. Clearly, the development of future short-range systems will focus on point-to-multipoint connectivity as well as on the interaction between short-range with other networks. One of the most exciting developments in short-range

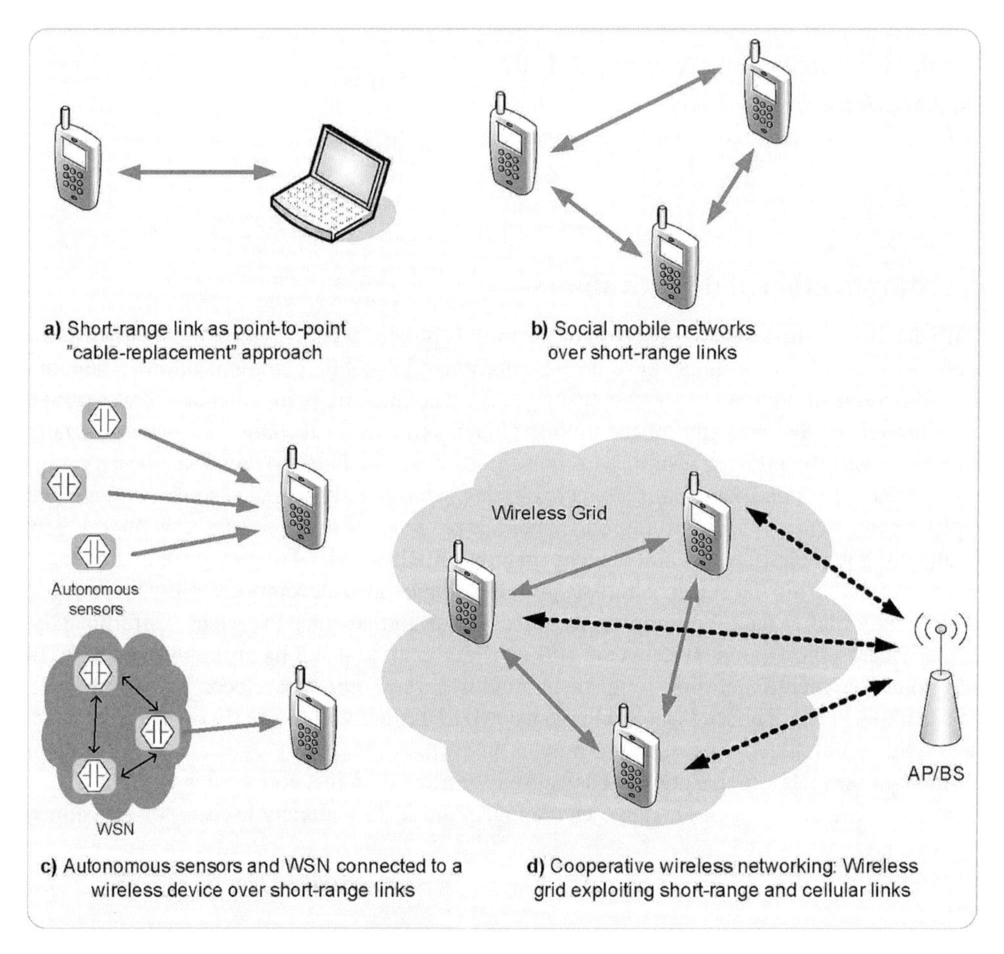

Figure 2.1 Evolutive aspects of short-range communication systems: from simple point-to-point systems to the concept of Wireless Grids.

communications results when such networks are connected to other networks, particularly cellular networks. In its simplest way, this could be realized by a network extension approach, where the short-range network is appended with the cellular network. A simple example of such a two-hop, dual-technology link would be the mobile phone in Figure 2.1(a, b or c) transmitting to the base station any information received over a short-range link. Figure 2.1(d) illustrates a more complex approach, exploiting rich and dynamic cooperation between the two networks, and aiming at improving communication performance as well as enhancing the efficiency in the utilization of resources. Such an approach, called *wireless grid*, has a lot of potential, and it will be discussed in the sequel.

2.2 Mobile Social Networks

Mobile social networks connect people that are within each other's proximity. The short-range technology is used to discover and connect other mobile phones in the range. This new kind of architecture is also referred to as mobile peer-to-peer. Multiple mobile peer-to-peer networks can coexist in the same place, as illustrated by Figure 2.2.

The first concrete applications of mobile social networks are on their way, and a very representative example of such is aka-aki [6], which enriches a conventional (web-based) social network such as Facebook [7], collecting information by short-range technology. Aka-aki users can access on the web at different profiles and add connections they made on their mobile life. The connections are collected by a mobile application running on the mobile phone. The application uses Bluetooth technology to identify known or new contacts using the same access technology. The mobile users can use the application either to get in contact right away or to contact those users later on the web. A similar approach is the Spider [8] application by Aalborg University. As illustrated in Figure 2.3, a mobile user can look at

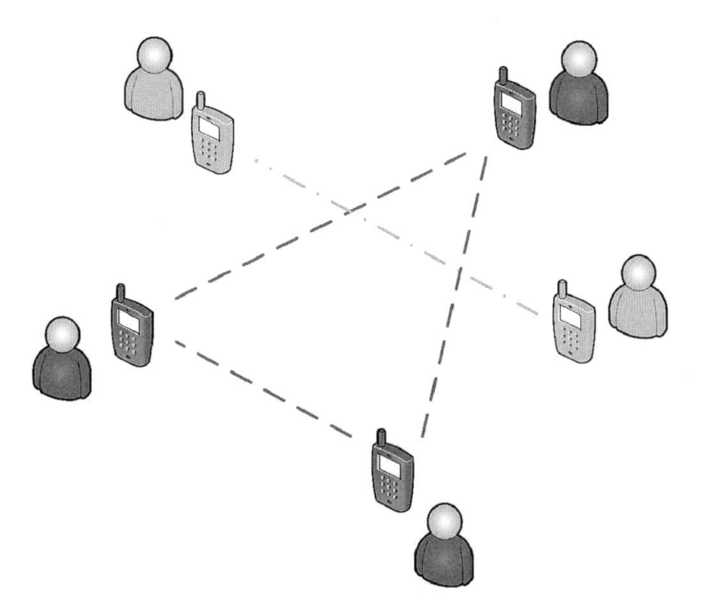

Figure 2.2 Mobile peer-to-peer architecture for mobile social networking.

Figure 2.3 Screenshot of a virtual meeting room in the Spider application [8].

a virtual world where its own virtual character is mobile. As soon as Bluetooth detects other close-by mobile phones that are also running the same application, more characters start to fill the virtual meeting room. Each mobile user can now steer its own character close to other characters and start some actions such as chatting, exchanging profiles or just looking at a real photo image of the actual user. The Spider approach is based on the concept of SMARTEX [9]. The idea behind SMARTEX is to exchange digital content among mobile phones, introducing a new concept of digital ownership [10]. Also here the Bluetooth technology is used to form the mobile peer-to-peer network.

Wireless technology brings an important extension to social networking. Users can be networked on the move, by cellular or short-range networks. The latter approach is particularly appealing as close-by users are likely to share some commonalities, interests or background.

2.3 Wireless Sensor Networks

In parallel with the evolution of cellular networks and mobile phones, wireless sensor networks (WSN) have attracted a lot of attention in recent years. The first research efforts were mainly focused on military applications, but wireless sensors have shown their great potential in civil environments too. A wireless sensor is composed of a wireless communication facility, sensor elements as well as a source of energy, typically a battery. A great array of different physical phenomena can be measured by the sensors, such as light, temperature, pressure, distance and many others. There are also some passive sensors such as RFID and UHF devices that do not need a battery as they are remotely powered by the electromagnetic waves from the requesting device.

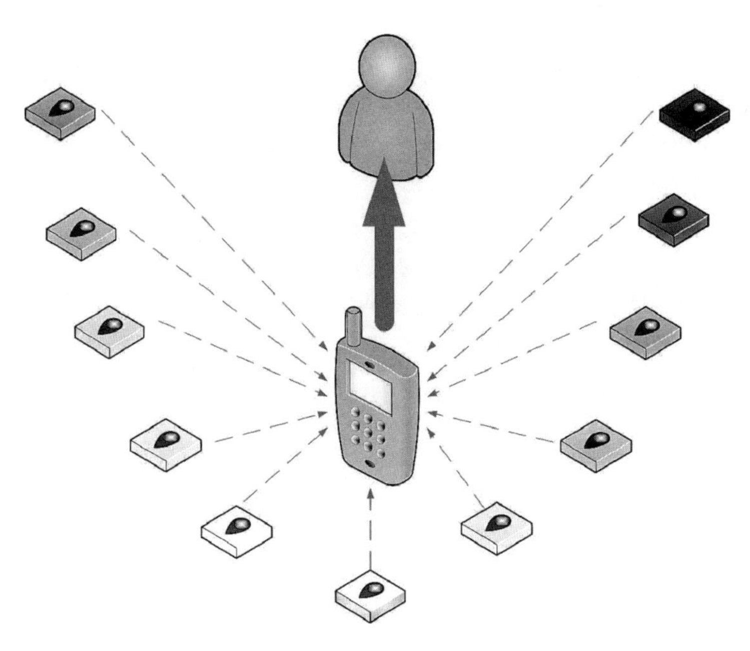

Figure 2.4 Mobile device gathering information from a Wireless Sensor Network.

Even though the combination of mobile phones with wireless sensors, see Figure 2.4, appears to be quite obvious and straightforward, its full potential has not been exploited yet. By linking a wireless sensor network to a wireless device such as a mobile phone, the user can control the networks as well as getting access to measured data. A wireless device equipped with cellular and short-range interfaces then becomes an extremely flexible tool for monitoring and controlling the sensor network, both remotely and locally. Some concrete examples of current developments in that direction include the joint development by Apple and Nike of sport shoes with integrated sensors measuring and transmitting relevant information (e.g., number of steps) to an access point held or worn by the user, for example an iPod using propriety technology. Other sensors are able to retrieve heart beat information or blood pressure and convey the data to a dedicated place such as a wrist watch. It is expected that the variety of external sensors is much larger than that of the embedded ones. In particular, external sensors will lead to new applications in the area of health care, intelligent housing and others. The advantage of omnipresent sensors is that complete measurement campaigns can be carried out, which are more meaningful than instant measurements points. An example is health parameters such as blood pressure or insulin levels, where the instantaneous value has almost no meaning, but values over a longer time period are more meaningful. Therefore sensors could help medical doctors to retrieve the value over a long time, without being dependent on the instantaneous value that they measureon the date they see the patient.

There are already some sensor platforms on the market. Nevertheless, in collaboration with the Technical University of Berlin, the mobile device group of Aalborg University (AAU) has designed their own platform, called *opensensor*. The main reason for this development was to have as much flexibility on the platform as possible. Flexibility refers to the number of connectable sensor parts as well as the wireless part. As the opensensor is to be used for

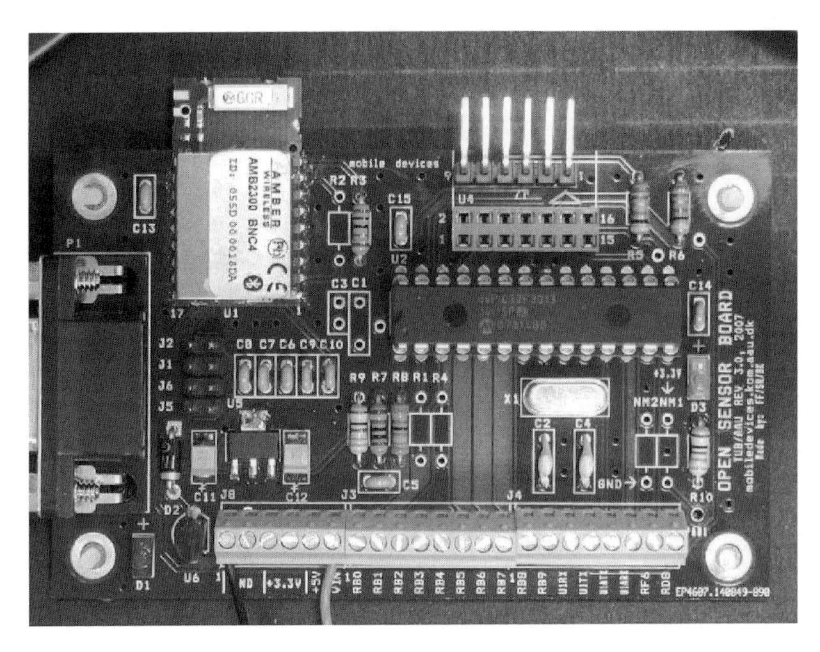

Figure 2.5 Opensensor hardware [10].

ongoing and future research projects as well as for teaching purposes, the design of the board was kept as simple as possible. Throughout the evolution of the opensensor, surface-mount devices (SMD) technology was deliberately avoided, designing the board for easy-to-mount standard components. In addition, this results in a low-cost solution, enabling researchers and students to implement their ideas in the field of wireless sensor networking on a real platform. As the convergence of wireless sensors with mobile phones is one of the design goals of the mobile device group at AAU, the opensensor offers the possibility to communicate with commercial mobile phone.

In Figure 2.5 a view of the opensensor hardware is depicted [11]. The opensensor initiative makes available to anyone detailed hardware specifications, software, as well as teaching material. The opensensor can be used for classical WSN implementation such as MAC design, routing, energy saving, and so on. The opensensor is powered up with a 9 V battery block and it has several interfaces to connect to the outer world. Two LEDs are on the board, the red one showing the operational phase of the opensensor while the green one, controlled by DSP, is used for monitoring purposes. The core of the opensensor is a 16 bit architecture dsPIC30f3013 processor produced by Microchip. It has low power consumption, features 24 KB of program memory, 2.048 B of RAM and has 28 pins, 20 of which can be used as in/output pins. For communication purposes the chip has two UARTS and one SPI port. The chip is capable of performing Analog to Digital conversion in 12 bit, 200 Ksps for measurement purposes, it has three timers and is also able to provide pulse-width modulation (PWM) which can be utilized for motor control. The opensensor board has up to five interfaces that allow communicating and programming the sensor. While the connection bar, the RS232 and the PICKIT pins are part of the board, the wireless communication interfaces nRF905 and Bluetooth are optional.

Figure 2.6 Practical implementation of an opensensor grid.

The opensensors can communicate among each other using the nRF905 or Bluetooth. In Figure 2.6 an opensensor grid with nine nodes is shown. This grid was used to demonstrate distributed storage approaches and network coding [12–14].

As mentioned before, the opensensor is also envisioned to ease the convergence of mobile phones and external wireless sensors, building up contextual information for the user, as previously mentioned. An example of external sensors and mobile phones, namely the *Parksensor* is shown in Figure 2.7. Here four external proximity sensors capable of measuring distance between objects are mounted at the corners of a car. The individual distance measurements are sent to the driver's mobile phone inside the car. The mobile phone can then display the situation in an appealing way, even supporting the driver with audio and sound information, ultimately resulting in a more comfortable and safer parking operation.

2.4 Cooperative Wireless Networking

Another field of application for short-range technology is cooperative wireless networking, where mobile phones connect with other phones in their proximity, forming so-called *wireless*

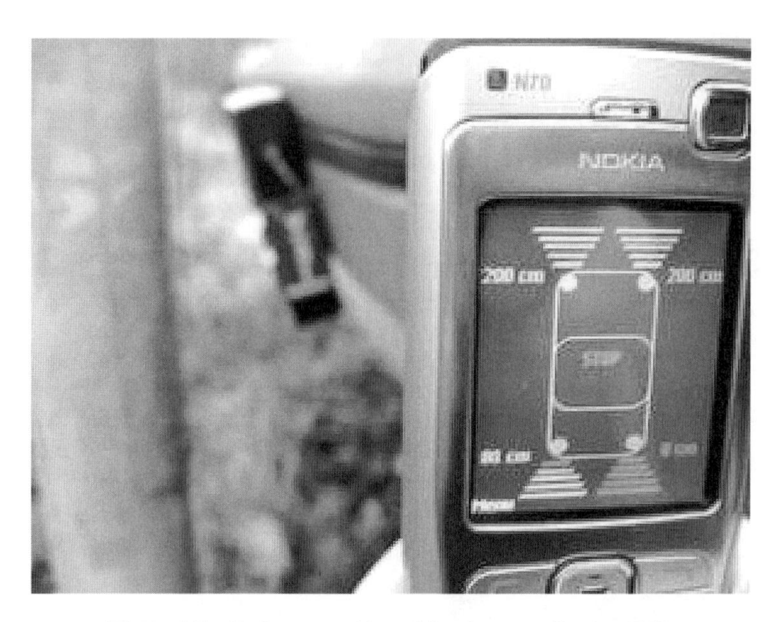

Figure 2.7 Parksensor with mobile phone application [15].

grids. Wireless grids can use their capabilities and resources in a much more efficient way than any stand alone device could ever do. In general, a wireless grid exploits not only the short-range links among the local cooperative cluster, but also the cellular links of the interacting wireless devices. This combination of access approaches with many complementary characteristics creates a novel framework for distributing information. The advantages of wireless grids include better utilization of radio and other shared resources, enhancement of communication capabilities such a data throughput and quality of service and a natural support for new types of cooperative services and applications. For instance, when the mobile phones are connected to the cellular overlay network as given in Figure 2.8, the wireless grid is able to provide high peak data throughput by accumulating the data rates of each single mobile phone. The wireless grid can use the accumulated data rate for better service support, less energy consumption and less cost for the user [16–18]. The concept of wireless grids is very flexible and has a lot of potential to become one of the paradigms of future wireless communications.

2.5 Top Ten Design Rules For Short-range Communications

As seen by the classification discussed in the previous sections, short-range technology can be used in various scenarios and for different applications. The demands for this technology differ from those in a cellular concept dramatically. In the following we identify and discuss the top ten design rules for short-range communications, particularly taking into account the emerging applications and novel scenarios for the future. The rules can be interpreted also as a wish list, where the need for each individual issue arises from a certain requirements for future communication architectures such as social-mobile networks, wireless grids and cooperative wireless networking. In many cases these individual issues are not totally separated and will overlap. Nevertheless, the following list aims at highlighting the key design issues to take into

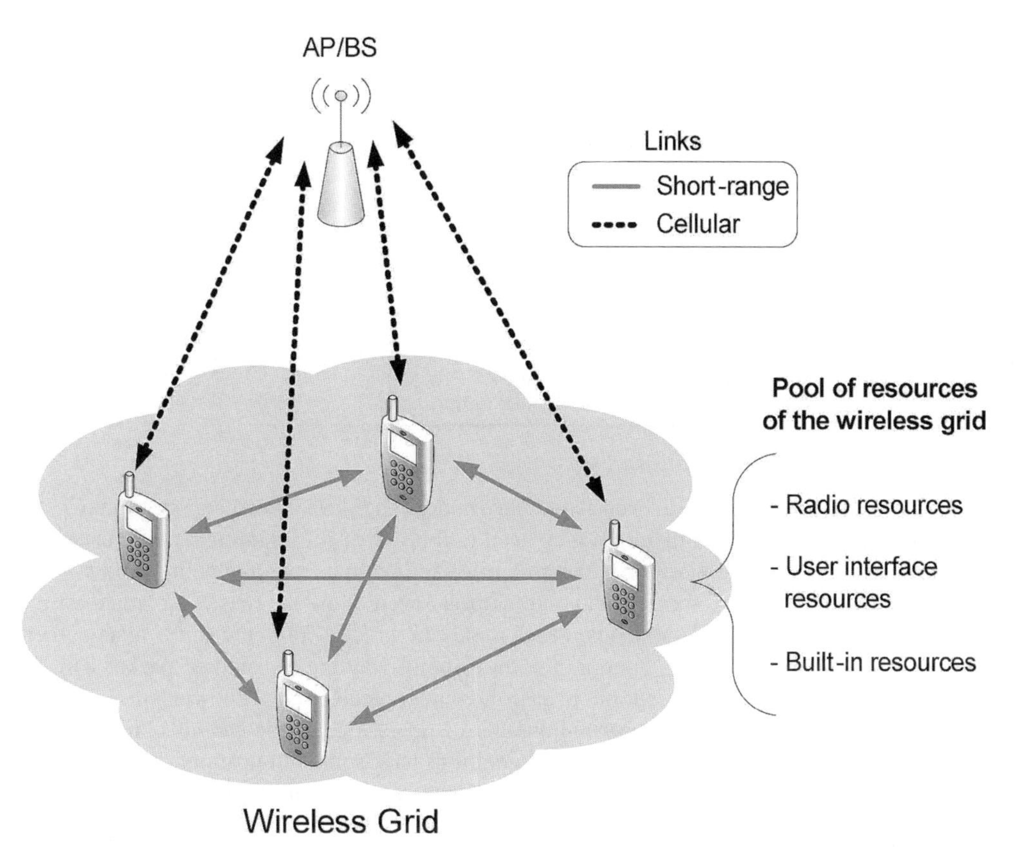

Figure 2.8 Wireless Grids: a cooperative cluster formed over short-range links, while connected to the cellular network.

account when designing short-range communication systems for future wireless networks. The top ten design rules are listed below in Table 2.1 and are explained in the following sections. Each listed topic is equally important.

2.5.1 Communication Architecture

As mentioned earlier the state-of-the-art communication architecture is dominated by point-to-point links. In the cellular environment the base station communicates with individual mobile devices and vice versa. Broadcast by the base station is envisioned, but rarely used. Even for multicast services, mostly unicast traffic channels are used. The only exceptions are technologies such as digital multimedia broadcasting (DMB) [19] and digital multimedia broadcasting-handheld (DVB-H) [20]. Those are in fact real broadcast communication systems without any feedback channel. Certainly, the mobile devices themselves use only unicast transmission towards the base station. It is convenient to extend this somewhat conservative thinking to the short-range world. With IrDA the first point-to-point optical short-range wireless communication was introduced. Even Bluetooth followed this path, being originally conceived as a wireless cable replacement. The Bluetooth architectures foresee one master

Table 2.1 Top ten design rules

Rank	Short-range Topic
1	Communication architecture
2	Energy awareness
3	Signaling and traffic channels
4	Scalability and connectivity
5	Medium access control and channel access
6	Self-organizing and presence
7	Service discovery
8	Security and privacy and authentication
9	Flexible spectrum
10	Software defined radio

node communicating with up to seven active slave nodes. The slave nodes cannot communicate directly with each other and therefore they need to relay their information via the master, in a centralized fashion. The master has the possibility to communicate in a point-to-point way with the slaves. Moreover, some Bluethooth chipsets even allow the master to broadcast its information. In the IEEE802.11 standard each node can directly communicate with any other node in its proximity. If this is done in a point-to-point way the transmitted packet will be acknowledged on successful reception. In case the sender receives no acknowledgment, it will do link level retransmission for a certain number of times. Each node can also broadcast or multicast its packet to other nodes, but such operations have some implications. First, the data rate is reduced on most chipsets to enable all nodes in the coverage area to receive the message. In addition, there is no feedback from the recipients, nor back-off in case of packet collisions.

It would be highly beneficial that future short-range communication systems have both point-to-point as well as point-to-multipoint communication capabilities. In particular point-to-multipoint capabilities are essential for service discovery, presence and self-organization. Furthermore, point-to-multipoint communications is the natural access paradigm for social and cooperative networking.

2.5.2 Energy Awareness

One of the fundamental challenges of battery-driven mobile devices is their energy consumption. The more energy is consumed the less operational time of the mobile device is available for the user. This is not desirable from the viewpoint of the users, but also from the service provider, as shorter operational times involve less use of the provided services and consequently less revenue. Furthermore, if too much energy is consumed in a short time, this will end up in a dramatic heating of the mobile device, particularly taking into account the small form factor of wireless devices. Nowadays the materials of a mobile device are chosen to distribute the generated heat over the whole mobile device in an attempt to dissipate it as much as possible. In Figure 2.9 temperature distributions of a commercial mobile phone are shown. The device heats up by nearly 8 °C just by changing the phone from offline mode to WLAN transmissions.

Manufacturers are certainly interested in reducing power consumption, as longer autonomy will make their products attractive, improving their competitive edge. Indeed, operational

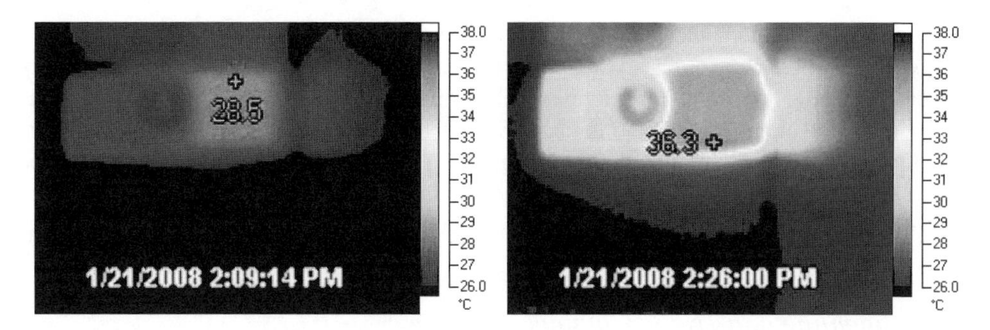

Figure 2.9 Internal temperature distribution in a commercial mobile phone: Maximum temperatures are 28.5 °C (offline mode) and 36.3 °C (WLAN on).

time is also a critical selling point for mobile phones and consequently manufacturers should continue developing techniques for achieving lower power consumption. The power drain resulting from the communicational functionalities of a typical mobile phone could be half or even two-thirds of the overall power consumption of the device [18, 21]. So, the designer of future air interfaces has a large margin of power to reduce. Consuming less energy can be achieved by using less complex system components, more efficient medium access schemes, and other issues addressed in the following. Most of the wireless technologies have four basic states, namely *sending*, *receiving*, *idle* and *sleep*. A wireless device can be characterized by a given average power consumptions associated with each of these states. The energy consumed in each state depends on the power level and the related time that state is active. For instance, if two sending data rates R1 and R2 are available to transmit a given data chunk of size D with sending power P1 and P2 respectively, R1 should be used for transmission if $R1 < R2 \times P2/P1$ and R2 otherwise. The power levels for the different states change according to the technology and the implementation of different distributors. In the following table the power levels for 3G, WLAN and Bluetooth are given for a typical commercial mobile phone. In general, the power levels for sending are larger than those of the receiving state which are in turn larger than the power during the idle state (Table 2.2).

A lesson learned from our previous research work [22–24] is that one of the dominating factors for the overall energy consumption is in fact the idle power state. This is particularly true for scenarios where the wireless link is not used in an on-off model as cable replacement. The idle power state could be eliminated easily if the incoming traffic could be known beforehand or somehow signaled in an energy-efficient way. For streaming services, where data is coming from an access point or a neighboring device with a predefined traffic pattern (e.g., video frames every 40 ms), the mobile devices should be able to switch off the RF chain and

Table 2.2 Power levels for different wireless technologies [22–24]

	Sending	Receiving (W)	Idle (W)
3G	—	1.3	0.6
WLAN	1.6 W	1.2	0.9
Bluetooth	0.4 W	0.3	0.1

baseband circuitry among the activity phases, a procedure that is known from DVB-H to save energy. More problematic are the cases where the traffic pattern is unknown and the inactivity phase is very long. For example, a mobile device using IEEE802.11 would significantly waste energy just by waiting for an incoming call, while the call itself may take only a short while compared with the inactivity phase. Switching off the RF/BB stages in this case would not be a good idea as incoming calls cannot be received. In this case other solutions should be found, and some possible ideas are discussed in the following section.

2.5.3 Signaling and Traffic Channels

In general, short-range communication systems employ the same physical and logical resource for signaling and traffic transportation. This has some serious drawbacks. The first is that normal traffic channels can block important and even critical signaling messages. The second is that the same amount of energy per bit is spent for the signaling and the traffic transportation. So, one design rule is to split the signaling from the main data exchange. Coming back to the previously presented example of the IEEE802.11 technology used for VoIP, instead of wasting energy for waiting an incoming call, the RF/BB should be switched off. Then, a special procedure to wake the mobile device up and listen for incoming calls is required. Two possible approaches are discussed next:

1. *Overlay Concept:* The originating SIP server calls the mobile device over the cellular network (e.g., 2G/3G). Next, the mobile device, spending only a small amount of energy by powering up the 2G/3G, will receive the incoming call, identifying the number, suppressing the ring tone and switching on the WLAN to receive the call over WLAN. Figure 2.10 depicts the underlying concept of overlay wake up.

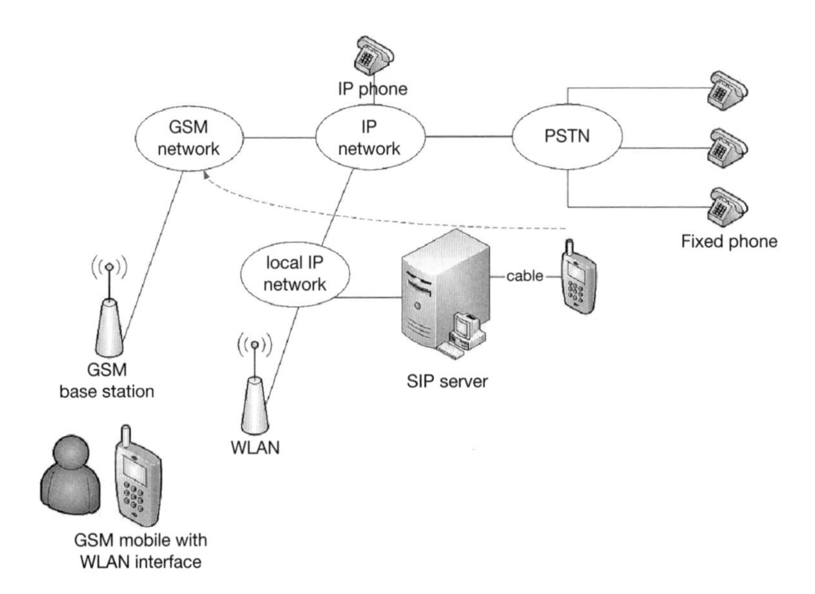

Figure 2.10 Overlay concept for energy saving with wake up calls.

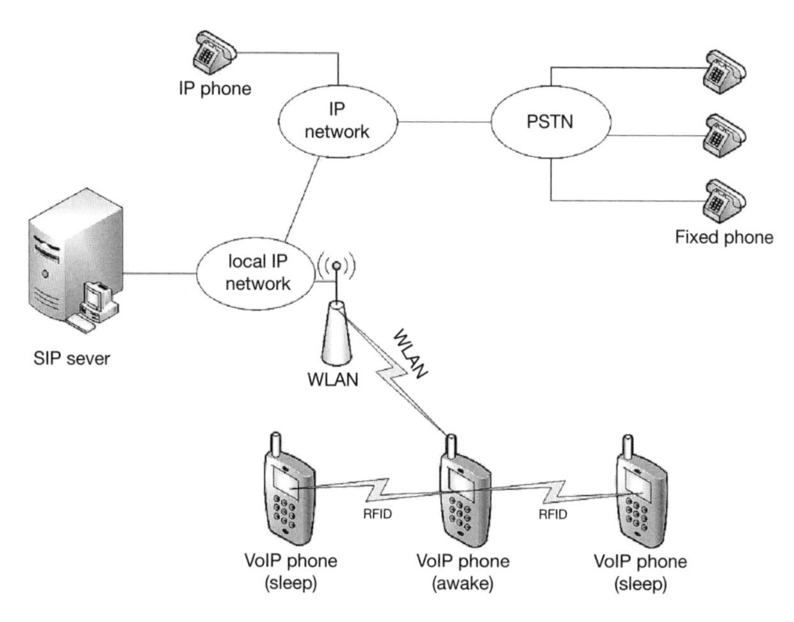

Figure 2.11 Cooperative concept for energy saving with wake up calls.

2. *Cooperative Concept:* A group of mobiles in an office scenario is waiting for incoming VoIP calls over WLAN. To save energy some of the mobile devices switch off their WLAN, while other mobile devices, referred to as watchdogs, will listen on the WLAN AP. Whenever there is an incoming call for a mobile under their responsibility, they need to wake them up. Here RFID or UHF technology comes into the picture. The watchdog will initiate a radio wave that carries enough energy to wake up the neighbouring mobile devices after receiving SIP messages over the WLAN AP. With coded RFID or UHF it might be even possible to wake up dedicated mobile devices. In order to balance the energy expenditure of all cooperating mobile devices, the role of the watchdog should be changed among the mobile devices in a round robin fashion. The concept of cooperative wake up is shown in Figure 2.11.

2.5.4 Scalability and Connectivity

It is highly important that the whole wireless communication system scales. In other words, scalability should not be limited to a certain number of participating devices, as we see it today in Bluetooth. As the foreseen scenarios are highly dynamic and heterogeneous, the number and type of participating devices in a short-range network can vary dramatically. In particular, wireless grids offer a wirelessly scalable architecture. Indeed, the more cooperating units, the larger is the resource pool that can be formed, as each wireless device can contribute with its resources. These resources, for example radio resources, built-in resources, user-interface resources, and so on can not only be shared by all cooperating devices, but they can also be moved or borrowed to particular node or nodes of the network that may need them.

The concept of network connectivity is related to network scalability. Connectivity here refers to the capability of the short-range network to be connected to another network. This is a highly desired feature in very many scenarios. In fact, a short-range network can be extended to another network, for example the cellular network. Information generated within the short-range network is made available widely, for instance for monitoring purposes. Moreover, the short-range network can be controlled remotely with this approach. Network connectivity to any complementary wireless/mobile network increases the capabilities and operating domain of the short-range network considerably. Such a connectivity, leading to cooperation between networks, is one of the underlying principles of wireless grids. One or more nodes of the short-range network can serve as the interface or gateway to another network.

2.5.5 Medium Access Control and Channel Access

The medium access control (MAC) has the biggest impact on the short-range performance, whereas performance refers to all other top ten requirements. From the energy saving perspective, a slotted system would be desirable such that mobile devices can go to a deep sleep more or even better switch-off their air interface if not needed to save the largest possible amount of energy. Even though the channel is slotted, the MAC should support high and low data rates equally efficiently. For the self-organization the amount of data to be exchanged is small, but the messages need to be exchanged very quickly to perform fast self organization. The MAC should be different for the signaling and data channel if there is such a separation. If the signaling is separated from the data channel, the requirement of being slotted can be dropped on the signaling domain.

2.5.6 Self-organizing and Presence

One important factor of short-range technology is that it has, most of the time, no structure and no continuity over time. Therefore short-range networks need the capability of self- organization. The first step towards self-organization is presence, where the device can announce its existence to the other devices in proximity to it. Though the concept of presence can be implemented now with existing technologies, it results, on the other hand, in a highly inefficient solution, particularly in terms of energy and spectrum usage. Note that presence need not necessarily be announced by means of signaling from the device willing to be taken into account. In fact, such devices, their resources and capabilities can be sensed by the network in many different ways, including in a centralized fashion by the access point or base station, or cooperatively, involving several nodes that sense relevant information. Future cognitive networks will have implemented a cognitive cycle, extending the sensing of spectrum (as in cognitive radios) to include the other shared resources, as discussed in [17].

2.5.7 Service Discovery

Short-range networks are set up with a certain goal in mind. For example if we assume a wireless grid with heterogeneous wireless devices, each device may want to use remote capabilities (e.g., possessed by other devices in the grid) whenever these are not implemented onboard or the available capabilities are not sufficient. To understand the available capabilities within the wireless grid, a given device needs somehow to announce the capabilities it is willing to share. Thus service discovery is an essential capability for upcoming short-range technologies.

2.5.8 Security and Privacy and Authentication

Security is nowadays already implemented in most wireless technologies and users are aware of the risks of not using it. But all implemented solutions so far have a centralized entity to assure security, mostly based in the core network. In the future, with upcoming mobile peer-to-peer networks becoming one of the key communication paradigms, security needs to be established among two entities with no external referee. Another trend in security is that it tends to be implemented at lower protocol layers. Security started with HTTP security (HTTPS) and solutions such as IPsec and WEP came later.

2.5.9 Flexible Spectrum

In order to have enough bandwidth and to assure coexistence among the large number of wireless technologies, the available spectrum needs to be assigned in a flexible manner. Spectrum is a limited and highly valuable radio resource, the scarcity and value of which will increase with the emergence of advanced bandwidth-eager services as well as the increasing number of wireless subscribers. Considerable efforts are being put into developing new techniques to enhance the spectral efficiency of the network. Cognitive radio is perhaps the most representative solution to that challenge. The concept of wireless grids is attractive from the standpoint of spectrum usage. In fact, the composite topology involving the combination of centralized and distributed access architectures can help to improve the spectral usage of the network. The cellular network uses licensed spectrum while the short-range networks typically exploits license-free spectrum. Cooperative strategies exploiting this synergy can be devised to maximize the use of the short-range network, aiming ultimately to minimize the amount of licensed spectrum.

2.5.10 Software Defined Radio Design

For some of the envisioned scenarios mobile devices might use multiple air interfaces at the same time. In accordance with the flexible spectrum section, the hardware of the mobile device needs to be flexible. Instead of just accumulating multiple wireless technologies, there should be a platform that can flexibly adjust to the needs of the mobile device and transform to the need technology. This can be done by software designed radio (SDR).

2.6 Conclusion

After having introduced the top ten design-rules (or wish list) for wireless short-range communications, we summarize these requirements in Table 2.3, for the three initial scenarios, namely social-mobile networks, wireless sensor networks and cooperative networking. For each individual scenario not all of the requirements are needed.

Table 2.3 Top ten design rules applied to different scenarios

	Social Mobile Networks	Wireless Sensor Networks	Cooperative Wireless Networks
Communication architecture	P2P, P2MP	P2P, P2MP	P2P, P2MP
Energy awareness	Yes	Yes	Yes
Signaling and traffic channels	Yes	Yes	Yes
Scalability and connectivity	Yes	Yes	Yes
Medium access control and channel access	SoA	Advanced for energy	Advanced
Self-organizing and presence	Yes	Yes	Yes
Service discovery	Must	Maybe	Must
Security and privacy and authentication	Must	Must	Must
Flexible spectrum usage	No	No	Yes
Software defined radio	No	No	Yes

2.7 Outlook

As mentioned beforehand the top ten design rules for short-range systems can also be seen as a wish list. Nevertheless, if we have a look at the development of Bluetooth technology, we will see that some of the points discussed are addressed there already. Indeed, besides the actual version of Bluetooth v2.x, two further developments are on their way. The ultra

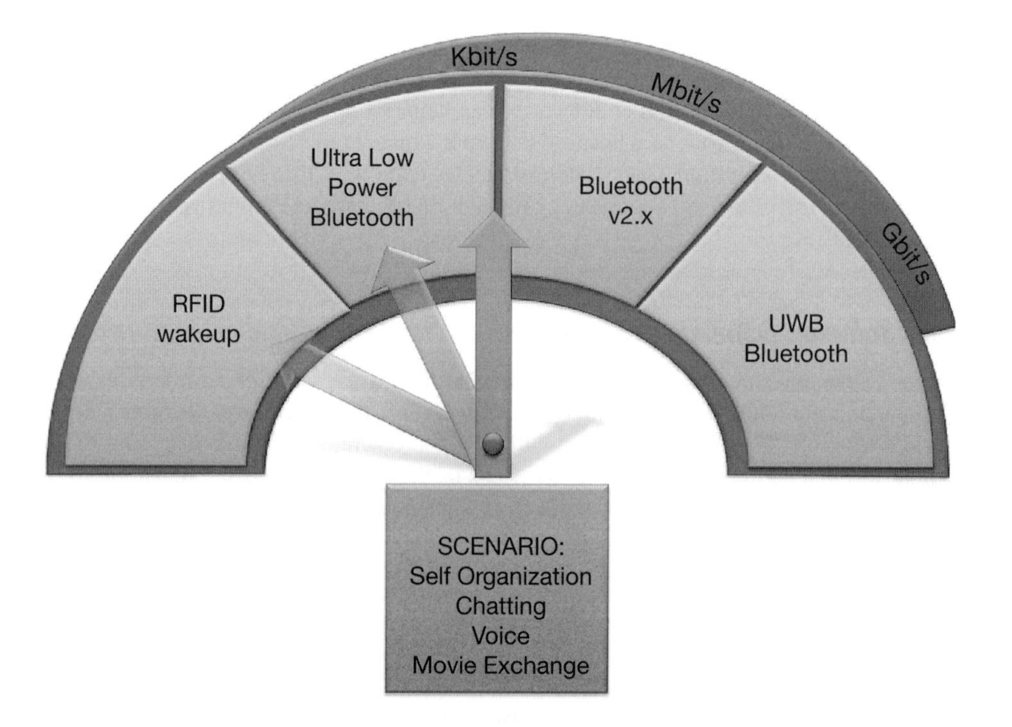

Figure 2.12 Composition of several Bluetooth technologies.

wide band (UWB) Bluetooth supporting very high data rates of up to 400 Mbit/s over short distances and the ultra low power (ULP) Bluetooth (formerly known as WiBree) with low data rates, but extremely low energy consumption. The ULP Bluetooth will be an integral part of the standard Bluetooth chip. This will help to penetrate the market very quickly. The UWB Bluetooth on the other hand will be realized by an additional chipset in the beginning. From a mobile device's point of view some of the top ten requirements are fulfilled already.

The ULP Bluetooth can be used to do the initial detection of neighboring devices or wireless sensors. In case of short data exchange or signaling conventional Bluetooth technology can be used. If larger data files need to be exchanged among the devices UWB Bluetooth can be then used. This *troika* can be extended by a simple RFID, UHF, or NFC entity to carry out rudimentary wake ups.

As given in Figure 2.12, this would end up in four short-range technologies appropriate to the needs and scenarios. These technologies are just like gears in an engine, to change the operating point of the system to the best possible solution.

Acknowledgements

We would like to thank Nokia for providing technical support as well as mobile phones to carry out the measurement campaign. Special thanks to Mika Kuulusa, Gerard Bosch, Harri Pennanen, Nina Tammelin and Per Moeller from Nokia. This work was partially financed by the X3MP project granted by Danish Ministry of Science, Technology and Innovation. Furthermore we thank all students and colleagues at the mobile device group and AAU who have participated in the ongoing projects, namely J. Rasmussen, P. Østergaard, J. Jensen, A. Grauballe, A. Sapuppo, G. Perrucci and B. Krøyer. We also thank VTT for partially funding our research activities within the framework of the project CHOCOLATE.

References

1. IRDA, http://www.irda.org/.
2. Bluetooth/WiBRee/ULP, http://www.bluetooth.org.
3. WLAN IEEE 802.11, http://www.ieee802.org/11/.
4. ZigBee, http://www.ieee802.org/15/pub/TG4.html.
5. Near Field Communications, http://www.nfc-forum.org.
6. aka-aki, http://www.aka-aki.com.
7. Facebook, http://www.facebook.com/.
8. SPIDER application, Aalborg University (2008) http://mobiledevices.kom.aau.dk/projects/student_projects/spring_2007/social_network/.
9. Pedersen, M. and Fitzek, F.H.P. (2007) *Mobile Phone Programming – SMARTEX: The SmartME Application*, Springer, pp. 271–4, ISBN 978-1-4020-5968-1 11.
10. Stini, M., Mauve, M. and Fitzek, F.H.P. (2006) Digital ownership: From content consumers to owners and traders. *IEEE Multimedia-IEEE Computer Society*, **13**(5), 4–6.
11. Grauballe, A., Perrucci, G.P. and Fitzek, F.H.P. (2008) *Introducing Contextual Information to Mobile Phones by External and Embedded Sensors*, in International Workshop on Mobile Device and Urban Sensing – MODUS08. St. Louis, Missouri, USA, http://mobiledevices.kom.aau.dk/opensensor.
12. Madsen Tatiana, T., Grauballe Anders, A., Jensen, M.G. *et al.* (2008) *Reliable Cooperative Information Storage in Wireless Sensor Networks*, in The 15th International Conference on Telecommunications – IEEE ICT.
13. Grauballe, A., Jensen, M., Paramanathan, A. *et al.* (2008) *Implementation of Cooperative Information Storage on Distributed Sensor Boards*, in IEEE International Conference on Communications (ICC 2008) - CoCoNet Workshop.

14. Pedersen, M.V., Fitzek, F.H.P. and Larsen, T. (2008) *Implementation and Performance Evaluation of Network Coding for Cooperative Mobile Devices*, in IEEE International Conference on Communications (ICC 2008) – CoCoNet Workshop.
15. Rasmusse, J., Østergaard, P., Jensen, J. *et al.* (2007) Chapter, in *Mobile Phone Programming – Parking Assistant Application*, Springer, http://mobiledevices.kom.aau.dk/projects/parksensor/.
16. Fitzek, F.H.P., Kyritsi, P. and Katz, M. (2006) *Cooperation in Wireless Networks – Power Consumption and Spectrum Usage Paradigms in Cooperative Wireless Networks*, Chapter 11, Springer, pp. 365–86, ISBN 1-4020-4710-x.
17. Fitzek, F.H.P. and Katz, M. (2007) *Cognitive Wireless Networks – Cellular Controlled Peer to Peer Communication*, Springer, pp. 31–59, ISBN 978-1-4020-5978-0 2.
18. Fitzek, F.H.P. and Katz, M. (eds) (2006) *Cooperation in Wireless Networks: Principles and Applications – Real Egoistic Behavior is to Cooperate!*, Springer, ISBN 1-4020-4710-X.
19. DMB, http://www.worlddab.org/.
20. DVBH, http://www.dvb-h.org/.
21. Fitzek, F.H.P. and Reichert, F. (2007) *Mobile Phone Programming and its Application to Wireless Networking*, Springer.
22. Perrucci, G.P., Fitzek, F.H.P. and Petersen, M.V. (2008) *Heterogeneous Wireless Access Networks: Architectures and Protocols – Energy Saving Aspects for Mobile Device Exploiting Heterogeneous Wireless Networks*, Springer.
23. Pedersen, M.V., Fitzek, F.H.P. and Larsen, T. (2008) *Implementation and Performance Evaluation of Network Coding for Cooperative Mobile Devices*, in IEEE International Conference on Communications (ICC 2008) – CoCoNet Workshop.
24. Petersen, M.V., Perrucci, G.P. and Fitzek, F.H.P. (2008) *Energy and Link Measurements for Mobile Phones Using ieee802.11b/g*, in The 4th International Workshop on Wireless Network Measurements (WiNMEE 2008) - in conjunction with WiOpt 2008, Berlin, Germany.

Part II

UWB Communications: State of the Art, Challenges and Visions

3

UWB Propagation Channels

Andreas F. Molisch
Mitsubishi Electric Research Labs, Cambridge, MA, USA, and also at The Department of Electrical and Information Technology, Lund University, Lund, Sweden

3.1 Introduction

The evaluation of the performance of UWB systems, including interference to and from legacy systems, requires knowledge of the propagation channels the systems are operating in. This chapter first describes techniques for measuring such channels and how to process the measured data. Deterministic and stochastic channel modeling techniques are then summarized. The main work in the future will have to lie in additional measurement campaigns, the development of high-resolution channel parameter extraction techniques, and a refinement of statistical channel models. More detailed reviews and extensive references can be found in [1–3].

3.2 State of the Art

3.2.1 Measurement Techniques

Any reasonable channel model has to be based on – or verified by – measurement results. Ideally, the measurement setup and parameter extraction technique should not have any impact on the results and the final model. For UWB channel impulse responses, there are essentially three techniques: impulse generators combined with sampling oscilloscopes, correlative channel sounders and network analyzers.

3.2.1.1 Pulse Generators

In this technique, the channel is excited by a short pulse (which can be seen as an approximation to a delta function), and the received signal is recorded, for example, by a sampling oscilloscope. Absolute delays can be measured if trigger signals, for example via a cable connection from TX to RX, are used. A major difficulty in this approach is the generation of

Short-Range Wireless Communications Rolf Kraemer and Marcos D. Katz

very short pulses. In order not to have noticeable noise enhancement in the analysis, the power spectrum of the pulse should be fairly constant in the frequency range of interest. Thus, the method seems suitable for low-frequency measurements [4, 5]. The technique also requires the generation of high-amplitude pulses, which is an added difficulty.

3.2.1.2 Correlative Channel Sounders

Correlative channel sounders emit a signal $s(t)$ that has desirable properties with respect to peak-to-average ratio. The receiver then correlates this with a signal $g(t)$. By choosing transmit signal and receive correlator so that $s(t) \otimes g(t) \approx \delta(t)$, the correlative sounder becomes equivalent to an excitation with a short pulse, assuming that the channel stays constant for the duration of $s(t)$. Typically, $s(t)$ is a pseudonoise sequence. The correlative sounder reduces the peak-to-average-power (compared with a pulse generator). The delay resolution of such a system is determined by the chip duration, that is the time variability of the signal $s(t)$. Generating a sequence of very short-duration chips can be even more difficult than generating one short pulse.

3.2.1.3 Network Analyzers

By far the most popular method for UWB channel sounding is based on vector network analyzers. Those devices perform the measurement in the frequency domain, by exciting the channel (and measuring at the receiver) with a slowly sweeping (or stepping) sinusoidal waveform. The major disadvantage is that each measurement sweep takes a long time (of the order of several seconds), and the channel has to stay completely static during that time. When modeling the impulse response, we need to either extract the parameters of the multipath components (MPCs) in the frequency domain, or to transform the signal back into the delay domain. The windowing for this back transformation can have a considerable impact on the results – it leads to additional delay dispersion, and can introduce artificial correlation between adjacent delay bins.

3.2.2 Deterministic Channel Description

3.2.2.1 Canonical Representation of the Impulse Response

In general, there are two different types of channel modeling approaches: deterministic and stochastic. Deterministic, or site-specific modeling, encompasses electromagnetic field computations that solve Maxwell's equations (or some approximation thereof) for specific boundary conditions (environments). Most deterministic channel modeling methods are based on the 'high-frequency approximation' of the wave equation, using ray launching or ray tracing. In ray launching, 'rays', representing plane waves, are emitted. Interaction with surrounding objects is taken into account by changing the direction and strength of the rays. At the receiver, the strength, delay and direction of the incoming rays is recorded, and gives the impulse response. Existing ray tracing programs assume that the impulse response can be written in the well-known form

$$h(\tau) = \sum_{i=1}^{N} a_i \delta(\tau - \tau_i) \tag{3.1}$$

where N is the number of multipath components, and the a_i and τ_i are the complex amplitudes and delays of the multipath components. In UWB channels, on the other hand, each ray undergoes dispersion (due, e.g., to the frequency-dependence of reflection- and diffraction coefficients). The correct channel description is then given by

$$h(\tau) = \sum_{i=1}^{N} a_i \chi_i(\tau) \otimes \delta(\tau - \tau_i) \tag{3.2}$$

where $\chi_i(\tau)$ denotes the distortion of a single pulse by the frequency selectivity of the interaction. As a consequence, the propagation characteristics at different frequencies are also different, and the WSSUS assumption breaks down.

3.2.2.2 Measurement Evaluations

For the analysis of pulsebased measurements, the CLEAN algorithm can be used [5]. This algorithm is essentially a serial interference cancellation: the parameters of the strongest multipath component (MPC, pulse echo) are determined, and subtracted from the received signal. Then, we search for the next strongest MPC, subtract its contribution, and so on. In order for this algorithm to work, the 'basic' pulse shape of a received echo (including distortions by the antennas) must be known; this can be obtained from back-to-back calibration, see below. An alternative technique is the SAGE algorithm, which performs maximum-likelihood estimation of the parameters of the MPCs in iterative form. SAGE has been used extensively in the analysis of narrowband array measurements; however, its common form is based on the assumption that the impulse response can be written in the shape specified by Equation (3.1).

3.2.3 Statistical Channel Modeling

We turn next to stochastic channel modeling. In many cases, it is not possible, or even desirable, to model the impulse response in a specific environment. When designing a new system, it is required that this new system works in the majority of all envisioned locations, not just in a single, specific place. For this reason, stochastic models are preferable for system design and testing. Those models correctly reproduce the *statistics* of the impulse response.

3.2.3.1 Pathloss

In a UWB channel, we define a *frequency dependent pathloss*

$$PL(f) = E\left\{\int_{f-\Delta f/2}^{f+\Delta f/2} \left|H(\tilde{f})\right|^2 \mathrm{d}\tilde{f}\right\} \tag{3.3}$$

where the expectation $E\{\}$ is taken over an area that is large enough to allow averaging out of the shadowing as well as the small-scale fading $E\{\} = E_{lsf}\{E_{ssf}\}\}$. Δf is chosen small enough so that diffraction coefficients, dielectric constants, and so on, can be considered constant within that bandwidth. The frequency dependence of the pathloss is usually modeled as [6]

$$\sqrt{PL(f)} \propto f^{-m}. \tag{3.4}$$

The pathloss also depends on the distance between TX and RX; it is common to assume that the frequency dependence and distance dependence are independent of each other. Shadowing is treated similar to narrowband systems.

3.2.4 Delay Dispersion, Angular Dispersion and Small-scale Fading

3.2.4.1 Arrival Statistics of Multipath Components

In UWB systems, the impulse response of the channel can become discontinuous, in other words, there are 'empty' delay bins (bins containing no energy) between the arriving multipath components. The Saleh–Valenzuela model is one approach to describing this effect [7], by using the following model for the impulse response

$$h(t) = \sum_{l=0}^{L} \sum_{k=0}^{K} a_{k,l} \delta(t - T_l - \tau_{k,l}), \tag{3.5}$$

where $a_{k,l}$ is the tap weight of the kth component in the lth cluster, T_l is the delay of the lth cluster, $\tau_{k,l}$ is the delay of the kth MPC relative to the lth cluster arrival time T_l. By definition, we have $\tau_{0,l} = 0$. The distributions of the cluster arrival times and the ray arrival times are given by a Poisson processes

$$\begin{aligned} p(T_l | T_{l-1}) &= \Lambda \exp[-\Lambda(T_l - T_{l-1})], \quad l > 0 \\ p(\tau_{k,l} | \tau_{(k-1),l}) &= \lambda \exp[-\lambda(\tau_{k,l} - \tau_{(k-1),l})], \quad k > 0 \end{aligned} \tag{3.6}$$

where Λ is the cluster arrival rate, and λ is the ray arrival rate.

Generalizations of the model include a mixed Poisson process for the ray arrival rate, and treating the number of clusters K as a random variable.

In addition to the deterministic components, a Rayleigh distributed 'clutter' describes the contributions that stem from diffuse scattering and other propagation paths that are not covered by the ray tracing. Kunisch and Pamp [6] suggest that each deterministic ray gives rise not only to a single MPC, but rather is associated with clusters whose power distributions are drawn from an exponential distribution.

3.2.4.2 Amplitude Statistics

We next turn to the variations of the MPC amplitudes, caused by the superposition of unresolvable components. In conventional wireless systems, it is well established that the amplitudes in delay bins $[(i - 0.5)\Delta, \ (i + 0.5)\Delta]$ where Δ is the delay bin width, are Rayleigh fading for $i > 0$, and possibly Rician fading for the LOS component $i = 0$. This is due to the superposition of many multipath components within each delay bin. In UWB systems, the number of components falling within each delay bin is much smaller, which leads to a change in the statistics. Amplitude distributions that have been suggested include the Nakagami distribution, Rice distribution, lognormal distribution, POCA and NAZU distributions, and Weibull distribution. Of these, the Nakagami and the lognormal distributions are the most popular.

3.2.4.3 Angular Dispersion

The different multipath components arrive at the receiver not only with different delays, but also with different angles. This fact is of importance for systems with multiple antennas, as well as for analyzing the impact of nonisotropic antenna patterns. The angular power spectrum, that is the power (averaged over the small-scale fading) coming from a certain direction, is often modeled as a Laplacian function, following the approach in narrowband systems. The few existing measurements indicate that the angular dispersion strongly depends on the measured delay.

3.2.5 Standardization Activities

Standard models for the mobile radio channel are important tools for the development of new radio systems. They allow a fair comparison of different system approaches, as well as an assessment of benefits for different transceiver and multiple-access structures without excessive effort. Such standardized models should be detailed enough to reflect all relevant properties of the channel but should also be simple enough to allow rapid implementation and fast simulation times. The trade-off between simplicity and accuracy in the modeling is a difficult one.

The first UWB standard channel model was developed for the IEEE 802.15.3a group. Establishment of the channel model started in May 2002, and finished in November of the same year. That short time was related to the timescale of the physical-layer standardization process; it also implied that the number of measurements on which this model could be based was fairly small. The goal was to establish a channel model for a range of 0–10 m, as this is the range of interest for high data-rate communications. The model distinguishes between four radio environments: LOS with a distance between TX and RX of 0–4 m (CM1), NLOS for a distance 0–4 m (CM2), NLOS for a distance 4–10 m (CM3), and a ‘heavy multipath’ environment (CM4). It is a straightforward Saleh-Valenzuela model, with lognormal fading of the MPCs, as well as of the clusters.

In 2003, the channel modeling group of another standard, the IEEE 802.15.4a group, was established. It had more than a year to complete its work, and came up with a model that showed a number of improvements:

- models for a larger number of environments;
- frequency dependence of the pathloss and thus implicitly the distortions of each separate MultiPath Components (MPCs);
- modelling of the number of clusters of MPCs in the Saleh–Valenzuela (S–V) model as a random variable;
- a power delay profile that models a ‘soft’ onset, so that the first arriving paths can be considerably weaker than later MPCs; this is critical for accurate assessment of ranging capabilities of UWB;
- a new model for body-area networks that includes correlated lognormal shadowing.

There is often the misconception that the 802.15.3a model should be used for the simulation of systems with high data rate, while the 802.15.4a model is suitable for systems with low data rate. This is wrong. The applicability of a channel model does not depend on the data rate.

Rather, the 802.15.4a model is a more refined description, and based on more measurements, and should thus preferentially be used for any system simulations.

3.3 Challenges

3.3.1 Measurement Techniques

In UWB measurements, the influence of the antenna on the frequency response of the system can be dominant. The simplest form of calibration (for SISO measurements) is to record the output signal with an antenna that is placed in close vicinity, and with a LOS connection, to the transmit antenna. The output from this antenna then provides the signal waveform that would be observed in a pure LOS situation. Alternatively, the conventional back-to-back calibration together with a measurement of the frequency dependent antenna patterns could be used. However, a true calibration requires that the received signal is calibrated, depending on direction and frequency. The situation is made more challenging by the fact that UWB antennas can have different antenna patterns at different frequencies. Thus, a complete calibration requires the knowledge of the frequency-and-direction dependent transfer function $H(f, \Omega_{TX}, \Omega_{RX})$, which should be combined with a directionally resolved measurement of all the MPCs.

When designing a measurement campaign, it is also important to plan the measurement points in such a way that the extraction of model parameters is facilitated. For example, almost all channel models distinguish between 'small-scale' and 'large-scale' fading. The statistics of those two types of fading can only be separated if multiple measurements are done in an area where the large-scale fading can be considered constant. From those multiple measurement points both the small-scale statistics, and the small-scale average can be determined. The distribution of small-scale averages over large areas then gives the large-scale fading statistics. Note that there are several measurement campaigns that did not use such a procedure; their results thus cannot be used to derive a channel model that distinguishes between small-scale and large-scale fading.

In an environment with significant propagation in three-dimensional space, the elevation angle of the MPCs has a strong impact on the MIMO channel matrix, even if the MIMO arrays are horizontal ULAs. The attenuation of an MPC depends on the elevation angle via the antenna elevation pattern. In addition, the elevation pattern is strongly frequency dependent. Thus, all measurements that aim to extract a two-directional channel description should – in principle – use planar arrays for the measurements. This implies, however, a much higher effort for the actual measurements.

3.3.2 Deterministic Channel Description

The generic description methods for UWB channels are by now well established. The key problem lies in the identification of the distortion functions $\chi(\tau)$. Electromagnetic computations that are not based on the high-frequency assumption seem to be a promising way to establish those values. Measurements of the frequency-dependent reflection coefficients and measurement or modeling of diffraction and scattering coefficients will be required as a basis of (frequency-dependent) ray tracing. The problems here do not lie so much in the fundamental formulations, but rather 'crunching the numbers' and performing a number of time-consuming measurements.

For the measurement evaluation, we see two main challenges:

1. Generalization of the SAGE algorithm to the UWB case, including the possibility of spherical (and not planar) waves. As mentioned previously, the SAGE algorithm assumes the validity of Equation (3.1). Attempts have been made to use it on subbands of the UWB spectrum, and then combine the results of the MPC parameters in the different subbands. However, this is not a completely general approach, as combining the information in the subbands suffers from certain arbitrariness.
2. All of the high-resolution algorithms such as CLEAN and SAGE assume that the pulse shape of each single MPC is known, and only the delay, directions and amplitudes are unknown. However, this is not fulfilled in UWB – the function $\chi(\tau)$ is unknown. Misestimation of this pulse shape can lead to significant errors of all the involved parameters. Both CLEAN and SAGE utilize a form of interference cancellation; that is the contribution of an identified MPC is subtracted from the total signal, before the next MPCs are estimated. If the pulse shape is misestimated, the subtracted contribution is not the true one, giving rise to a wrong 'residual' signal. Investigations of this effect on the evaluation of UWB measurements seem required.

Another important problem is the estimation of possible diffuse background radiation, that is signal energy that cannot be explained by the discrete-components model (3.2).

3.3.3 Stochastic Channel Modeling

For stochastic channel modeling, there are enormous challenges. The key problem arises from the scarcity of measurements that are currently available. While good generic stochastic models have been established, they are parameterized only in a few selected environments (indoor residential, indoor office, industrial), and even then the parameterization is often based only on a few (or even just a single) measurement campaign. Thus, a lot of new measurements and evaluations will be required in the future. Other problems that remain are the following:

1. Are distance- and frequency dependence of the pathloss really separable? Up to now, this assumption has always been made, but it is only based on convenience, not on measurements. In other words, is the pathloss exponent independent of the frequency?
2. Is the shadowing variance independent of frequency?
3. More measurements are needed so that the Erceg–Ghassemzadeh model for the pathloss (treating the pathloss exponent as random variable, not as constant) can be fully parameterized in different types of environments.
4. Models for the angular dispersion. It can be anticipated that the angular dispersion will show a dependence on the frequency band, as well as the delay range (long delays versus small delays).

Another area of stochastic channel modeling is the description of the temporal evolution of the channel. Traditionally, this has been described by the Doppler spectrum. However, the Doppler spectrum is a concept related to the WSSUS model of Bello. But UWB channels are not WSSUS, so that completely new description methods might be required. It seems also clear that different models are required for different sources of time variance (i.e. moving transceiver, or moving, shadowing, objects).

3.3.4 Impact of Channels on Geolocation

The propagation channels also have an important impact on the ranging and geolocation capabilities of UWB devices. The channels and channel models for ranging are, in principle, the same as for communications. However, different effects are important for ranging. Most significant is the identification of the first multipath component in a measured impulse response or power delay profile. There are three major effects of the PDP on the identifiability of the first MPC:

1. The shape of the power delay profile: It is a common assumption of narrowband channel models that the power delay profile (averaged over the small-scale fading) shows a sharp onset, followed by an exponential decay. Such a PDP allows for easy identification of the first MPC. However, recent measurements in office and industrial environments have a different PDP, namely a soft onset (PDP increasing with delay) until a maximum value is reached (typically several tens of nanoseconds after the first MPC, followed by a further decay. Identification of the first MPC is more difficult.
2. The fading statistics of the first components: the larger the fading variability, the larger the probability that the first MPC is in a fading dip, and cannot be identified.
3. The interpath and inter-ray arrival statistics: The farther the clusters and MPCs are separated in the delay domain, the more difficult it is to distinguish MPCs from random noise spikes.

3.4 Vision

The vision is to have a complete stochastic channel model, based on extensive measurements, for all environments of interest for UWB communications. Furthermore we want to have efficient deterministic channel prediction methods.

A large number of measurements will be required in the future in order to gain a better understanding, and statistically viable models, of UWB propagation channels. As a rough estimate, we anticipate that five person-years of work are required for a model of each environment, if such a model is to include directional information as well as models for the temporal variability. A total of ten environments are anticipated to be relevant for future UWB applications. It is obvious that the effort of such extensive measurements cannot be borne by a single industrial or academic entity. It would therefore be advisable to have a coordinated effort by various interested parties. With such a concerted effort, it would be realistic to have a highly reliable model in about 4–5 years time. Faster completion is not anticipated, since much of the effort cannot be parallelized.

For faster measurements, sophisticated time-domain measurement equipment (pulse generators and sampling scopes) with bandwidths of >7 GHz will be required. However, the current state of the art seems to be limited to about 5 GHz. Thus, further development will be required.

References

1. Molisch, A.F. (2005) Ultrawideband propagation channels – theory, measurement and models. *IEEE Transactions on Vehicular Technology*, invited paper, **54** (5), 1528–45.
2. Molisch, A.F. (2007) *Ultrawideband Propagation Channels and Their Impact on System Design*, in Proceeding IEEE International Symposium on Microwave, Antenna, Propagation and EMC Technologies for Wireless Communications (MAPE), pp. K4-1–K4-5.

3. Molisch, A.F. (2008) *Ultrawideband Propagation Channels*, IEEE Proceeding, Special Issue on UWB.
4. Cassioli, D., Win, M.Z. and Molisch, A.F. (2002) The ultrawide bandwidth indoor channel: from statistical model to simulations. *IEEE Journal on Selected Areas in Communications*, **20** (6), 1247–57.
5. Cramer, J.M., Scholtz, R.A. and Win, M.Z. (2002) Evaluation of an ultrawideband propagation channel. *IEEE Transactions on Antennas and Propagation*, **50** (5), 561–70.
6. Kunisch, J. and Pamp, J. (2002) *Measurement Results and Modeling Aspects for the UWB Radio Channel*, in Proceeding IEEE Conference on Ultra Wideband Systems and Technologies, pp. 19–23.
7. Saleh, A. and Valenzuela, R.A. (1987) A statistical model for indoor multipath propagation. *IEEE Journal on Selected Areas in Communications*, **5** (2), 128–37.

4

Pulse Shaping and Diversity

Lorenzo Mucchi
Department of Electronics and Telecommunications, University of Florence, Italy

4.1 Pulse Shaping

Unlike conventional wireless communications systems that are carrier-based, UWB systems transmit information using extremely short pulses that spread the energy from near DC to a few gigahertz without using a frequency carrier. This carrier-less technique will greatly reduce the complexity and cost of the transceiver. In an UWB system, the choice of the impulse (pulse shape) will strongly affect the design of the filters, the choice of receiver bandwidth, the bit error rate performance in general, and the performance in multipath propagation environments. Impulse radio signals in UWB systems utilize the extremely broad bandwidth for transmission and coexist with other radio applications in the same frequency spectrum. It is desirable for UWB signals to spread the energy as widely in frequency as possible to minimize the power spectral density and hence the potential for interference to other user systems.

Impulse radio is based upon transmitting and receiving very short pulses. Pulse generation and pulse shaping are among the most fundamental problems in UWB systems. Fundamentally, three types of UWB pulses have often appeared in the recent UWB research literature [1, 2], that is Gaussian pulse, monocycle pulse and doublet pulse. These pulses are usually employed to conduct basic theoretic analysis and simulation study. Mainly the pulses are used to:

- Efficient Spectral Utilization – Operating below the noise floor, UWB radios must emit at low power. But as any other communication system, the performance of a UWB system heavily relies on the received signal-to-noise-ratio (SNR), which is proportional to the transmit power. Maximization of the latter, however, can be achieved only if the spectral shape of the FCC mask is exploited in a power-efficient manner.
- Flexible Interference Avoidance – To avoid interference to (and from) coexisting narrowband systems, their corresponding frequency bands must be avoided. Since the nature and number

Short-Range Wireless Communications Rolf Kraemer and Marcos D. Katz

of coexisting services may change depending on the band used, the avoidance mechanism should also allow for sufficient flexibility.
- Multiple Orthogonal Pulses – Traditionally, UWB multiple access is achieved by employing time hopping (TH) codes [3, 4]. User capacity of UWB radios can be further improved by partitioning the ultrawide bandwidth into sub-bands, allowing users to hop among these sub-bands according to user-specific hopping patterns. Since hopping takes place over sub-bands centered around different frequencies, similar to narrow-band systems, frequency hopping (FH) can also enhance system capacity and reinforce the low probability of interception/detection (LPI/LPD) of UWB radios.

Unfortunately, the widely adopted Gaussian monocycle is not flexible enough to meet these challenges [5]. To design pulse shapers with desirable spectral properties, two approaches can be employed: Carrier-modulation and/or baseband analog/digital filtering of the baseband pulse shaper. The former relies on local oscillators at the UWB transmitter and receiver, which being prone to mismatch give rise to carrier frequency offset/jitter. In multi-band UWB systems with FH, multiple carrier frequency offset/jitter's emerge with this approach. Although passing the (Gaussian) pulse through a baseband analog filter can reshape the pulse without introducing carrier frequency offset/jitter, it is well known that analog filters are not as flexible as digital filters, which are accurate, highly linear and perfectly repeatable [6].

The information bits are embedded in the transmitted pulse sequence. Many UWB modulation types, such as time hopping pulse position modulation, time hopping binary phase and amplitude modulation, direct sequence spread spectrum and orthogonal frequency division modulation, are commonly used in the UWB communication systems.

Due to its wide bandwidth, UWB can convey greater data throughput than conventional narrow band signals. In the IEEE802.15.3a standard, which adopts the UWB techniques, the data rate can achieve up to 480 Mbps. With other merits such as flexible implementation, base band processing (for impulse radio only), and relatively simple hardware, UWB has been shown to be promising for short-range wireless communications, such as wireless video playing, home entertainment, and so on. Since UWB signal bandwidth is ultra wide, how to control its interference to existing narrow-band systems has drawn increasing research interest.

Pulse shaping filters [7] are a useful means to control the signal spectrum and to avoid interference of UWB to other legacy narrow band signals. In [8], pulse shaping is performed in frequency domain to produce a carrier interferometer waveform such that the receiver can fully exploit frequency diversity by decomposition and recombination of the received signal. In [9], a carrier interferometer chip shaping technique is developed in order that UWB DS-CDMA can operate over nonadjacent frequency bands.

Paper [10] designs a transmission pulse shaping filter using the Chebyshev approximation, and also addresses several other issues such as single band UWB, multi-band UWB, clock and frequency effects, and narrow band interference. This paper introduces an optimal finite impulse response (FIR) filter design approach, which can be easily adapted to devise either transmitting or receiving pulse shaping filters for the UWB systems. The pulse-shaping filter is treated as an individual filter design problem. The advantage of this kind of method is that the filter design is isolated from other system factors (e.g., multiple access interference), and so the design problem is relatively easy to analyze. However, since the filter optimization is either at the transmitter or at the receiver side, this method also has its own disadvantage that it may not be optimal from a systematic perspective.

4.1.1 Alternative Methods

One approach to the design of digital pulse shapers that comply with the FCC spectral masks is to employ prolate spheroidal wave functions to generate pulses from the dominant eigenvectors of a channel matrix that is constructed by sampling the spectral mask. Pulses generated from different eigenvectors are mutually orthogonal, but require a high sampling rate that could lead to implementation difficulties. Other pulse shaping methods include exploiting the properties of Hermite orthogonal polynomials, and fine-tuning higher-order derivatives of the Gaussian pulse, the latter of which is not flexible in fitting FCC spectral mask changes as well as other regional regulations. All these pulses do not achieve optimal spectral utilization. For flexible pulse shaping and convenient use of off-the-shelf hardware components, digital FIR filter design solutions may be more appropriate, as indicated in [11], where some algebraic transformations facilitated the formulation of various single pulse design problems as convex optimization problems, from which globally optimal solutions can be efficiently obtained.

4.1.2 Challenge/Vision

The best approach in this case is probably to jointly design the transmitter and the receiver pulse shaping filters.

In UWB system, the sampling rate and symbol rate are usually ultra high. If the pulse shaping filter is implemented in digital signal processors, for example with FIR digital filter, it would require extremely high speed analog-to-digital converter (ADC) and extremely powerful digital signal processors. All of these will result in expensive and complex hardware.

The key advantage of IIR filters over FIR filters is that they can usually meet the same specifications with a lower filter order, and therefore have lower computational complexity and are cheaper to implement than FIR filters. Furthermore, the IIR filter can be implemented not only in digital forms, but also in analog forms. Analog IIR filters are inexpensive and don't require digital computation power at all. These are the essential merits of IIR filters, which are very attractive to UWB pulse shaping filter designers.

However, IIR filters also have their own disadvantages, that is IIR filters have nonlinear phases, they are more sensitive to coefficient quantization, they may have limit cycle problems, and so on. Due to the nature of infinite response of IIR filters, the signal energy tends to spread out in the time domain when it passes through an IIR filter. All of these may result in SNR degradation of correlation receivers with a time-limited template.

Pulse shapers respecting the FCC spectral mask were proposed recently in [12, 13]. Targeting multiple orthogonal pulses that are FCC mask compliant, the resulting pulses in digital form correspond to the dominant eigenvectors of a matrix, which is constructed by sampling the FCC mask. In [14] a method was developed to shape the transmitted pulse, based on combination of a set of base waveforms. A Gaussian pulse, which is the most commonly adopted waveform for UWB, is assumed as the initial waveform. Two different ways of modifying the waveform are then analyzed: pulse width variation and differentiation.

Recently, in [15] an efficient and exact method that permits accurate and tractable computation of UWB system BERs was provided. A set of Gaussian monocycles was the original proposal for UWB communications systems and has been widely adopted in the investigation of UWB applications. An orthogonal pulse set based on modified Hermite polynomials (HP) was introduced in [16]. More recently, a novel pulse design algorithm utilizing the concepts of prolate spheroidal (PS) functions was proposed in [17]. Compared with the Gaussian monocycle and the PS pulse, the HP pulse needs frequency shifts and bandpass filters to meet FCC masks, thus requiring much more complicated implementation. Moreover, the multiple access performance of a single HP pulse is worse than the performance of the other two pulses when all pulses are constrained to meet the FCC spectral emission requirements.

Finally, in [18] the design of the pulse shaper is proposed to not only offer optimality in *meeting* the FCC mask, but also *optimally exploit* the allowable bandwidth and power. Moreover, it can be implemented without requiring expensive ADC circuitry and without modifying the analog components of existing UWB transmitters, whereas converting the digital designs in [19] into analog form entails digital-to-analog (D/A) operations at 64 GHz rate.

ADC speed is increasing by a factor of 10 every 5 years. Digital FIR pulse shaping design will have the possibility to exploit properly sampled signals compared with current practice. Will it be possible to design a digital shaping filter that implies low complex hardware low power consumption and assures a non-rippled limited-in-time template at the same time?

Pulse shaping filters should be able essentially to assure a narrow pulse template without oscillations that fits the power limit per frequency, that does not need a bandpass filter, that is able to mitigate multiple access interference (orthogonal basis design), that eases the acquisition process and that allows the best energy recovery at the receiver, at the same time.

Moreover, the quantization error effect could still be a problem [20], and further research is needed in this area.

4.2 Diversity

In spite of the numerous advantages of UWB, the transmitted power, at most −2.6 dBm, but likely several dB less, will thus tend to limit applications to relatively short ranges or to moderate data rates. It is therefore crucial to develop solutions that make the best possible use of the radiated and received power, for the feasibility and future commercial success of UWB communications systems.

The challenge is that these ultrawide bandwidth systems are still subject to the major impairment and constraints which previous generations of mobile communication systems had to face. First, they have to operate over a wide variety of complex and harsh time-varying channels whose quality will quite often be insufficient to support the quality of service required by a specific telecommunication application. Due to this major impairment, efficient *fade mitigation* techniques are required to improve radio link performance. In addition, handheld and portable terminals have limitations on their power and size and must be cheap enough for users to afford, but sophisticated enough to communicate reliably with their base stations. This constraint puts a fundamental *complexity and power consumption limit* on these terminals which has to be kept in mind in the design of these systems. To realize systems which overcome the aforementioned impairment and constraints, many of these proposed technologies make use, in one form or another, of *diversity techniques*. These techniques improve the system performance by transmission and/or reception as well as appropriate combining of differently faded replicas of the same information-bearing signal. In general, the more available diversity paths, the larger the potential diversity benefit, and as a consequence a wide range of important emerging wireless communication systems tend to rely on physical layer solutions, operating in diversity rich environments, that is with a large number of diversity paths, as the UWB channel does.

On another front, in designing such systems to achieve a specified quality of service, one is also faced with the practical reality of satisfying a predetermined degree of complexity. Thus, instead of using optimum (from a performance standpoint) diversity combining schemes such as maximal-ratio combining (MRC) (for coherent receive diversity systems) which rely on the estimation and coherent combining of all available diversity paths, one often opts for less complex and/or power consuming combining schemes at a sacrifice in a certain degree of system performance.

Because they already possess rich diversity inherently, multiple transmit antennas do not provide diversity gain in the strict sense, that is the slope of bit error rate (BER) versus signal-to-noise ration (SNR), but can reduce the complexity of the RAKE receiver. As for the effect of receive diversity, multiple receive antennas can improve the performance of the UWB system by providing higher antenna array combining gain, even without providing the diversity gain in the strict sense.

Consequently, although fading may not be crucially serious in the pulsed mode UWB system, receive antenna diversity is suggested for the UWB system to improve energy capture [21, 22]. In the literature, not many papers have been reported to address the issue of employing transmit diversity for the pulsed UWB system, for example [23, 24]. The former, evaluates the performance of the pulse-amplitude modulation (PAM) signals in the UWB multiple-input multiple-output (MIMO) channel. The latter, proposes an STBC scheme for the PPM-based UWB system in the flat fading real channel, where the received pulses through the radio channel are assumed to be orthogonal with each other. In general, spatial diversity order increases with the number of transmitting and receiving antennas, although the greater the number of antennas involved, the smaller is the diversity gain, due to the presence of multipath.

Although the UWB channel is defined as dense multipath and the diversity exploited by recombining these paths partially or selectively could be seen to be enough for the reception, often this is not true. If the antenna shape is considered in the system model, the channel frequency response could be very different, depending on the polarization of the transmitted pulse waveform. This means that diversity is available and can be used (cross-polarization antenna) to assure the QoS of the link independently on the antenna shape and the orientation between transmitting and receiving antenna [25].

In [26] a good study on the tradeoff between the intrinsic diversity of UWB signals and the diversity gain due to coding is presented. Moreover, in the paper two new space-time codes for pulsed UWB are proposed in order to have full-rate full-diversity at the receiver.

Multiband UWB, that is broadband OFDM, is normally designed not to have multipath phenomena in each subcarrier, but it can easily make use of frequency diversity by allowing the same symbol to be sent by different subcarriers in different macrobands.

4.2.1 Challenge/Vision

Among the various forms of diversity techniques, most of them have been studied by assuming perfect channel knowledge. The design of diversity techniques is still able to yield advantages with practical channel estimation. Moreover, diversity, when talking about ultrawide bandwidth impulse radio signals, can be mainly exploited in the spatial, time, pulse and polarization domain. Basically, the goal for searching for diversity is increasing the system capacity, data rate, application range and the multiple standards coexistence. In other words, diversity is nowadays mainly seen as a way to optimize a specified system. In the future diversity could change this meaning. Diversity will go over just the PHY layer and will mean more a way to make a communication system more robust against unknown channel state information, network morphology and user distribution.

References

1. Win, M.Z. and Scholtz, R.A. (2000) Ultra-wide bandwidth time-hopping spread-spectrum impulse radio for wireless multiple-access communications. *IEEE Transactions on Communications*, **48** (4), 679–91.
2. Chen, X. and Kiaei, S. (2002) *Monocycles Shapes for Ultra Wideband System*, in Proceeding IEEE Conference on Ultra Wideband Systems and Technologies, pp. 597–600.
3. Scholtz, R.A. (1993) *Mutiple Access with Time Hopping Impulse Modulation*, in Proceeding IEEE Military Communications Conference (MILCOM), pp. 447–50.
4. Choi, J.D. and Stark, W.E. (2002) Performance of ultra-wideband communications with suboptimal receivers in multipath channels. *IEEE Journal on Selected Areas in Communications*, **20** (9), 1754–66.
5. Withington, P. (1998) *Impulse Radio Overview*, Time Domain Corporation, http://leitl.org/docs/uwb/pulse-radio-overview.pdf.
6. Oppenheim, A.V. *et al.* (1999) *Discrete-Time Signal Processing*, 2nd edn, Prentice Hall.
7. Wu, Z., Zhu, F. and Nassar, C.R. (2002) *High Performance Ultra-Wide Bandwidth Systems via Novel Pulse Shaping and Frequency Domain Processing*, in Proceeding IEEE Conference on Ultra Wide-band Systems and Technologies, pp. 53–8.
8. Wu, Z., Zhu, F. and Nassar, C.R. (2002) *Ultra Wideband Time Hopping Systems: Performance and Throughput Enhancement via Frequency Domain Processing*, in Proceeding IEEE Asilomar Conference on Signals, Systems and Computers, Vol. **1**, pp. 722–7.
9. Wu, Z., Nassar, C.R. and Shattil, S. (2001) *Ultra Wideband DS CDMA via Innovations in Chip Shaping*, in Proceeding IEEE Vehicular Technology Conference (VTC), Vol. **4**, pp. 2470–4.
10. Luo, X., Yang, L. and Giannakis, G.B. (2003) *Designing Optimal Pulse-Shapers for Ultra-Wideband Radios*, in Proceeding IEEE Conference on Ultra Wideband Systems and Technologies, pp. 349–53.
11. Wu, X., Tian, Z., Davidson, T.N. and Giannakis, G.B. (2006) Optimal waveform design for UWB radios. *IEEE Transactions on Signal Processing*, **54** (6), 2009–21.
12. Parr, B. *et al.* (2003) A novel ultra-wideband pulse design algorithm. *IEEE Communications Letter*, **7** (5), 219–21.
13. Parr, B., Cho, B. and Ding, Z. (2003) *A New UWB Pulse Generator for FCC Spectral Masks*, in Proceeding IEEE Vehicular Technology Conference (VTC), Vol. **3**, pp. 1664–6.
14. De Nardis, L., Giancola, G. and Di Benedetto, M.-G. (2004) *Power Limits Fulfilment and MUI Reduction Based on Pulse Shaping in UWB Networks*, in Proceeding IEEE International Conference on Communications (ICC), Vol. **6**, pp. 3576–80.
15. Hu, B. and Beaulieu, N.C. (2003) Exact bit error rate analysis of TH-PPM UWB systems in the presence of multiple access interference. *IEEE Communications Letter*, **7** (12), 572–4.
16. Michael, L.B., Ghavami, M. and Kohno, R. (2002) *Multiple Pulse Generator for Ultra-Wideband Communication Using Hermite Polynomial Based Orthogonal Pulses*, in Proceeding IEEE Conference on Ultra Wideband Systems and Technologies, pp. 47–51.
17. Parr, B., Cho, B., Wallace, K. and Ding, Z. (2003) A novel ultra-wideband pulse design algorithm. *IEEE Communications Letter*, **7** (5), 219–21.
18. Luo, X., Yang, L. and Giannakis, G.B. (2003) Designing optimal pulse-shapers for ultra-wideband radios. *Journal of Communications and Networks*, **5** (4), 344–53.
19. Parr, B. *et al.* (2003) A novel ultra-wideband pulse design algorithm. *IEEE Communications Letter*, **7** (5), 219–21.
20. Franz, S. and Mitra, U. (2006) *UWB Receiver Design for Low Resolution Quantization*, in Proceeding European Signal Processing Conference (EUSIPCO).

21. Sibille, A. and Bories, S. (2003) *Spatial Diversity for UWB Communications*, in Proceeding European Personal Mobile Communications Conference (EPMCC), pp. 367–70.
22. Tan, S.S., Kannan, B. and Nallanathan, A. (2003) *Ultra-Wideband Impulse Radio Systems with Temporal and Spatial Diversities*, in Proceeding IEEE Vehicular Technology Conference (VTC), Vol. **1**, pp. 607–11.
23. Weisenhorn, M. and Hirt, W. (2003) *Performance of Binary Antipodal Signaling over the Indoor UWB MIMO Channel*, in Proceeding IEEE International Conference on Communications (ICC), pp. 2872–8.
24. Yang, L. and Giannakis, G.B. (2004) Analog space-time coding for multiantenna ultra-wideband transmissions. *IEEE Transactions on Communications*, **52**, (3), 507–17.
25. Argenti, F., Bianchi, T., Mucchi, L. and Ronga, L.S. (2006) Ultra-wideband transmission with polarization diversity, in *UWB Communication Systems – A Comprehensive Overview* (eds M.-G. Di Benedetto *et al.*), Hindawi Publishing Corporation.
26. Mucchi, L. and Puggelli, F. (2008) *Proposal of Two New Space-Time Codes for Extending the Range of UWB Systems*, accepted for publication at ICUWB Conference.

5

Noncoherent Detection

Mike Wolf and Nuan Song

Communications Research Laboratory, Ilmenau University of Technology, PO Box 100565 D-98684 Ilmenau, Germany

Due to the broadband nature of UWB-signals, many distinct propagation paths are resolvable at the receiver. To efficiently capture the energy contained in the multipath arrivals, the receiver has to combine the multipath components which are spread over time. This problem grows with the signal bandwidth B, and the potential difficulties from the channel estimation and receiver-complexity point of view are by far not unique to impulse radio only. The same happens to UWB-transmission based on chirp-, on direct-sequence or on other spread-spectrum signals.

A coherent receiver, which is able to utilize the energy of dozens of multipath arrivals by means of a RAKE, cf. [1], will be very complex and costly. The hardware itself may consume a lot of power, and the time needed to estimate the corresponding amplitudes and phases wastes capacity.

Therefore, suboptimal noncoherent receivers have regained popularity. There are two basic possibilities to establish noncoherent detection:

1. envelope detection and
2. differential detection.

With both types of receivers, it is possible to capture the multipath energy easily, since in both cases multipath combining does not require any knowledge with respect to the phases or polarities. Unfortunately, this advantage can only be obtained at the expense of a higher sensitivity to inband interference, to noise, and to intersymbol interference (ISI). For this reason, such suboptimal detectors can not be treated as a replacement of coherent solutions at all, but they are interesting alternatives, especially for low-power systems operating at low or at medium data rates.

Short-Range Wireless Communications Rolf Kraemer and Marcos D. Katz

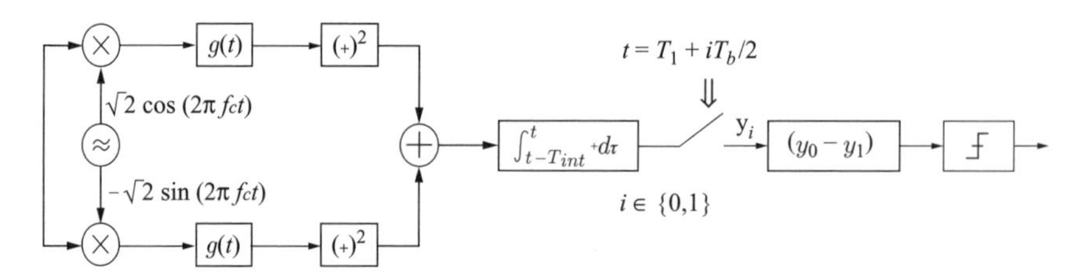

Figure 5.1 2-PPM envelope detector: $g(t)$ is the impulse response of a lowpass filter (matched to the baseband-pulse, if applicable). Instead of quadrature downconversion, the received signal could also be bandpass filtered and squared afterwards.

In the simplest case, path-diversity combining can be achieved by means of a single, analog integration device, as shown in Figure 5.1 for envelope detection. The integrator effectively provides a binary weighting of the multipath arrivals: all components inside the integration window of size T_{int} are weighted with '1', while all the others are weighted with '0'. Both, for envelope detection as well as for differential detection, the E_b/N_0-performance (where E_b is the bit energy and N_0 is the noise power density) can be improved, if a number of N_{int} subintervals, each of size T_b/N_{int}, is used instead of a single integration window (T_b: bit interval). Then, weighted combining according to the N_{int} different energies could be applied. Noncoherent detection with weighted combining is widely discussed in the literature [2–4].

Although envelope detection has also been suggested [5, 6] for UWB-communications, the majority of the UWB-literature about noncoherent detection deals with differential detection. It seems to be straightforward to combine differential detection with DPSK (binary Differential Phase-Shift Keying). This is the usual solution in fibre optics [7] and promises a 3 dB advantage with respect to the required E_b/N_0-ratio compared with envelope detection of OOK (On-Off Keying) or orthogonal 2-PPM (Pulse-Position Modulation). However, if the multipath combining takes place by means of an analog integrator, located before the AD-conversion (AD: Analog-to-Digital), the delay (or the delays, if a quadrature down-conversion is used) has to be realized in the analog domain. This is quite complicated, since for differential phase modulation formats the delay has to be equal to the bit/symbol interval. Since any spread spectrum signal is characterized by a large product T_bB with B being the bandwidth, the delay-bandwidth product is as large as T_bB. The realization of an all-pass filter having a constant group delay T_b within a bandwidth $>>1/T_b$ is quite challenging.

To admit shorter delays, TR (Transmitted Reference) approaches have been suggested. In this case, the auto-correlation of the current signal takes place with an unmodulated reference signal (impulse) introduced prior to the modulated symbol. Usually, TR-signalling is suggested for impulse-radio systems in conjunction with antipodal modulation (2-PSK). Many authors, for example [8, 9] propose combining TR-signalling with time- and delay-hopping to enable a special kind of CDMA. The TR-approach also has two disadvantages. First, the reference pulse wastes 3 dB of the total power, if every modulated pulse is assigned to its own reference pulse. Second, the autocorrelation process is disturbed, if the delay between the reference pulse and the modulated pulse is smaller than the excess delay of the channel.

The E_b/N_0-performance of noncoherent transmission *with single window* combining can be well estimated, even if no concrete realizations of the multipath channel are considered. Such a theoretical prediction provides very useful insights into basic characteristics of suboptimal receivers, such that general design rules can be derived. This is illustrated for orthogonal 2-PPM impulse radio transmission combined with envelope detection in Figure 5.1. However, if DPSK-differential detection with single window combining is considered instead, it can be shown that the statistic of the decision variable is identical to the statistic of the random variable $(y_0 - y_1)$ shown in Figure 5.1 – except for a 3 dB difference in favour of DPSK.

For an ideal AWGN-channel, binary orthogonal transmission exhibits a bit error rate [10]

$$p_b = \frac{1}{2}e^{-\frac{E_b}{2N_0}}, \tag{5.1}$$

if noncoherently detected. In this case, the receiver shown in Figure 5.1 would not require (and not contain) the integrator. The binary decision at the receiver is based on the two samples y_0 and y_1. If $y_0 \geq y_1$, the pulse was probably transmitted at the first position, if $y_0 < y_1$, the pulse was probably transmitted $T_b/2$ seconds later. Since the distributions of y_0 and y_1 are known, the derivation of Equation (5.1) is straightforward. Supposing that the pulse was transmitted at position #1, y_0 exhibits a noncentral χ^2-distribution with two degrees of freedom. The noncentrality parameter is equal to the bit energy E_b. The random variable y_1 is χ^2-distributed with two degrees of freedom, whereas the variance is determined by the noise power density N_0.

If a multipath channel is additionally introduced, and the diversity combining is achieved using a single analogue integrator, as shown in Figure 5.1, a number of $N = T_{\text{int}}B$ paths could be resolved within the integration interval of size T_{int}. If the pulse is transmitted at position #1, y_1 is approximately χ^2-distributed with $2N$ degrees of freedom[1], whereas y_0 exhibits a noncentral χ^2-distribution with $2N$ degrees of freedom. The noncentrality parameter is equal to the energy $E \leq E_b$ contained in the window of size T_{int}.

The performance loss estimation due to the described noncoherent combining is shown in Figure 5.2. In the case of envelope detection, an ideal coherent 2-PPM receiver acts as the reference. For any argument N it is assumed that the window of size T_{int} contains the whole bit energy.

A lot of general statements can be obtained:

- Noncoherent detection performs best, if the major part of the received energy is concentrated in a very short time interval of size $1/B$. For impulse radio, this may nearly be the case, if the received signal contains a strong LOS-component, or, if noncoherent detection is combined with time-reversal [11]. Such a predistortion could be applied at the base station transmitter. The corresponding terminal receiver utilizes only the small time-interval within the energy is focused.
- For a direct-sequence signal, the signal energy is already spread over the time at the transmitter. If one combines this 'chip'-energy noncoherently, an additional performance loss is introduced which depends on the code length, or more precisely, on the number of nonzero code-elements. It is therefore much more advantageous to combine either impulse radio

[1] It should be noted that this is an approximation, since $2N = 2T_{\text{int}}B$ fully uncorrelated Gaussian noise terms are supposed prior to the squaring operation.

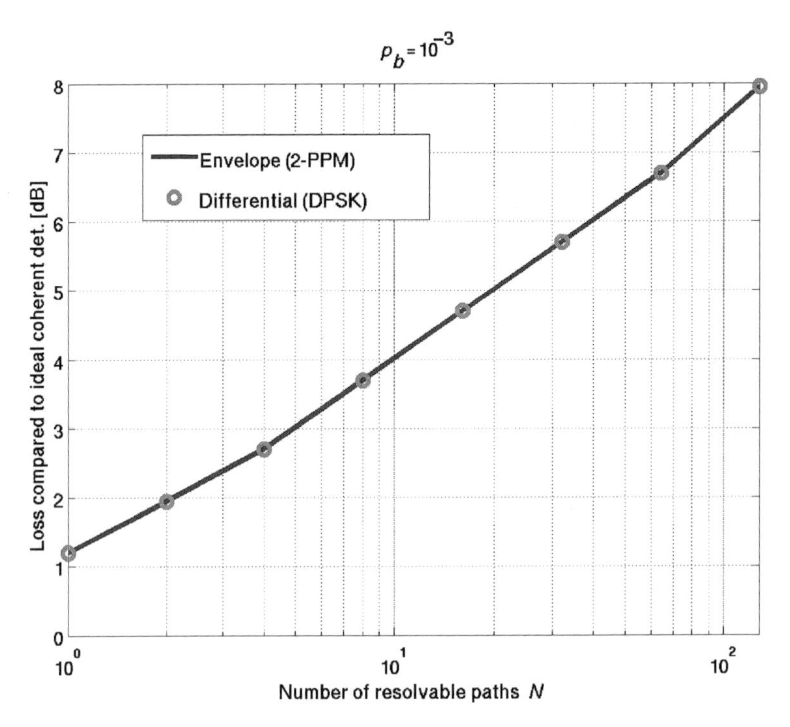

Figure 5.2 Estimation of the performance loss due to noncoherent detection compared with ideal coherent detection. It is assumed, that the window of size $T_{\text{int}} = N/B$ always contains the whole energy E_b.

concepts with noncoherent detection or to apply a filter matched to the whole transmitted direct-sequence signal prior to the noncoherent part of the receiver. At the output of this filter, which is possibly realized by means of a surface acoustic filter, we have again an impulse-like signal enabling a short integration interval.

- If T_{int} is increased, the energy E contained within the interval T_{int} increases as well, until $E = E_b$ (not contained in Figure 5.2). However, with T_{int} increasing the noise-power is increased simultaneously. It is not easy to find an optimum value for T_{int} adaptively during the transmission, since the dependency of the required E_b/N_0 on T_{int} is not necessarily a monotonic function due to the fact that the impulse response may exhibit clusters.
- A large bandwidth B always increases the diversity order and therefore the performance loss compared with (ideal) coherent detection. However, to determine an optimal value for B, the gain with respect to the multipath fading has to be weighed against the loss due to the increased noise.

The estimation used to calculate the loss displayed in Figure 5.2 does not consider any intersymbol or intrasymbol interferences. This topic is discussed in [12]. It should be noted that the performance of TR-systems will also depend on the delay between the reference pulse and the modulated pulse [3], as long as the delay is smaller than the excess delay of the channel.

References

1. Choi, J.D. and Stark, W.E. (2002) Performance of ultra-wideband communications with suboptimal receivers in multipath channels. *IEEE Journal on Selected Areas on Communications*, **20** (9), 1754–66.
2. Tian, Z. and Sadler, B.M. (2005) *Weighted Energy Detection of Ultra-Wideband Signals*, in Proceeding IEEE Workshop on Signal Processing Advances in Wireless Communications, pp. 1068–72.
3. Song, N., Wolf, M. and Haardt, M. (2007) *Low-Complexity and Energy Efficient Non-coherent Receivers for UWB Communications*, in Proceeding IEEE International Symposium on Personal Indoor and Mobile Radio Communications (PIMRC), pp. 1–4.
4. Wu, J., Xiang, H. and Tian, Z. (2006) Weighted noncoherent receivers for UWB PPM signals. *IEEE Communications Letters*, **10** (9), 655–7.
5. Weisenhorn, M. and Hirt, W. (2004) *Robust Noncoherent Receiver Exploiting UWB Channel Properties*, in Proceeding Joint International Workshop on Ultra Wideband Systems and IEEE Conference on Ultrawideband Systems and Technologies, pp. 156–60.
6. Oppermann, I., Stoica, L., Rabbachin, A. *et al.* (2004) UWB wireless sensor networks: UWEN – a practical example. *IEEE Communications Magazine*, **42** (12), S27–32.
7. Winzer, P. and Essiambre, R.J. (2006) Advanced optical modulation formats. *Proceedings of the IEEE*, **94** (5), 952–84.
8. Hoctor, R.T. and Tomlinson, H.W. (2001) *An Overview of Delay-Hopped, Transmitted-Reference RF Communications*, G.E. Research and Development Center, Technical Information Series.
9. Witrisal, K., Leus, G., Pausini, M. and Krall, C. (2005) Equivalent system model and equalization of differential impulse radio UWB systems. *IEEE Journal on Selected Areas in Communications*, **23** (9), 1851–62.
10. Proakis, J.G. (2001) *Digital Communications*, McGraw-Hill.
11. Guo, N., Qiu, R.C. and Sadler, B.M. (2005) An ultra-wideband autocorrelation demodulation scheme with low-complexity time reversal enhancement. *IEEE Military Communications Conference (MILCOM)*, **6**, 3066–72.
12. Romme, J. and Durisi, G. (2004) *Transmit Reference Impulse Radio Systems Using Weighted Correlation*, in Proceeding Joint International Workshop on Ultra Wideband Systems and IEEE Conference on Ultrawideband Systems and Technologies, pp. 141–5.

6

Transmit Reference UWB Systems

A.-J. van der Veen and Geert Leus

Delft University of Technology, Department of Electrical Engineering, Mekelweg 4, 2628 CD Delft, The Netherlands

Impulse Radio Ultra Wideband (IR-UWB) is a promising alternative technology for ultra wideband wireless communications. Instead of modulating information on a carrier, in current wireless communication technologies, data is transmitted using a coded series of very narrow pulses (of less than a nanosecond duration). The initially predicted advantages of IR-UWB over conventional communication techniques are significant: the devices are expected to be small and cheap, the foreseen data rates and user concentrations are large, the method is robust against multipath reflections, the power efficiency is high, and security is inherently present. Besides communications, the pulse-based transmissions of IR-UWB are similar to those used in radar, and can be used for precise localization at accuracy. Hence, IR-UWB also is an enabling technology for position-aware devices, with applications such as tagging and tracking of assets and personnel.

Although IR-UWB is promising in theory, the state-of-the art in IR-UWB can be considered rather immature. Many of the predicted theoretical advantages have not yet been demonstrated. The main reason for this is that the standard RAKE receiver structure that is envisioned in IR-UWB is too complex. A RAKE receiver consists of a bank of so-called fingers, where each finger correlates the incoming pulse with a locally generated template synchronized to a specific delay. All finger outputs are then combined in an optimal fashion, which depends on the power that is received on the different fingers. However, there are a few problems related to this receiver structure. First of all, it is not clear which template we should use. Due to the ultra wideband nature of the pulses, we cannot assume that the received pulse is a superposition of several delayed and scaled copies of the transmitted pulse. In fact it is a superposition of delayed and *distorted* copies of the transmitted pulse, where each copy is related to a different propagation path with different propagation characteristics. Hence, it is unlikely that we can recognize the transmitted pulse in the received pulse. Next, the delays that are used in the different fingers should be perfectly synchronized to the main reflection

Short-Range Wireless Communications Rolf Kraemer and Marcos D. Katz

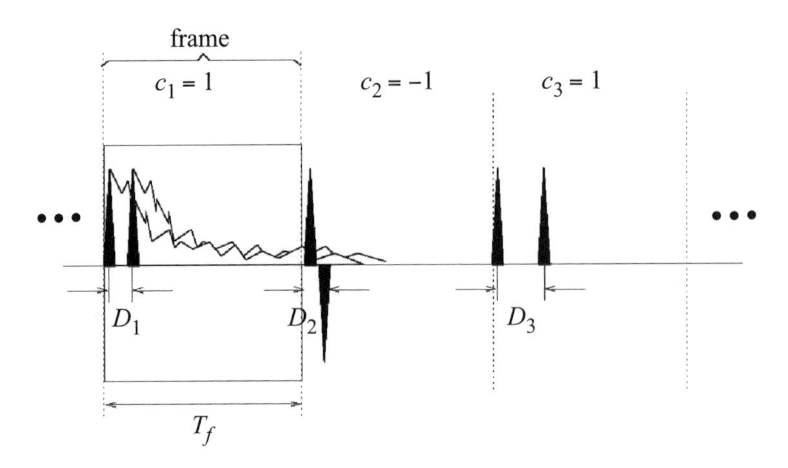

Figure 6.1 Transmit-reference modulation scheme.

paths in the environment, which is difficult to realize. To solve these two problems, one could consider an arbitrary template that spans the bandwidth of the signal and uses a large number of fingers with fixed delays. Such a system is still too complex, compared the data rates that can be achieved. Further, it is unclear how the correlations with the template signal can be implemented. If this is done in the analog domain, then a linear filter has to be constructed that has the template signal as its impulse response - difficult if the template is not constant but extracted from the received signal. Other receivers that have been proposed for IR-UWB are all-digital receivers, but they require high-rate high-precision A/D converters, which are currently too costly in terms of price, size and power.

An important breakthrough that could possibly save IR-UWB is the so-called transmit reference ultra wideband (TR-UWB) approach. This basically is an old idea that was reintroduced by Hoctor and Tomlinson in the UWB context [1]. The approach boils down to sending a pulse pair or doublet instead of a single pulse, where each doublet consists of a reference pulse followed by an amplitude-modulated data pulse (see Figure 6.1). At reception, this reference pulse shows how the transmitted pulse has been distorted by the propagation environment. Hence, the received reference pulse can be used as a noisy template to collect the data by a very straightforward autocorrelation mechanism. We simply have to correlate the received signal with a delayed version of itself, where the delay is the same as the one between the data and the reference pulse. A major advantage of such a TR-UWB system is that the analog processing is simple and data-independent, and does not require synchronization nor channel estimation. The data rates after correlation are much lower (related to the frame rate), and easily captured by current A/D conversion technology. Multi-user systems can be created by combining with CDMA at the frame level (the user code has two dimensions and consists of the usual chip code, plus a delay code). Nonetheless, there are still a few major issues that have to be dealt with to make TR-UWB a viable UWB candidate, as discussed below.

6.1 Challenges

The crucial component of any TR-UWB system is the delay of the autocorrelation receiver. It can be viewed as the dual of the oscillator in a narrowband communication system. However, building an ultra wideband delay with a high accuracy and a small size and power is a major challenge. Several advances have already been reported, such as analog delay filters or quantized analog delays, but it appears that the range of the delays should be limited to a small multiple of the pulse width (a few nanoseconds) in order to satisfy the requirements. Since the environment may spread a transmitted pulse over more than 50 ns, adopting a small delay will introduce a significant overlap between the received reference pulse and the received data pulse, leading to so-called inter pulse interference (IPI). Moreover, if a high data rate is required the doublets should be closely spaced, possibly leading to additional so-called inter frame interference (IFI). Digital signal processing comes into the picture to combat this IPI and IFI in the digital domain [2–7]. Note that, compared the all-digital receivers proposed for IR-UWB, which operate at the Nyquist rate, digital equalizers proposed for TR-UWB only operate at a multiple of the frame rate, which generally is much smaller than the Nyquist rate. This also means that the number of unknown propagation parameters required to construct a digital equalizer for TR-UWB is much smaller than for all-digital receivers for IR-UWB. It is clear that the IPI and IFI will influence the performance, even with an optimal digital equalizer. This is the price we have to pay for the advantages that come with TR-UWB. Another price lies in the 'squaring' of the signal in the correlation process, which essentially doubles the noise, introduces many cross-terms and increases the required dynamic range of the A/D converter.

To avoid spectral peaks, it is required that the delay between the reference pulse and data pulse changes according to some random code, known at the receiver. This means that we need some capability to implement different delays in the autocorrelation receiver. A first approach to realize this is to consider a bank of autocorrelation receivers, each one implementing a different delay, where the integration length W equals the frame duration T_f, or possibly a fraction thereof (see Figure 6.2). This has the additional advantage that the digital equalizer could combine the outputs of all these branches instead of just using the output of the branch with the correct delay. Due to correlation in the propagation environment, not only the branch with the correct delay will be excited, also the other branches contain some useful information about the data that could be exploited to improve the performance, including possibilities for blind synchronization [7]. Especially when there is a delay mismatch, this architecture has significant advantages. A drawback is of course the additional size and power needed to implement the different branches. Alternatives to this structure are envisioned in the near future. One possibility is to use an adaptive delay that can be changed according to the random

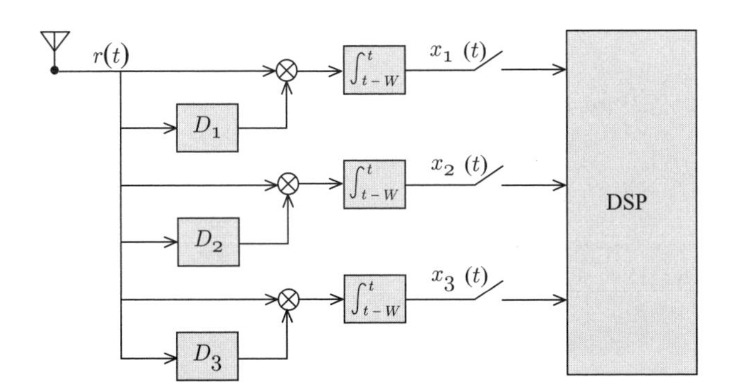

Figure 6.2 TR-UWB receiver structure.

code. Another possibility is to use a cyclic delay, that is a small delay with an attenuated feedback loop from the output to the input. Such a device allows capturing all delays that are a multiple of the implemented small delay, at the cost of additional IPI and IFI.

A major problem in TR-UWB, and IR-UWB in general, is the influence of narrowband interference (NBI), for example generated by WiFi systems. Due to the generated cross-terms in the autocorrelation process, TR-UWB generally suffers more from NBI than IR-UWB. However, solutions have been proposed or are envisioned to solve this problem. First of all, ultra wideband antennas and LNAs are proposed that contain notches at the known NBI frequencies. Further, if a bandwidth from around 3–10 GHz is used, frequency downconversion techniques are envisioned that fold the 5 GHz close to DC and fold the 2.4 GHz out of the desired band. Hence, a simple band-pass filter can then be used to remove the WiFi NBIs. The autocorrelation principle is then applied to the in-phase and quadrature components, instead of to the original incoming signal. This has the additional advantage that the delays can be designed for frequencies up to 3 GHz instead of for frequencies up to 10 GHz, with savings on size and power. Finally, NBI can be further reduced in the digital domain, by applying special NBI cancellation schemes, with or without exploiting the knowledge of the NBI frequency [5].

6.2 Vision

Among IR-UWB systems, TR-UWB currently seems the only viable candidate. Major advantages are its simple and data-independent analog processing, and relatively modest sampling rate (not much higher than the data rate). A disadvantage is the noise enhancement. The maximal data rate is limited by the amount of tolerable ISI in relation to the complexity of the digital receiver. Using current technology, it should be possible to achieve at least 100 Mbps.

References

1. Hoctor, R. and Tomlinson, H. (2002) *Delay-Hopped Transmitted Reference RF Communications*, in Proceeding IEEE Conference on Ultra Wideband Systems and Technologies, pp. 265–70.
2. Dang, Q.H., Trindade, A., Van Der Veen, A.J. and Leus, G. (2006) Signal model and receiver algorithms for a transmit-reference ultra-wideband communication system. *IEEE Journal on Selected Areas in Communications*, **24**(4), 773–9.
3. Witrisal, K., Leus, G., Pausini, M. and Krall, C. (2005) Equivalent system model and equalization of differential impulse radio UWB systems. *IEEE Journal on Selected Areas in Communications*, **23**(9), 1851–62.
4. Dang, Q.H. and Van Der Veen, A.J. (2006) *Resolving Inter-Frame Interference in a Transmit-Reference UWB Communication System*, in Proceeding IEEE International Conference on Acoustics, Speech, and Signal Processing (ICASSP), Vol. **4**, pp. 481–4.
5. Dang, Q.H. and Van Der Veen, A.J. (2006) *Narrowband Interference Mitigation for a Transmit-Reference Ultra-Wideband Receiver*, in Proceeding European Signal Processing Conference (EUSIPCO).
6. Xu, Z. and Sadler, B.M. (2006) Multiuser transmitted reference ultrawideband communication systems. *IEEE Journal on Selected Areas in Communications*, **24**(4), 766–72.
7. Djapic, R., Leus, G. and Van Der Veen, A.J. (2004) *Blind Synchronization in Asynchronous UWB Networks Based on the Transmit-Reference Scheme*, in Proceeding IEEE Asilomar Conference on Signals, Systems, and Computers, Vol. **2**, pp. 1506–10.

7

Multiband Modulation in UWB Systems

Antti Anttonen and Aarne Mämmelä
VTT Technical Research Centre of Finland

7.1 Introduction

Due to its extremely large bandwidth UWB technology provides many well-known advantages compared with a narrowband system, such as robustness to multipath fading, the possibility to increase data rate, and so on. However, achieving very high data rates (i.e. above 1 Gbit/s) with feasible and robust UWB technology is still a significant challenge, mainly due to strict transmission power masks to enable coexistence with other systems, channel impairments and high required sampling rate. Currently, it is not very clear which transceiver technology, in terms of modulation, multiple access and receiver structure, gives the optimal cost and performance trade-off with very high data rates for different application scenarios. One of the most significant system design decisions includes how to use the maximum allocated 7.5 GHz bandwidth in practice with a UWB radio. A data symbol can be transmitted using a single signal which occupies the whole allocated band. On the other hand, several parallel signals, each occupying only a small part of the whole bandwidth, can be employed. The latter method, recently proposed for UWB, is called a multiband approach in which data is transmitted in multiple subbands (SB) each occupying at least 500 MHz bandwidth [1]. Consequently, each SB becomes a miniature UWB system. In the following, the state of the art, challenges and visions of a generalized UWB multiband modulation approach are briefly discussed.

7.2 State of the Art

The multiband approach has been proposed for use in conjunction with several UWB modulation schemes which have traditionally been monoband solutions. The terms multiband communications and multicarrier communications are often used interchangeably in the

Short-Range Wireless Communications Rolf Kraemer and Marcos D. Katz

literature. Here multiband transmission is defined as having SBs with nonoverlapping bandwidths and multicarrier transmission having SBs with overlapping bandwidths. In a classical parallel data system (i.e. a multiband system), the total signal frequency band is divided into a number of nonoverlapping frequency subchannels. Each subchannel is modulated with a particular modulation method, resulting in a frequency-multiplexed transmission system. The nonoverlapping frequency subchannels ensure the elimination of interchannel interference (ICI). This means that different SBs are orthogonal to each other. Furthermore, to obtain optimal diversity gain the channels should be uncorrelated. However, there is a trade-off between the ICI and spectral efficiency. To improve the spectral efficiency the use of overlapping subchannels, defined here as a multicarrier system, is a natural approach. It is also possible to preserve orthogonality between overlapping subchannels with an orthogonal frequency division multiplexing (OFDM) scheme, which is a special case of the multicarrier transmission. Therein a number of lower rate subcarriers are overlapping, but experience no ICI due to the orthogonal frequency spacing of all subcarriers.

Instead of using a transmission signal which occupies the whole spectrum, it can be shaped such that it occupies only a narrower SB from the allocated band. This leads to a multisignal transmission along adjacent bands in order to use the whole available bandwidth more efficiently. The main design parameters leading to different performance complexity trade-offs include the total available bandwidth (B_{tot}), the bandwidth of SB (B_{sb}), the number of SBs (K), and the applied time-frequency coding strategy. The generalized multiband transceiver system is illustrated in Figures 7.1 and 7.2. The transmitter consists of three main parts, namely demultiplexer, data modulation and time-frequency multiplexer. The demultiplexer divides the data into different SBs to increase the data rate, or alternatively copies the data in the time or frequency domain to get diversity gain. The modulator part includes data symbol mapping, spectral shaping, subcarrier modulation, or spreading operations, depending on the target system. The time-frequency multiplexer employs a time-frequency code, which specifies the SBs that are used for a selected time slot. For each time slot the frequency utilization can remain the same or change after a certain time period and the latter resembles the frequency-hopping spread spectrum principle. In general, either only one subband can be hopped or alternatively at the other end all subbands can be hopped. The time-frequency multiplexer can be implemented using an oscillator and a mixer bank or a filter bank. The receiver then performs the functions in reverse order, as seen in Figure 7.1.

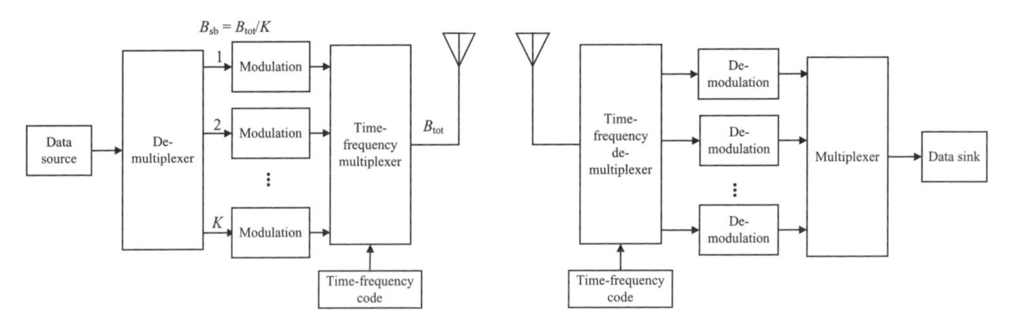

Figure 7.1 Block diagram of a generalized multiband transceiver.

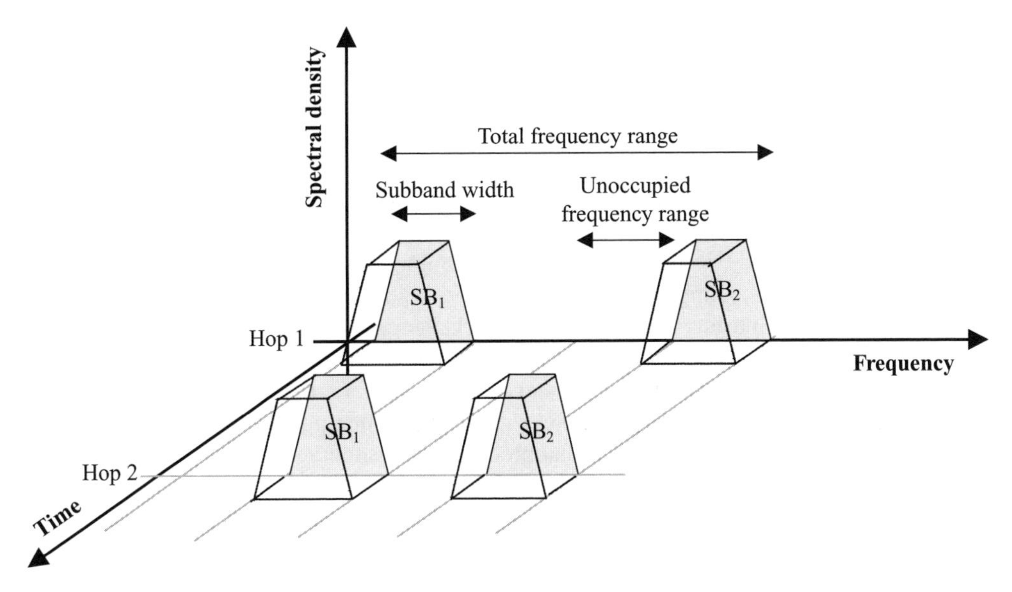

Figure 7.2 Illustration of a frequency-hopped multiband strategy with two subbands.

The first laboratory prototype based on UWB multibands was built in 2001 by General Atomics [1]. Several variations have been proposed for multiband UWB. In addition to detailed band plan, these variations differ in terms of how the modulation and spreading are done in each SB. These include, for example multiband OFDM system [2] and multiband impulse radio system [3].

The motivation to use multiband compared to monoband is many-fold. The multiband approach provides means for coordinated multiple access as different users can be allocated for different bands. Some additional degrees of freedom can be achieved to avoid significant intersymbol interference or to enable higher transmission power, but still respecting the average regulator mask values by adjusting the time-frequency code. Further advantages include fewer collisions from adjacent uncoordinated piconets due to the lower duty cycle of each SB, ability to adaptively select the SB which provides good interference and coexistence robustness and good bit rate scalability. Finally, an increased degree of freedom to find compromises between receiver complexities is achieved since lower bandwidth eases the implementation of the analog-to-digital converter.

7.3 Challenges

While enabling several benefits, the multiband approach has some disadvantages compared with monoband UWB occupying the same bandwidth. Multiband schemes often require a complex multiband signal generator which is able to quickly switch between frequencies [4]. Pulse shaping of individual bands needs to be designed to enable maximization of transmission power and minimization of ICI and complexity of the signal generator. Synchronization of frequency-hopped multiband signals introduces additional complexity increase in the receiver

[5–7]. Finally, parallel signal processing of multiband signals results in more expensive digital and analog parts for the transceivers.

7.4 Visions

It is foreseen that the multiband modulation approach will play an important role, especially in high data rate UWB radios of the future. This is suggested by the inherent benefits that this technique provides. Energy efficiency will be an important aspect, including transmitted energy and signal processing energy [8]. However, according to the above discussion the multiband technique introduces some additional challenges which require special attention. Generally, by using less bandwidth per information bit with multiband modulation, the risk of losing the benefits of traditional UWB transmission increases. Furthermore, a rapid frequency-switching multiband signal generation along with precise switching offset synchronization and low complex techniques for parallel signal processing of simultaneously received signals in different subbands must be carefully addressed. Finally, the design of time-frequency codes to minimize multiuser interference while maintaining reasonable complexity is an important research area.

References

1. Aiello, G.R. and Rogerson, G.D. (2003) Ultra-wideband wireless systems. *IEEE Microwave Magazine*, **4** (2), 36–47.
2. Batra, A., Balakrishnan, J., Aiello, G. *et al.* (2004) Design of a multiband OFDM system for realistic UWB channel environments. *IEEE Transactions on Microwave Theory and Techniques*, **52** (9), 2123–37.
3. Paquelet, S., Aubert, L.-M. and Uquen, B. (2004) *An Impulse Radio Asynchronous Transceiver for High Data Rates*, in Proceeding Joint International Workshop on Ultra Wideband Systems and IEEE Conference on Ultrawideband Systems and Technologies, pp. 1–5.
4. Mishra, C. *et al.* (2005) Frequency planning and synthesiser architectures for multiband OFDM UWB radios. *IEEE Transactions on Microwave Theory and Techniques*, **53** (12), 3744–56.
5. Berens, F., Dimitrov, E., Kaiser, T. *et al.* (2007) *The PULSER II View Towards Very High Data Rate OFDM Based UWB Systems*, in Proceeding IST Mobile and Wireless Communications Summit, pp. 1–5.
6. Anttonen, A., Siltala, S. and Mämmelä, A. (2007) *Timing and Phase Offset Sensitivity of Autocorrelation Based Frequency Synchronization in an FH-OFDM System*, in Proceeding IEEE Vehicular Technology Conference (VTC 2007-Spring), pp. 2379–83.
7. Hahtola, P., Anttonen, A. and Mämmelä, A. (2007) *Acquisition of an Unknown Hopping Code in an Ultra-Wideband FH-OFDM System*, in Proceeding IEEE International Symposium on Wireless Communication Systems (ISWCS), pp. 577–81.
8. Mämmelä, A., Saarinen, I. and Taylor, D. (2005) *Transmitted Energy as a Basic System Resource*, in Proceeding IEEE Global Telecommunications Conference, pp. 3456–60.

8

Design of Synchronization Algorithms for UWB Systems

Claudio R.C.M. da Silva[1] and Laurence B. Milstein[2]

[1]*Wireless@Virginia Tech, The Bradley Department of Electrical and Computer Engineering, Virginia Polytechnic Institute and Sate University, Blacksburg, VA, USA*
[2]*The Center for Wireless Communications, Department of Electrical and Computer Engineering, University of California, San Diego, La Jolla, CA, USA*

The very large bandwidth of UWB signals is what makes this technology unique and of great potential (large channel capacity and lack of significant fading), but it also brings technical challenges that greatly increase the complexity of UWB devices. More specifically, due to the fine time resolution of UWB signals, the received signal is composed of a large number of low-energy multipath components that need to be acquired, have their channel coefficients estimated, and raked so that the UWB device can achieve a performance close to theoretical limits.

In particular, it is known that a major practical implementation challenge of UWB receivers is the design of high-accuracy, rapid synchronization algorithms. The main reason for the acquisition complexity is the lack of sufficient channel state information, as the receiver has to first synchronize with the received symbols before it can perform channel estimation [1, 2]. The affects of having no or little knowledge of the state of the channel can be seen in Figure 8.1, where the probabilities of false lock of three estimators that differ in the amount of *a priori* knowledge of the state of the channel are plotted. It is seen that there is a significant gap in performance between a theoretical synchronization algorithm assumed to have *a priori* (and complete) knowledge of the channel and a practical algorithm (based on a maximum-likelihood approach) that corresponds to the case in which the receiver has no knowledge of the channel. It is also seen in this figure that an algorithm (based on a maximum *a posteriori* formulation) assumed to have knowledge of the channel *statistics* (obtained, e.g., from *a priori* channel measurements) has a performance only a few dB away from the theoretical estimator [1, 2].

Short-Range Wireless Communications Rolf Kraemer and Marcos D. Katz

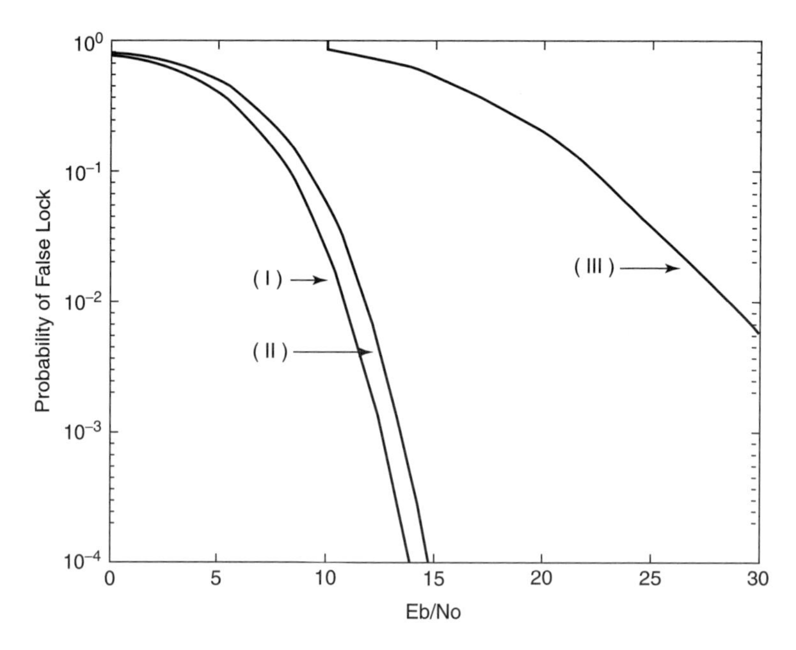

Figure 8.1 Probability of false lock for a UWB communication system: (I) maximum-likelihood estimator with perfect channel estimation (*a priori* knowledge of the channel realization); (II) maximum *a posteriori* estimator (*a priori* knowledge of the channel statistics); (III) maximum-likelihood estimator (no *a priori* knowledge of the channel realization or statistics) [2]. Reproduced by Permission of © 2007 IEEE.

Among different technologies that have been considered for implementing UWB communication systems, impulse-radio and OFDM schemes have been receiving most of the academic and commercial interest. Although the difficulty in performing signal acquisition is inherent to UWB communication systems (due to the large transmitted bandwidth), each implementation technology has its own peculiarities. In particular, a significant amount of effort has been focused on synchronization for impulse radio-based systems. Due to the use of an extremely short, low duty cycle UWB pulse (frame duration much larger than the pulse duration), the delay uncertainty region contains a large number of potential timing offsets compared with narrowband systems [3]. In addition, timing requirements are stringent because even minor misalignments may result in lack of sufficient energy capture to make data detection possible [4]. Conventional sliding correlation-based algorithms developed for narrowband systems would require a long search time and an unreasonably high sampling rate [5]. Recent algorithms proposed for UWB systems include coarse bin search, subspace spectral estimation, generalized likelihood ratio test, and cyclostationarity-based approaches (see [3–5] and references therein). As a consequence of the high complexity of the synchronization and channel estimation stages of impulse radio-based UWB radios, different suboptimum schemes have recently been proposed. Among such schemes, significant effort has recently focused on transmitted-reference techniques in which channel information is 'extracted' from reference symbols (see, among many others, [6–8]). However, these schemes usually suffer in performance (both data rate and error rate) compared with UWB implementations that have explicit synchronization and channel estimation stages.

Regarding OFDM-based UWB systems, the UWB PHY and MAC standard approved by Ecma International (led by the WiMedia Alliance) defines a time-domain PLCP (Physical Layer Convergence Protocol) preamble sequence for packet/frame synchronization with either 24 (standard mode) or 12 (burst mode) symbols (7.5 and 3.75 μs, respectively), followed by a frequency domain sequence (6 symbols, 1875 μs) for channel estimation [9]. Although OFDM is a mature technology used in many communications systems such as ADSL and 802.11a/g, there are still unanswered questions on the capability of this technology to cope with some of the unique characteristics (e.g., large frequency selectivity and possible presence of strong in-band interference) of UWB systems.

The degree of difficulty of accurate synchronization of UWB receivers increases even more with the possible presence of narrow-band interference (NBI) resulting from the spectral overlay. As the UWB signals have low power and broad spectrum, it is possible, at least in principle, to allow for such signals to overlay narrow-band systems with no noticeable interference. Based on this concept, the FCC allocated in early 2002 a 7.5 GHz bandwidth for UWB communication systems that force its coexistence with narrow-band systems, in an attempt to better utilize the spectrum. Because of the critical importance of the acquisition stage, the affect of NBI on the UWB receiver during this stage might be especially harmful, since, if acquisition fails, the desired signal cannot be successfully detected. The probability of false lock when acquisition is performed in the presence of NBI is shown in Figure 8.2. It is seen that the performance severely degrades when an interference mitigation technique is not

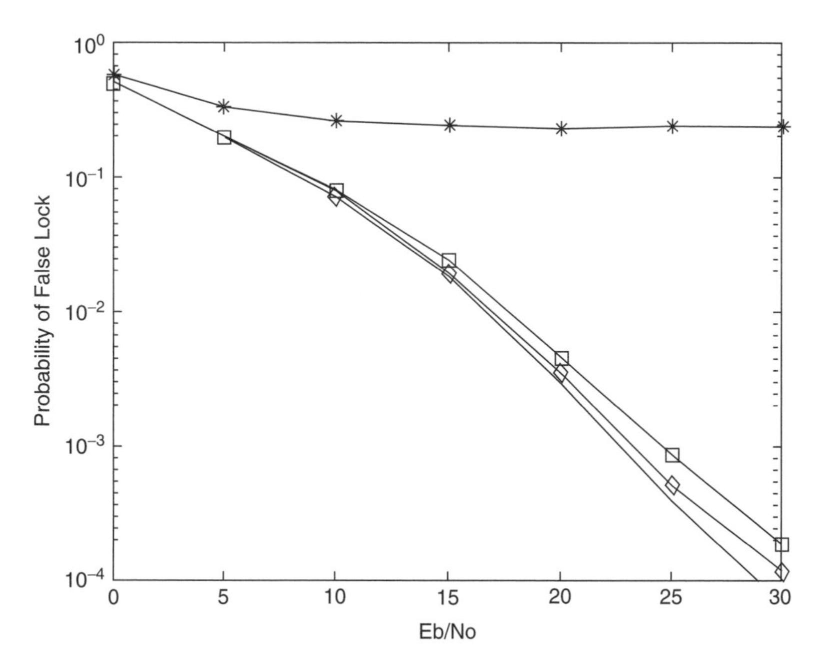

Figure 8.2 Probability of false lock for a UWB communication system in the presence of narrow-band interference (using a maximum-likelihood estimator). Symbols: (dot) absence of NBI; (star) no interference mitigation; (square) and (diamond) interference mitigation method ('covariance matrix estimation' and 'spectral encoding', respectively) is employed. SIR = −15 dB [2]. Reproduced by Permission of © 2007 IEEE.

used. However, the performance when either of two mitigation techniques (covariance matrix estimation and 'spectral encoding') is used closely approaches the performance corresponding to the absence of NBI [1, 2].

Spectral overlay is a major concern with UWB systems that currently deters its world wide adoption. Concerns about possible interference of UWB to GPS systems, for example, led the FCC to protect the GPS band by limiting UWB transmissions to the 3.1–10.6 GHz range. More recently, the European Commission regulators have issued a set of draft rules to define regulations that would require UWB devices to implement certain mitigation techniques (such as 'detect and avoid' and a cap on device activity) in order to reduce the likelihood of interference with other systems [10].

Synchronization is an important issue that should be further explored (research into fast and highly accurate acquisition algorithms) in order for UWB to become a practical communications technology. The viability of the UWB technology depends on significant advances in both system level algorithm design and circuit implementation methods in areas such as synchronization, channel estimation and multipath combining algorithms. The determining factor, however, that will dictate the viability and performance of UWB systems is the ability of these systems to cope with the spectral overlay. The reader is referred to [1, 2, 11, 12] for further results on the effects of the spectral overlay on the design of the multipath combining and channel estimation of UWB signals.

References

1. da Silva, C.R.C.M. and Milstein, L.B. (2007) Coarse acquisition performance of spectral-encoded UWB communication systems in the presence of narrow-band interference. *IEEE Transactions on Communications*, **55**(7), 1374–84.
2. da Silva, C.R.C.M. (2005) *Performance of Spectral-Encoded Ultra-Wideband Communication Systems in the Presence of Narrow-Band Interference*, Ph.D. dissertation, University of California, San Diego.
3. Ibrahim, J. and Buehrer, R.M. (2006) Two-stage acquisition for UWB in dense multipath. *IEEE Journal on Selected Areas on Communications*, **24**(4), 801–7.
4. Liuqing, Y. and Giannakis, G.B. (2004) Ultra-wideband communications: An idea whose time has come. *IEEE Signal Processing Magazine*, **21**, 26–54.
5. Tian, Z. and Giannakis, G.B. (2005) A GLRT approach to data aided timing – Part I: algorithm. *IEEE Transactions on Wireless Communications*, **4**(6), 2956–67.
6. Choi, J.D. and Stark, W.E. (2002) Performance of ultra-wideband communications with suboptimal receivers in multipath channels. *IEEE Journal on Selected Areas on Communications*, **20**(9), 1754–66.
7. Zhang, H. and Goeckel, D.L. (2003) *Generalized Transmitted-Reference UWB Systems*, in Proceeding IEEE Conferense on Ultra Wideband Systems and Technologies, pp. 147–51.
8. Carbonelli, C., Franz, S., Mengali, U. and Mitra, U. (2004) *Semi-Blind ML Synchronization for UWB Transmitted Reference Systems*, in Proceeding IEEE Asilomar Conference on Signals, Systems, and Computers, pp. 1491–5.
9. Ecma International (2007) *High Rate Ultra Wideband PHY and MAC Standard*, 2nd edn, Standard ECMA-368.
10. Wood, S. (2006) *UWB Standards*, WiMedia Alliance white paper.
11. da Silva, C.R.C.M. and Milstein, L.B. (2006) The effects of narrow-band interference on UWB communication systems with imperfect channel estimation. *IEEE Journal on Selected Areas on Communications*, **24**(4), 717–23.
12. da Silva, C.R.C.M. and Milstein, L.B. (2005) Spectral-encoded UWB communication systems: Real-time implementation and interference suppression. *IEEE Transactions on Communications*, **53**(7), 1391–401.

9

An Overview of UWB Systems with MIMO

Thomas Kaiser, Feng Zheng, Bamrung Tau Sieskul
and Kiattisak Maichalernnukul

The Institute of Communications Technology, Faculty of Electrical Engineering and Computer Science, Gottfried Wilhelm Leibniz University Hannover, Appelstrasse 9A, 30167 Hanover, Germany

9.1 Introduction

It has been observed for more than two decades that data rates of wireless and wireline communication systems increase exponentially [1]. Today's peak data rates for short-range communications are 480 Mbit/s, we expect a peak rate of 1 Gbit/s for 2008 and a peak rate of 10 Gbit/s for 2015. Major applications are high-quality video streaming or ultrahigh speed data exchange; for instance, almost 100 Gbyte capacity is expected in 2010 for *mobile* storages, meaning around 40 min copying time with today's state-of-the-art wireless devices.

The final breakthrough of UWB suffers from severe implementation challenges, for example compact antennas, efficient amplifiers and low-power AD converters, and so on. Hence, alternatives for boosting data rates are generally welcome. UWB represents a short-range and therefore mainly indoor communication technique, so the environment is marked by dense multipath propagation – or, in other words, 'rich scattering'. Fortunately, for such type of environments, MIMO-systems (multiple input and multiple output) permit a significant increase of spectral efficiency. Therefore, 'MIMO' is a promising 'add-on' candidate for enabling future ultrahigh speed wireless communications. Simply speaking, a two-transmit and two-receive antenna system requires doubled hardware efforts at both transmitter and receiver sides and additional digital signal processing, but doubles, in average, the data rate. Thus, MIMO basically offers a technological trade-off. Around two years ago such a trade-off caused the breakthrough of MIMO-WLANs, mainly because the analog radio frequency front end becames too expensive for modulation orders higher than 64-QAM. A similar situation

Short-Range Wireless Communications Rolf Kraemer and Marcos D. Katz

will occur for WiMAX and also for cellular communication systems (3G LTE, IEEE 820.20) soon, where MIMO is meanwhile considered as mandatory on the technical roadmap. In conclusion, 'UWB MIMO' is worth further investigation and seems to be a feasible approach for highest data rates on the horizon.

9.2 State of the Art

There are a few studies on the UWB-MIMO, compared to volumes of literature in the narrowband MIMO research. Basically, these studies can be categorized into four fields: UWB-MIMO channel measurement and modeling, channel capacity, space-time coding and beamforming. The channel measurement and modeling provide some guidelines for the practical system design. The channel capacity describes a limit of the benefits in some sense for a UWB system employing multiple antennas, while the space-time coding provides a realization tool towards reaching the limit. UWB beamforming is of great importance for indoor localization, which has become a hot topic in UWB applications.

For the channel measurements and characterization, several reports have been published, see, for example [2–4]. The full characterization of the spatial correlation of the UWB channels is provided in [3], where it is found that in the range of 2.5 times the coherence distance (about 4 cm), the antenna correlation follows a pattern of the first kind zeroth-order Bessel function of the distance, while an almost constant correlation coefficient (smaller than 0.4) is observed when the antenna distance is larger than the 2.5 times the coherence distance. It is particularly interesting that, as shown in [3–7], the antenna angular orientation and the signal polarization can be used to decrease the correlation of the spatial channels or to improve the system performance. Another approach describing the spatial channel correlation property is from the deterministic view point [8]. This is defined as the average value of all the cross-correlation functions of different spatial channels normalized by the autocorrelation functions of corresponding spatial channels. A possible unfavorable electromagnetic coupling between UWB antenna elements has been proven to be small, even for marginal antenna separations [9].

Regarding the channel capacity of UWB-MIMO systems, the research results can be found in [10–13]. In [12, 13], it is shown that for N_{T} transmit and N_{R} receive antennas (for simplicity, it is assumed that $N_{\mathrm{T}} = N_{\mathrm{R}}$ there), the UWB MIMO ergodic channel capacity linearly increases with N_{R}. However, for the MISO (multiple transmit antennas and single receive antenna) case, it is not always beneficial to employ more transmit antennas. It is shown in [12, 13] that the outage probability decreases with the number of transmit antennas when the communication rate is lower than the critical transmission rate, but increases when the rate is higher than another value. This critical transmission rate is determined by the fading power and the SNR of the system at the transmitter side. We can roughly say that it is not beneficial to use multiple transmit antennas if the required transmission rate (normalized by system bandwidth) is higher than the critical transmission rate or equivalently when the available power at the transmitter side is too low. In [10], a fixed region of scattering environments for UWB-MIMO systems is considered. Thus, the number of spatial degrees of freedom of the scattered field, denoted by η, is limited. It is shown in [10] that the system capacity is fundamentally limited by the three numbers: N_{T}, N_{R} and η. This is not strange since η will place a limit on the rank of the UWB-MIMO channel matrix, and hence it will affect the number of the independently separate channels. In [11], it is shown that if several different antennas (a loop antenna and two orthogonal bow-tie antennas) are placed in the same place instead of separately in different places at sufficient distance, as in the traditional spatial antenna array, the spectral efficiency of such a system approaches that of the traditional spatial array. This is due to the fact that the rank of the involved channel matrix is well maintained to be equal for both kinds of systems.

Since a UWB system is often required to work at low power by relevant regulation bodies, it is important to investigate the system capacity at low power or low SNR regime. For wideband systems, it is shown in [14, 15] that very large bandwidths yield poor performance for systems that spread the available power uniformly over time and frequency. In [16], it is shown that the input signals needed to achieve capacity must be 'peaky' in time or frequency for a wideband fading channel composed of a number of time-varying paths. We can witness this phenomenon for UWB-MIMO systems, as shown in Figures 9.1 and 9.2 [17]. In both the figures, the uniform power spectrum allocation (UPSA) and optimal power spectrum allocation (OPSA) policies are investigated respectively for the transmitted UWB signals, where for the OPSA, a water-filling algorithm is applied to adjust the power distribution across both frequency and antenna domains according to the channel multi-path fading, and for the UPSA case, the transmitted power is uniformly distributed across both frequency domain and antenna domain. We use $C_{\mathrm{uni}}/C_{\mathrm{opt}}$ as an index for the efficiency of the UPSA relative to the OPSA, where C_{uni} and C_{opt} denote the ergodic channel capacity under the UPSA and OPSA policies, respectively. It is demonstrated that the efficiency of the UPSA is lower than 0.61 when the SNR is lower than -20 dB. This means that the transmission rate can be increased more than about 1.6 times if

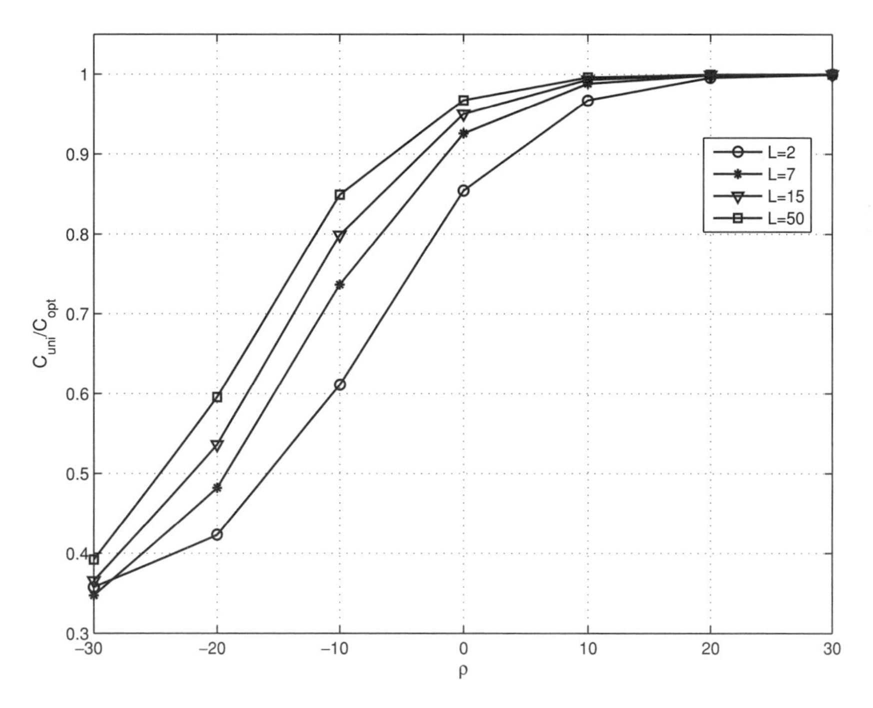

Figure 9.1 Efficiency of the UPSA relative to the OPSA for the 4 × 4 MIMO system, where ρ is the SNR at the transmitter side and L is the numbers of the resolvable multi-paths.

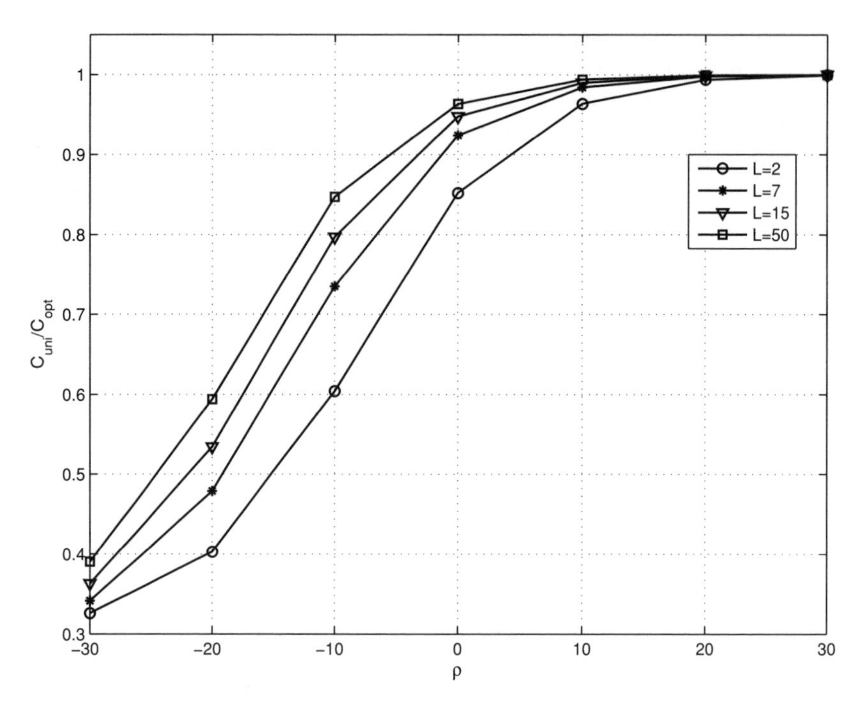

Figure 9.2 Efficiency of the UPSA relative to the OPSA for the 8 × 8 MIMO system, where ρ is the SNR at the transmitter side and L is the numbers of the resolvable multi-paths.

the OPSA, instead of the UPSA, is adopted. But when $\rho \geq 10\,\text{dB}$, $C_{\text{uni}}/C_{\text{opt}}$ approaches one. Therefore, the optimal power distribution algorithm, such as the water-filling approach, should be considered if the SNR is rather low, while the water-filling algorithm is just to make the transmitted signals to be 'peaky' in both frequency and antenna domains. When the channel fading information is unknown at the receiver, the system performance is characterized in [18] for MIMO wideband systems.

For the space-time coding (STC), the first result was reported in [19] for impulse-radio UWB systems, where it is shown that the receive diversity order is equal to the product of the number of receive antennas and the number of RAKE fingers. Note that a larger number of antennas promises only a limited diversity gain because of the distinct UWB multipath diversity [20]. In [8], a spatial multiplexing coherent scheme for UWB-MIMO (2 × 2) system is experimentally investigated. In [21], the performance of a space-time trellis code for a 2 × 2 UWB-MIMO system is evaluated. For general MIMO impulse UWB systems, the STC method was provided in [1, 22–25]. For OFDM based UWB systems [26], presented a STC method, essentially similar to the STC for the wideband OFDM. The report [27] showed an approach to increasing the spatial diversity via antenna selection across data frames. In [28], a space-time selective RAKE receiver is proposed considering the presence of narrowband interference and multiple access interference. In [29], a time interleave multiple transmit antenna UWB system is proposed, where N_f monocycle pulses per information symbol are transmitted discontinuously through the time interleaver to get more temporal diversity. Spatial diversity and temporal diversity are compared therein. In [30], a zero-forcing scheme is proposed to remove the interference among the multiple data streams in UWB-MIMO systems. A space-time trellis coding scheme is proposed in [31]. In [20, 32, 33], the multiple access performance of the UWB-MIMO systems is investigated. Spatial multiplexing is proposed in [34], where the VBLAST (Vertical Bell Laboratory Layered Space-Time) algorithm has been applied to UWB systems and a significant multiplexing gain could be proven.

For UWB-MIMO space-time coding, it is found in [35] that there is a fundamental compromise among the available SNR, coding interval and the number of transmit antennas. The basic conclusion is: only when the available SNR is sufficiently high, can the coding gain be obtained by deploying the transmit power into more transmit antennas and by using a longer coding interval, if the total power over all the transmit antennas and the coding interval is fixed. In other words, if the available SNR is too low, it is better to use fewer transmit antennas and shorter space-time codes. Therefore, we can see another kind of 'peaky' phenomenon. It is highly expected to give some quantitative characterization for how high the SNR should be so that the space-time coding can indeed provide rewarding gains. However, this is not available for general UWB-MIMO systems.

About UWB beamforming, systematic studies for the problem have been presented in [36–38] with several fundamental differences being found. In [39], a digital UWB beamforming was proposed. The effect of multipaths, simulated by the ray-tracing technique, on the beamformer output was illustrated in [40]. In [41–43], the UWB beamformer was used to find the location of the source. In [44], an adaptive beamformer for multi-band UWB wireless systems is proposed, where it is shown that the signal bandwidth has little impact on the beamwidth or direction, and hence the beam-focusing capability will not be sensitive to the signal bandwidth used. In [45], the measured transient response of a uniform linear UWB array shows a peaked output without sidelobe. In [46], an interesting algorithm for calculating the weighting coefficients for wideband beamformers is developed.

Beamforming for UWB impulse signals has some peculiar properties that are quite different from the narrowband beamformers. For example, the use of unequal weighting filters for the individual antenna branch will increases the sidelobe level in UWB beamforming, and thus optimal beamformers, as shown in [38], are those in which the weighting filters for all the antenna branches are identical. A basic difficulty in UWB beamformer is how to deal with multipaths of the inherited UWB signal propagations. In narrowband array processing technology, this problem can be ignored, but we cannot ignore it in the UWB array since the multi-path is one of the most pronounced characteristics of UWB channels. In this research field, ranging [43, 47] and sensing are promising applications of multiple antenna UWB technology. Ultra short pulses allow spatial resolution even down to subdezimeter range and are therefore another current UWB research topic [48]. By means of multi-antenna techniques, additional spatial parameters can be generally extracted (e.g., direction of arrival or direction of departure) leading to an enhanced ranging accuracy. Further applications cover the detection of breast cancer [49] or mine localization [50] as well as the detection of fires by active UWB radiation [51].

Even though a typical power increment (array gain) by multiple antennas is noticed, a general investigation on bandwidth dependence substantiated by quantitative results is still lacking. In general, it is evident that a boost of bandwidth is accompanied by a diminished small-scale fading. Hence, a threshold region will exist, but not yet be quantified. Exceeding this region makes UWB SIMO (single transmit antenna and multiple receive antennas) less promising because only the array gain persists.

This literature overview is not complete, and the number of scientific contributions to UWB-MIMO is continuously increasing. However, compared with the immense research on narrowband and wideband MIMO, it is reliable to say that the research on UWB-MIMO is still in its infancy.

9.3 Challenges

The challenges certainly lie in the research and not in the development domain. In principle, it is rather unclear from several perspectives on how bandwidth affects the performance of multiantenna systems. For instance, how do fading models or diversity gain depend on the bandwidth? Note that array gain may even double on a logarithmic scale (6 dB per doubling the number of antennas) compared with wideband systems, since the inverse bandwidth of an UWB signals may become less than the travel time through an array [38]. Spatial channel measurements and spatial channel modeling are of particular importance for a thorough investigation of UWB-MIMO systems. For more details about this challenge, see the contribution by Andreas M. Molisch in Chapter 1. A large number of new algorithm challenges can also be foreseen. For example, in multiband-OFDM, multiple antennas can be exploited to improve the crucially bad packing detection. Hardware challenges differ only slightly from the UWB SISO case, namely the synchronization of the mixers (if needed) for downconversion and upconversion, and for antenna array topologies.

For localization problems, multipath propagation and non-line-of-sight situations cause major challenges; multiple antennas provide spatial information and are therefore promising for new solutions.

9.4 Visions

The visions are apparent: UWB-MIMO provides a possible solution for highest data rates for short-range communication and accurate indoor localization under non-line-of-sight conditions in the subdecimeter or even subcentimeter range. In order to achieve them, several scientific disciplines will be involved, that is performance analysis, channel modeling, algorithm development and hardware development.

References

1. Cherry, S. (2004) Edholm's law of bandwidth. *IEEE Spectrum*, **41**(7), 58–60; http://ieeexplore.ieee.org/stamp/stamp.jsp?arnumber=01309810.
2. Keignart, J., Abou-Rjeily, C., Delaveaud, C. and Daniele, N. (2006) UWB SIMO channel measurements and simulations. *IEEE Transactions on Microwave Theory Techniques*, **54**(4), 1812–9.
3. Malik, W.Q. (2008) Spatial correlation in ultrawideband channels. *IEEE Transactions on Wireless Communications*, **7**(2), 604–10.
4. Malik, W.Q. and Edwards, D.J. (2007) *UWB Impulse Radio with Triple-Polarization SIMO*, in Proceeding IEEE Global Telecommunications Conference (GLOBECOM), pp. 4124–8.
5. El-Hadidy, M. and Kaiser, T. (2007) *An UWB Channel Model Considering Angular Antenna Impulse Response and Polarization*, in Proceeding European Conference on Antennas and Propagation.
6. El-Hadidy, M. and Kaiser, T. (2007) *A Signal Processing Framework for MIMO UWB Channels with Real Antennas in Real Environments*, in Proceeding IEEE International Conference on Ultra-Wideband, pp. 111–6.

7. Malik, W.Q. and Edwards, D.J. (2007) Measured MIMO capacity and diversity gain with spatial and polar arrays in ultrawideband channels. *IEEE Transactions on Communications*, **55**(12), 2361–70.
8. Tran, V.P. and Sibille, A. (2006) Spatial multiplexing in UWB MIMO communications. *Electronics Letters*, **42**(16), 931–2.
9. Sibille, A. and Bories, S. (2003) *Spatial Diversity for UWB Communications*, in Proceeding European Personal Mobile Communications Conference (EPMCC), pp. 367–70.
10. Martini, A., Franceschetti, M. and Massa, A. (2007) *Capacity of Wide-Band MIMO Channels via Space-Time Diversity of Scattered Fields*, in Proceeding IEEE Asilomar Conference on Signals, Systems, and Computers, pp. 138–42.
11. Rajagopalan, A., Gupta, G., Konanur, A.S. *et al.* (2007) Increasing channel capacity of an ultrawideband MIMO system using vector antennas. *IEEE Transactions on Antennas and Propagation*, **55**(10), 2880–7.
12. Zheng, F. and Kaiser, T. (2006) Channel capacity of MIMO UWB indoor wireless systems, in *UWB Communication Systems – A Comprehensive Overview* (eds M.-G.D. Benedetto and T. Kaiser *et al.*), Hindawi Publishing Corporation, pp. 376–409.
13. Zheng, F. and Kaiser, T. (2004) *On the Evaluation of Channel Capacity of Multi-Antenna UWB Indoor Wireless Systems*, in Proceeding IEEE International Symposium on Spread Spectrum Techniques and Applications, pp. 525–9.
14. Gallager, R.G. and Médard, M. (1997) *Bandwidth Scaling for Fading Channels*, in Proceeding IEEE International Symposium on Information Theory, p. 471.
15. Médard, M. and Gallager, R.G. (2002) Bandwidth scaling for fading multipath channels. *IEEE Transactions on Information Theory*, **48**(4), 840–53.
16. Telatar, I.E. and Tse, D.N.C. (2000) Capacity and mutual information of wideband multipath fading channels. *IEEE Transactions on Information Theory*, **46**(4), 1384–1400.
17. Zheng, F. and Kaiser, T. (2008) On the evaluation of channel capacity of UWB indoor wireless systems. *IEEE Transactions on Signal Processing*, to appear.
18. Ray, S., Zheng, L. and Médard, M. (2007) On noncoherent MIMO channels in the wideband regime: Capacity and reliability. *IEEE Transactions on Information Theory*, **53**(6), 1983–2009.
19. Yang, L. and Giannakis, G.B. (2004) Analog space-time coding for multiantenna ultra-wideband transmissions. *IEEE Transactions on Communications*, **52**(3), 507–17.
20. Weisenhorn, M. and Hirt, W. (2003) *Performance of Binary Antipodal Signaling over the Indoor UWB MIMO Channel*, in Proceeding IEEE International Conference on Communications, pp. 2872–8.
21. Tyagi, A. and Bose, R. (2007) A new distance notion for PPAM space-time trellis codes for UWB MIMO communications. *IEEE Transactions on Communications*, **55**(7), 1279–82.
22. Abou-Rjeily, C., Daniele, N. and Belfiore, J.-C. (2006) *Diversity-Multiplexing Tradeoff of Single-Antenna and Multi-Antenna Indoor Ultra-wideband Channels*, in Proceeding IEEE International Conference on Ultra-Wideband, pp. 441–6.
23. Abou-Rjeily, C., Daniele, N. and Belfiore, J.-C. (2005) *Differential Space-Time Ultra-Wideband Communications*, in Proceeding IEEE International Conference on Ultra-Wideband, pp. 248–53.
24. Siriwongpairat, W.P., Olfat, M. and Liu, K.J.R. (2005) Performance analysis and comparison of time-hopping and direct-sequence UWB-MIMO systems. *EURASIP Journal on Applied Signal Processing*, **2005**(3), 328–45.
25. Abou-Rjeily, C., Daniele, N. and Belfiore, J.-C. (2006) Space-time coding for multiuser ultrawideband communications. *IEEE Transactions on Communications*, **54**(11), 1960–72.
26. Siriwongpairat, W.P., Su, W., Olfat, M. and Liu, K.J.R. (2006) Multiband-OFDM MIMO coding framework for UWB communication systems. *IEEE Transactions on Signal Processing*, **54**(1), 214–24.
27. Wang, L.-C., Liu, W.-C. and Shieh, K.-J. (2005) On the performance of using multiple transmit and receive antennas in pulse-based ultrawideband systems. *IEEE Transactions on Wireless Communications*, **4**(6), 2738–50.
28. Chang, T.-H., Chang, Y.-J., Peng, C.-H. *et al.* (2005) *Space Time MSINR-SRAKE Receiver with Finger Assignment Strategies in UWB Multipath Channels*, in Proceeding IEEE International Conference on Ultra-Wideband, pp. 242–7.
29. Ezaki, T. and Ohtsuki, T. (2003) *Diversity Gain in Ultra Wideband Impulse Radio (UWB-IR)*, in Proceeding IEEE Conference on Ultra Wideband Systems and Technologies, pp. 56–60.
30. Liu, H., Qiu, R.C. and Tian, Z. (2005) Error performance of pulse-based ultra-wideband MIMO systems over indoor wireless channels. *IEEE Transactions on Wireless Communications*, **4**(6), 2939–44.
31. Ohtsuki, T. (2004) *Space-Time Trellis Coding for UWB-IR*, in Proceeding IEEE Vehicular Technology Conference (VTC), pp. 1054–8.

32. Tan, S.S., Kannan, B. and Nallanathan, A. (2006) Multiple access capacity of UWB M-ary impulse radio systems with antenna array. *IEEE Transactions on Wireless Communications*, **5**(1), 61–6.
33. Tan, S.S., Kannan, B. and Nallanathan, A. (2006) Performance of UWB multiple-access impulse radio systems with antenna array in dense multipath environments. *IEEE Transactions on Communications*, **54**(6), 966–70.
34. Kumar, N.A. and Buehrer, R.M. (2002) *Application of Layered Space-Time Processing to Ultra-Wideband Communication*, in Proceeding IEEE Midwest Symposium on Circuits and Systems, pp. 597–600.
35. Kaiser, T. Zheng, F. and Dimitnov, E. (2008) An overview of ultra wideband systems with MIMO. *Proceedings of the IEEE*, to appear.
36. Hussain, M.G.M. (2002) Principles of space-time array processing for ultrawide-band impulse radar and radio communications. *IEEE Transactions on Vehicular Technology*, **51**(3), 393–403.
37. Hussain, M.G.M. (2005) Theory and analysis of adaptive cylindrical array antenna for ultrawideband wireless communications. *IEEE Transactions on Wireless Communications*, **4**(6), 3075–83.
38. Ries, S. and Kaiser, T. (2006) Ultra wideband impulse beamfoming: it is a different world. *Signal Processing*, **86**, 2198–207.
39. Kaiser, T., Ries, S. and Senger, C. (2006) UWB beamforming and DoA – estimation, in *UWB Communication Systems - A Comprehensive Overview* (eds M.-G.D. Benedetto and T. Kaiser *et al.*), Hindawi Publishing Corporation, pp. 330–52.
40. Neinhüs, M., El-Hadity, M., Held, S. *et al.* (2006) *An Ultra-Wideband Linear Array Beamforming Concept Considering Antenna and Channel Effects*, in Proceeding European Conference on Antennas and Propagation.
41. Eltaher, A., Ghalayini, I.I. and Kaiser, T. (2005) *Towards UWB Self-Positioning Systems for Indoor Environments Based on Electric Field Polarization, Signal Strength and Multiple Antennas*, in Proceeding International Symposium on Wireless Communication Systems, pp. 389–93.
42. Eltaher, A. and Kaiser, T. (2005) *A Novel Approach Based on UWB Beamforming for Indoor Positioning in None-Line-of-Sight Environments*, in Proceeding Radio Techniques and Technologies for Commercial Communication and Sensing Applications.
43. Kaiser, T., Senger, C. and Sieskul, B.T. (2006) *Antenna Arrays for UWB Indoor Localization in Non-Line of Sight Environments*, in Proceeding IEEE MTT-S International Microwave Symposium.
44. Malik, W.Q., Allen, B. and Edwards, D.J. (2006) *A Simple Adaptive Beamformer for Ultrawideband Wireless Systems*, in Proceeding IEEE International Conference on Ultra-Wideband, pp. 453–7.
45. Sörgel, W., Sturm, C. and Wiesbeck, W. (2005) *Impulse Responses of Linear UWB Antenna Arrays and the Application to Beam Steering*, in Proceeding IEEE International Conference on Ultra-Wideband, pp. 275–80.
46. Ghavami, M. (2002) Wideband smart antenna theory using rectangular array structures. *IEEE Transactions on Signal Processing*, **50**(9), 2143–51.
47. Kaiser, T. and Sieskul, B.T. (2006) *An Introduction to Multiple Antennas for UWB Communication and Localization*, in Proceeding Conference on Information Science and Systems (CISS), pp. 638–43.
48. Zetik, R., Sachs, J. and Peyerl, P. (2003) *UWB Radar: Distance and Positioning Measurements*, in Proceeding International Conference on Electromagnetics in Advanced Applications (ICEAA).
49. Bond, E.J., Li, X., Hagness, S.C. and Van Veen, B.D. (2003) Microwave imaging via space-time beamforming for early detection of breast cancer. *IEEE Transactions on Antennas and Propagation*, **51**(8), 1690–705.
50. Sachs, J. *et al.* (2002) *Integrated Digital UWB-Radar*, in Proceeding American Electro-Magnetics Conference (AMEREM), pp. 1–8.
51. Willms, I., Sachs, J. and Kaiser, T. (2004) *Active UWB Fire Detection*, in Proceeding Internationale Konferenz über Automatische Brandentdeckung (AUBE).

10

UWB Localization Algorithms

Sinan Gezici[1] and H. Vincent Poor[2]
[1] *Department of Electrical and Electronics Engineering, Bilkent University, Bilkent, Ankara 06800, Turkey*
[2] *Department of Electrical Engineering, Princeton University, Princeton, NJ 08544, USA*

10.1 Introduction

In addition to communications applications, UWB signals are also well-suited for location-aware applications due to their high time resolution and penetration capability [1, 2]. Since a UWB signal occupies a broad frequency spectrum that includes low frequencies as well as high frequencies, it has a higher probability of passing through or around obstacles. Moreover, the large bandwidths of UWB signals imply very high time resolution, which makes accurate time-of-flight, and hence range estimation possible.

The accurate localization capability of UWB signals facilitates a number of applications, such as noninvasive patient monitoring, personnel tracking, search and rescue operations and through-the-wall health monitoring of hostages [2]. In addition, the IEEE Task Group 4a designed an alternate PHY specification for the already existing IEEE 802.15.4 standard for wireless personal area networks (WPANs), which employs UWB as one of the signaling formats [3]. With additional features provided by the 15.4a amendment, the standard can provide high-precision localization capability, high aggregate throughput and low power consumption.

Although UWB signals can potentially provide very accurate location information, even in harsh environments, there are a number of challenges in practical systems related to technology limitations and nonideal channel conditions. Specifically, multipath and/or nonline-of-sight (NLOS) propagation, multiple-access interference (MAI) and high time resolution of UWB signals present practical difficulties for accurate location estimation. Therefore, special attention should be paid to mitigation of errors due to those kinds of error sources.

In this study, we first investigate, in Section 10.2, conventional localization algorithms, such as angle-of-arrival (AOA) and time-of-arrival (TOA)-based schemes, from a UWB perspective, and conclude that time-based schemes are more suitable for UWB localization systems.

Short-Range Wireless Communications Rolf Kraemer and Marcos D. Katz

Therefore, we focus on time-based systems for the remainder of the chapter. In Section 10.3, we study the theoretical limits for time-based localization and a number of TOA estimation algorithms. Then, in Section 10.4, we investigate challenges for accurate UWB localization, and present some of the potential solutions from the literature. Finally, we make some concluding remarks in Section 10.5.

10.2 Localization Techniques and UWB

Localization of a node[1] in a wireless network commonly involves extraction of information from radio signals traveling between that node and a number of reference nodes. The information extracted from radio signals can be in the form of AOA, TOA, time-difference-of-arrival (TDOA) and/or signal strength (SS) estimates. In the following, we first investigate localization techniques based on angle, power and time information, and assess their suitability for UWB applications. After observing the advantages of time-based techniques for UWB localization systems, we then study the relationship between TOA estimation and optimal localization.

10.2.1 Angle of Arrival

AOA-based localization involves measuring angles of the target node as seen by reference nodes. For two-dimensional localization, it is sufficient to obtain two AOA measurements between the target node and two reference nodes for localization via triangulation [4].

Commonly, antenna arrays are employed either at the target node (for *self-localization*), or at reference nodes (for *remote localization*) in order to obtain AOA information.

The Cramer–Rao lower bound (CRLB) for estimating AOA φ of a UWB signal by using a uniform linear antenna array with K elements can be expressed as[2] [5]

$$\sqrt{Var(\hat{\varphi})} \geq \frac{\sqrt{3}c}{\sqrt{2}\pi\sqrt{\mathrm{SNR}}\beta\sqrt{K(K^2-1)}d_s \sin\varphi}, \tag{10.1}$$

where $\hat{\varphi}$ represents an unbiased estimate of the AOA, c is the speed of light, d_s is the spacing between the antenna elements, SNR is the signal-to-noise ratio and β is the effective signal bandwidth given by

$$\beta = \left[\int_{-\infty}^{\infty} f^2\,|S(f)|^2\,\mathrm{d}f \Big/ \int_{-\infty}^{\infty} |S(f)|^2\,\mathrm{d}f\right]^{1/2} \tag{10.2}$$

with $S(f)$ denoting the Fourier transform of the UWB signal.

Although UWB signals can provide accurate AOA estimation, use of antenna arrays make the system costly, which annuls one of the main advantages of a UWB radio equipped with low-cost transceivers. In addition, the large bandwidths of UWB signals result in large numbers of multipath components, which make accurate angle estimation quite challenging/costly.

[1] In this chapter, a node refers to any device that is being located, or participating in localization of other devices.
[2] The angle φ is measured from the axis along the antenna array.

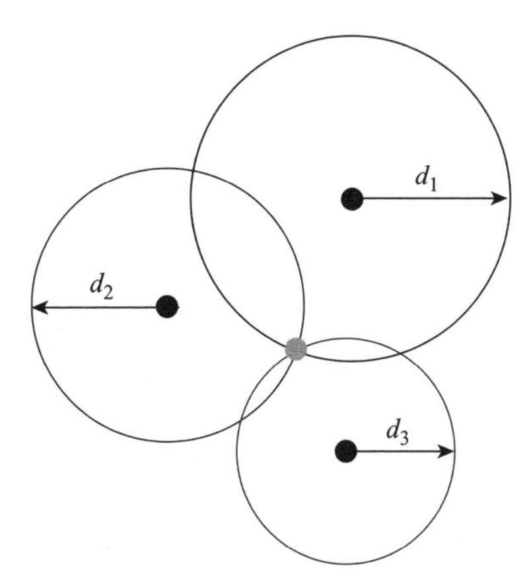

Figure 10.1 Distance-based localization. The distances can be obtained from SS or TOA measurements. The black nodes are the reference nodes.

10.2.2 Received Signal Strength

Distance between two nodes can be estimated by measuring the strength/power of the received signal at one of the nodes. For precise estimation, an accurate path-loss and shadowing model must be available, which makes SS-based localization algorithms very sensitive to the estimation of channel parameters. From distance estimates between the target node and at least three reference nodes, the location of the target node can be obtained by the well-known trilateration approach depicted in Figure 10.1.

The CRLB for a distance estimate $\hat{d}$ obtained from SS measurements can be expressed as [6]

$$\sqrt{Var(\hat{d})} \geq \frac{\ln(10)}{10} \frac{\sigma_{\mathrm{sh}}}{n_{\mathrm{p}}} d, \tag{10.3}$$

where d is the distance between the two nodes, n_{p} is the path loss factor and σ_{sh} is the standard deviation of the zero mean Gaussian random variable representing the log-normal channel shadowing effect [2]. From Equation (10.3), it is observed that the best achievable limit depends on the channel parameters and the distance between the two nodes. Therefore, the unique characteristic of a UWB signal, namely the very large bandwidth, is not exploited by the SS approach to increase the best achievable accuracy.

10.2.3 Time-Based Approaches

Time-based localization relies on measuring travel times of signals between nodes. For two nodes with a common timing reference, the node receiving the signal can determine the time-of-flight of the incoming signal from its TOA signal. For a single-path additive white Gaussian noise (AWGN) channel, the best achievable accuracy of a distance estimate $\hat{d}$ derived from TOA estimation satisfies the following inequality [7, 8]:

$$\sqrt{Var(\hat{d})} \geq \frac{c}{2\sqrt{2}\pi\sqrt{\mathrm{SNR}}\beta}. \tag{10.4}$$

Unlike the SS-based approach, the accuracy of a time-based approach can be enhanced by increasing the SNR or the effective signal bandwidth. Since UWB signals have very large bandwidths, this property facilitates extremely accurate localization of UWB devices using time-based techniques. As an example, for the second derivative of a Gaussian pulse with around 1 GHz bandwidth, an accuracy of less than a centimeter can be achieved at SNR = 5 dB [2].

Comparison of various localization techniques reveals that time-based schemes provide very good accuracy due to the high time resolution (large bandwidth) of UWB signals. Moreover, they are less costly than the AOA-based technique. Although SS estimation is simpler than TOA estimation in general, the distance (range) information obtained from SS measurements is very coarse compared with that obtained from TOA measurements. Due to the inherent suitability and accuracy of time-based approaches for UWB systems, we will focus our discussion on time-based UWB localization for the remainder of this discussion.

10.3 Time-Based Localization: Theoretical Limits and Practical Algorithms

In this section, we consider localization based on TOA information related to a target node and a number of reference nodes. First, we investigate the optimality of TOA-based localization and present theoretical limits for localization accuracy. After that, we consider a number of TOA estimation algorithms for UWB localization and study their properties.

10.3.1 Ranging and Optimal Localization

Ranging refers to estimation of distance ('range') between two nodes. In this context, TOA estimation is equivalent to ranging. On the other hand, localization is the estimation of position (location) of a node in a given network. Conventionally, localization is performed in two steps [2]. First, ranges between the target node and a number of reference nodes are estimated. Then, these range estimates are used to estimate the location of the target. However, this two-step approach is suboptimal in general. An optimal approach can be thought of as estimating the position of the target node directly from the received signals (*direct localization*) [9]. However, it can be shown that the two-step approach is asymptotically optimal under certain conditions, as we investigate below [10, 11].

We consider a synchronous UWB system with a target node and N reference nodes[3]. Let M of the reference nodes have NLOS to the target node, while the remaining nodes have line-of-sight (LOS). We assume that the identities of NLOS or LOS reference nodes are known; this information can be obtained by NLOS identification techniques [12–14]. Without this information, all first arrivals can be considered as NLOS signals.

The received signal related to the ith reference node can be expressed as

$$r_i(t) = \sum_{j=1}^{L_i} \alpha_{ij} s(t - \tau_{ij}) + n_i(t), \tag{10.5}$$

[3] Note that practical UWB systems are not usually synchronous. However, two-way ranging protocols [20] are commonly used in order to compensate for the timing offset.

for $i = 1, \ldots, N$, where L_i is the number of multipath components at the ith node, α_{ij} and τ_{ij} are, respectively, the fading coefficient and the delay of the jth path of the ith node, $s(t)$ is the UWB signal, and $n_i(t)$ is zero mean white Gaussian noise with spectral density $N_0/2$. The noise components related to different reference nodes are assumed to be independent. The delay of the jth path component at node i can be expressed, for two-dimensional positioning, as

$$\tau_{i,j} = \frac{1}{c}\left[\sqrt{(x_i - x)^2 + (y_i - y)^2} + l_{ij}\right], \tag{10.6}$$

for $i = 1, \ldots, N$, $j = 1, \ldots, L_i$, where c is the speed of light, $[x_i\ y_i]$ is the location of the ith node, l_{ij} is the NLOS propagation induced path length, and $[x\ y]$ is the location of the target node.

We assume, without loss of generality, that the first M nodes have NLOS, while the remaining $N - M$ have LOS. Then, $l_{i1} = 0$ for $i = M + 1, \ldots, N$, as the signal directly reaches the related node in an LOS situation. Hence, the parameters to be estimated are the NLOS delays and the location of the node, $[x\ y]$ which can be expressed as $\boldsymbol{\theta} = [xy\boldsymbol{l}_{M+1} \ldots \boldsymbol{l}_N \boldsymbol{l}_1 \ldots \boldsymbol{l}_M]$, where

$$\boldsymbol{l}_i = \begin{cases} (l_{i1}\ l_{i2}\ \ldots\ l_{iL_i}) & \text{for } i = 1,\ \ldots,\ M, \\ (l_{i2}\ l_{i3}\ \ldots\ l_{iL_i}) & \text{for } i = M+1,\ \ldots,\ N, \end{cases} \tag{10.7}$$

with $0 < l_{i1} < l_{i2} \ldots < l_{iN_i}$ [11]. Note that the first delay is excluded from the parameter set for LOS signals, since these are known to be zero.

From Equation (10.5), the likelihood function for $\boldsymbol{\theta}$ can be expressed as

$$\Lambda(\boldsymbol{\theta}) \alpha \prod_{i=1}^{N} \exp\left\{-\frac{1}{N_0}\int\left|r_i(t) - \sum_{j=1}^{L_i}\alpha_{ij}s(t - \tau_{ij})\right|^2 \mathrm{d}t\right\} \tag{10.8}$$

Then, the lower bound on the variance of any unbiased estimator for the unknown parameter $\boldsymbol{\theta}$ can be obtained from $\mathrm{E}_\theta\left\{(\hat{\boldsymbol{\theta}} - \boldsymbol{\theta})(\hat{\boldsymbol{\theta}} - \boldsymbol{\theta})^T\right\} \geq \mathbf{J}_{\boldsymbol{\theta}}^{-1}$, where $\mathbf{J}_\theta$ is the Fisher information matrix (FIM) [7]. Since the first two elements of $\boldsymbol{\theta}$ are the main parameters of interest, only the first 2 × 2 block of the FIM should be calculated. In the absence of any statistical information about the NLOS delays, it can be obtained, from Equation (10.8), after some manipulation, that [11]

$$[\mathbf{J}_{\boldsymbol{\theta}}^{-1}]_{2\times2} = \left[c^2\left(\mathbf{H}_{\mathrm{LOS}}\boldsymbol{\Psi}_{\mathrm{LOS}}\mathbf{H}_{\mathrm{LOS}}^T\right)^{-1}\right]_{2\times2}, \tag{10.9}$$

where $\mathbf{H}_{\mathrm{LOS}}$ is related to the LOS nodes only, and depends on the angles between the target node and the reference nodes, and $\boldsymbol{\Psi}_{\mathrm{LOS}} = \mathrm{diag}\{\boldsymbol{\Psi}_{M+1},\ \boldsymbol{\Psi}_{M+2},\ \ldots,\ \boldsymbol{\Psi}_N\}$, with

$$[\boldsymbol{\Psi}_i]_{jk} = 2\frac{\alpha_{ij}\alpha_{ik}}{N_0}\int\frac{\partial}{\partial\tau_{ij}}s(t - \tau_{ij})\frac{\partial}{\partial\tau_{ik}}s(t - \tau_{ik})\mathrm{d}t, \tag{10.10}$$

for $j \neq k$, and $[\boldsymbol{\Psi}_i]_{jj} = 8\pi^2\beta^2\mathrm{SNR}_{ij}$. Note that $\mathrm{SNR}_{ij} = |\alpha_{ij}|^2/N_0$ is the SNR of the jth multipath component for the ith node's signal, assuming that $s(t)$ has unit energy, and β is given as in Equation (10.2).

Note that Equation (10.9) proves that the best accuracy can be achieved by using the signals only from the LOS nodes. Moreover, the numerical examples in [11] show that, in most cases, the CRLB is approximately the same whether all the multipath components from the LOS nodes or just the first arriving paths of the LOS nodes are employed. Therefore, processing of the multipath components other than the first path does not significantly increase the accuracy, but adds computational load to the system.

Furthermore, the maximum likelihood (ML) estimate of the node location based on the delays of the first incoming paths from LOS nodes achieves the CRLB as the SNR and/or the effective bandwidth goes to infinity [11]. This result implies that for UWB systems, the first arriving signal paths from the LOS nodes are sufficient for an approximately optimal positioning receiver design. In other words, asymptotically optimal localization can be considered in two-steps as ranging (TOA estimation) and position estimation, as shown in Figure 10.2.

For a numerical example on the achievable accuracy of UWB localization systems, we consider a localization scenario in which the target node is in the middle of six reference nodes located uniformly around a circle. In Figure 10.3, the minimum positioning error, defined as

$$\sqrt{\mathrm{trace}\left(\left[J_\theta^{-1}\right]_{2\times 2}\right)},$$

is plotted versus the effective bandwidth for various numbers of NLOS nodes, M, at SNR = 0 dB. The channels between the target and the reference nodes have ten taps that are independently generated from a log-normally distributed fading model with random signs and exponentially decaying tap energy [15]. It can be observed that the large bandwidths of UWB signals make it possible to obtain location estimates with very high accuracy.

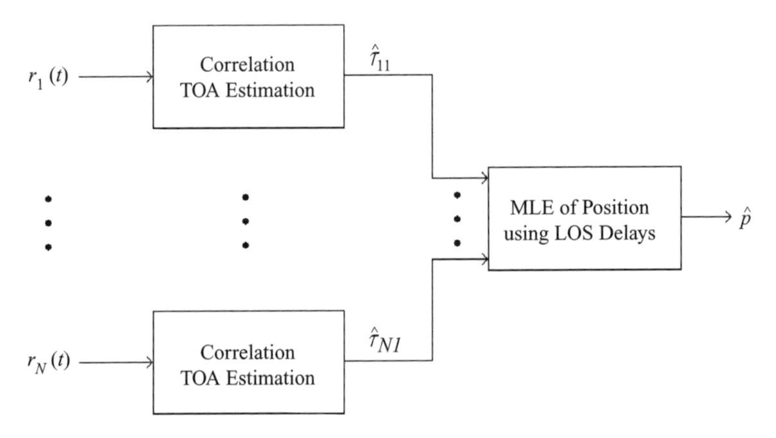

Figure 10.2 An asymptotically optimum receiver structure for localization. No information about the statistics of the NLOS delays is assumed.

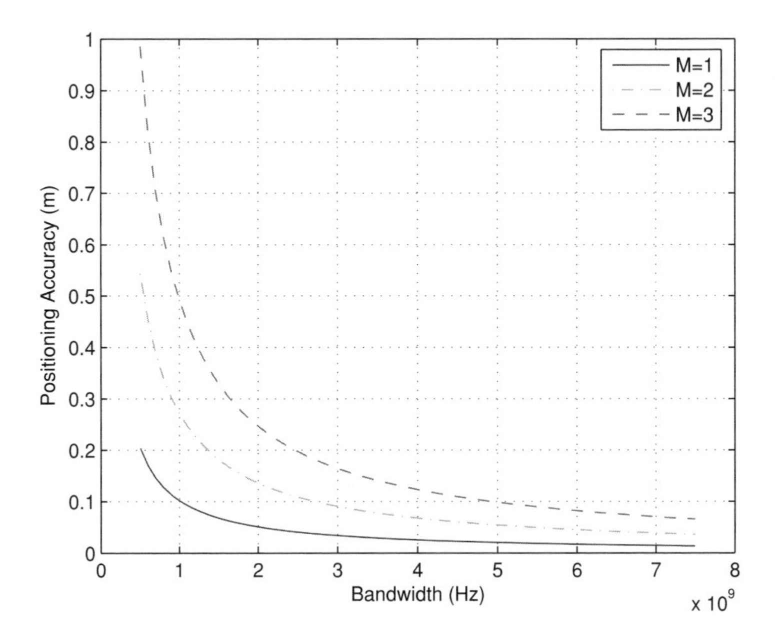

Figure 10.3 Minimum positioning error versus bandwidth for different numbers of NLOS nodes. For $M = 1$, node 1; for $M = 2$, nodes 1 and 2; and for $M = 3$, nodes 1, 2 and 3 are the NLOS nodes. The UWB channels are modeled as in [15] with $L = 10$, $\lambda = 0.25$ and $\sigma^2 = 1$.

Note that in the previous scenario, no information is assumed about the statistics of the NLOS delays. If the probability density function (PDF) of NLOS delays is available, it can be shown that the maximum *a posteriori* probability (MAP) estimate of the node location using the estimates of the delays of all the multipath components from all the nodes achieves asymptotic optimality [11]. However, in practice, the distribution of the NLOS delays is usually not available. Moreover, estimation of additional multipath delays increases the computational complexity of the localization algorithm. Therefore, only the TOA estimation of the first signal path will be considered in the sequel.

10.3.2 Time-of-Arrival Estimation Algorithms

Estimation of the first incoming signal path is the crucial step in TOA-based localization. An optimal TOA estimate can be obtained using a correlation receiver with the received waveform as the template signal[4], and choosing the time shift of the template signal corresponding to the maximum correlation with the received signal [16]. However, due to multipath propagation, the received waveform has many unknown parameters to be estimated. Hence, the optimal correlation based TOA estimation, considered in Section 10.3 (Ranging and Optimal Localization), is not practical. Therefore, the transmitted waveform is used in a 'conventional' correlation-based receiver as the template signal. However, this is obviously suboptimal in a

[4] Equivalently, a matched filter matched to the received waveform can be used.

multipath environment. Also, selection of the delay corresponding to the peak of the correlation does not necessarily yield the true TOA, since the first multipath component can be weaker than the others in some cases. Therefore, first-path detection algorithms are studied for the suboptimal correlation-based schemes [17]. One of the most challenging issues in UWB TOA estimation is to obtain a reliable estimate in a reasonable time interval under sampling rate constraints. In order to have a low-power and low-complexity receiver, one should assume symbol-rate (or, sometimes frame-rate) sampling at the output of the correlators. However, when symbol-rate samples are employed, the TOA estimation can take a very long time. To address this problem, a two-step TOA estimation algorithm that can perform TOA estimation from symbol-rate samples in a reasonable time interval is proposed in [18]. In order to speed up the estimation process, the first step provides a rough TOA estimate of the received signal based on SS measurements. Then, in the second step, the arrival time of the first signal path is estimated by employing a statistical change detection approach [19].

Another approach for the first-path detection is to use the generalized maximum likelihood (GML) estimation principle and to obtain iterative solutions after some simplifications [20, 21]. The advantage of the GML-based algorithm is that it is a recursive algorithm that can perform very accurate TOA estimation [20]. However, the main drawback is that it requires high-rate sampling, which is not practical in many applications.

An alternative to the GML-based approach is a low-complexity timing offset estimation technique using *symbol-rate* samples based on the idea of *dirty templates* [22–25]. In this approach, symbol-length portions of the received signal are used as noisy ('dirty') templates, the cross-correlations of which are employed to estimate the TOA of the received signal. This technique does not require estimation of parameters related to multipath components, except the delay of the first path. Also, significant multipath energy is collected due to the template structure. However, the noise in the template results in noise–noise cross-terms, which causes some performance loss. This effect can be mitigated to some extent by averaging operations [25]. Finally, TOA estimates obtained by the dirty template approach can have an ambiguity equal to the extent of the noise-only region between consecutive symbols. Therefore, another algorithm needs to be implemented after the dirty template based algorithm in order to obtain the exact TOA. In other words, this scheme can be used as the first step of a two-step TOA estimation algorithm.

In addition to the TOA estimation algorithms mentioned above, other approaches include use of coded beacon sequences to speed up the acquisition process by enabling searches over larger intervals [26], a frequency domain approach based on sub-Nyquist uniform sampling [27], a TOA estimation technique that tries to estimate the breakpoint between the noise-only and signal part of the received signal process [28], and a blind timing offset estimation based on the cyclostationarity of impulse radio UWB signals [29]. Although each algorithm has its advantages and disadvantages, the main issues for TOA estimation schemes for UWB systems are the following:

- A low sampling rate is required in order to have low power and practical designs. Therefore, algorithms using symbol-rate or frame-rate samples are preferable.
- For a given accuracy, TOA estimation should be performed using as few training symbols as possible; that is the time it takes to estimate the TOA should be short.
- Related to the previous issue, for a given time interval or a given number of training symbols, the TOA estimation should provide sufficient accuracy.

Considering these criteria, the dirty template based algorithm combined with a statistical change detection or a conventional correlation-based first-path detection scheme is a reasonable scheme for UWB TOA estimation. This is because the dirty template based algorithm can reduce the uncertainty about the TOA to a small region quickly, using symbol-rate samples. Then, a higher resolution algorithm based on correlation outputs of the received signal with a template signal matched to the transmitted symbol[5] can be used to estimate the TOA of the incoming signal.

Design of UWB TOA estimators that provide a trade-off between complexity and performance is still an active research area. Designing an optimal TOA estimator within the constraints discussed above, such as the maximum sampling rate and estimation interval, remains an open problem.

10.4 Challenges

Extremely accurate localization is possible in a single user, LOS and single-path environment. However, in a practical setting, multipath propagation, MAI, NLOS propagation and high timing resolution make accurate localization challenging. Multipath propagation results in a number of signal components arriving at the receiver. This has two adverse effects for accurate TOA estimation [2]. First, multipath propagation can result in partial overlap of multiple replicas of the transmitted signal. Therefore, the correlation peak may not yield the true TOA. Second, multipath propagation can cause scenarios in which the first multipath component is not the strongest one. Therefore, first-path detection algorithms, considered in Section 10.3 (Time-of Arrival Estimation Algorithms), need to be applied.

TOA estimation accuracy can also degrade in the presence of other nodes in the environment due to MAI. A technique for reducing the effects of MAI is to use different time slots for transmissions from different nodes (time-division multiplexing). However, even with such time multiplexing, there can still be MAI from neighboring networks (piconets). In order to reduce the effects of MAI, time-hopping (TH) codes with low cross-correlation properties can be employed [30], and pulse-based polarity randomization can be introduced [31, 32]. With known training patterns at the transmitter and the receiver, template signals consisting of a number of pulses matched to both TH and polarity codes can be used to mitigate the effects of MAI. Moreover, training sequences can be designed in order to facilitate TOA estimation in the presence of MAI [25].

One of the most important sources of errors in TOA estimation is NLOS propagation, which blocks the direct line-of-sight path between the transmitter and the receiver. One way to mitigate NLOS errors is to use nonparametric (pattern recognition/fingerprinting) techniques [33, 34]. The main idea behind nonparametric localization algorithms is to gather a set of TOA measurements from all the reference nodes at *known* locations beforehand, and use this set as a reference database to estimate the location when new measurements from a node are obtained. In the absence of a database, certain statistical information on NLOS can be used to mitigate NLOS errors. For example, [35] uses a simple variance test to identify NLOS situations and employs an LOS reconstruction algorithm. Moreover, by assuming a scattering model for the environment, the statistics of TOA measurements can be obtained, and then

[5] Symbol-rate samples can be obtained by using a template signal matched to the transmitted UWB symbol.

well-known techniques, such as MAP and ML, can be employed to mitigate the effects of NLOS propagation [36, 37]. In case of tracking a mobile user in a wireless system, biased and unbiased Kalman filters can be employed in order to provide accurate localization [34, 38].

Finally, high time resolution of UWB signals can pose certain challenges in practical systems. First, clock jitter becomes an important factor in evaluating the accuracy of UWB positioning systems [39]. Another consequence of high time resolution is that a large number of bins needs to be searched for true TOA. Finally, the high time resolution, or equivalently large bandwidth, of UWB signals makes it quite impractical to sample the received signal at or above the Nyquist rate, which is typically of the order of tens of GHz.

10.5 Conclusions

Theoretical analysis of UWB signals promises very accurate localization capability. Although it is not always possible to get close to the theoretical limits of localization accuracy with the current technology under practical constraints, it is expected in the near future that UWB localizers will be able to perform location estimation with subcentimeter accuracy for LOS scenarios and subdecimeter accuracy for NLOS scenarios while consuming around a few tens of milliwatts. One of the main technology improvements that would facilitate such low power and accurate localization is related to the design of low power analog-to-digital converters (ADCs) operating at the order of tens of GHz. In addition, ranging and localization algorithms that perform joint optimization of power consumption, localization accuracy and time constraints should be developed.

References

1. Gezici, S., Tian, Z., Giannakis, G.B. *et al.* (2005) *Localization via ultra-wideband radios. IEEE Signal Processing Magazine*, **22**(4), 70–84.
2. Sahinoglu, Z., Gezici, S. and Guvenc, I. (2008) *Ultra-Wideband Positioning Systems: Theoretical Limits, Ranging Algorithms, and Protocols*, Cambridge University Press.
3. IEEE P802.15.4a/D6 (Amendment of IEEE Std 802.15.4) (2006) *Part 15.4: Wireless Medium Access Control (MAC) and Physical Layer (PHY) Specifications for Low-Rate Wireless Personal Area Networks (LRWPANs): Amendment to Add Alternate PHY*.
4. Caffery, J. Jr. (2000) *Wireless Location in CDMA Cellular Radio Systems*, Kluwer Academic Publishers.
5. Mallat, A., Louveaux, J. and Vandendorpe, L. (2007) *UWB Based Positioning in Multipath Channels: CRBs for AOA and for Hybrid TOA-AOA Based Methods*, in Proceeding IEEE International Conference on Communications (ICC), pp. 5775–80.
6. Qi, Y. and Kobayashi, H. (2003) *On Relation Among Time Delay and Signal Strength Based Geolocation Methods*, in Proceeding IEEE Global Telecommunications Conference (GLOBECOM), Vol. **7**, pp. 4079–83.
7. Poor, H.V. (1994) *An Introduction to Signal Detection and Estimation*, Springer-Verlag.
8. Cook, C.E. and Bernfeld, M. (1970) *Radar Signals: An Introduction to Theory and Applications*, Academic Press.
9. Weiss, A.J. (2004) *Direct position determination of narrowband radio frequency transmitters. IEEE Signal Processing Letters*, **11**(5), 513–6.
10. Qi, Y., Kobayashi, H. and Suda, H. (2006) *On time-of-arrival positioning in a multipath environment. IEEE Transactions on Vehicular Technology*, **55**(5), 1516–26.
11. Qi, Y., Kobayashi, H. and Suda, H. (2006) *Analysis of wireless geolocation in a non-line-of-sight environment. IEEE Transactions on Wireless Communications*, **5**(3), 672–81.

12. Borras, J., Hatrack, P. and Mandayam, N.B. (1998) *Decision Theoretic Framework for NLOS Identification*, in Proceeding IEEE Vehicular Technology Conference (VTC), Vol. **2**, pp. 1583–7.
13. Gezici, S., Kobayashi, H. and Poor, H.V. (2003) *Non-Parametric Non-Line-of-Sight Identification*, in Proceeding IEEE Vehicular Technology Conference (VTC), Vol. **4**, pp. 2544–8.
14. Venkatraman, S. and Caffery, J. (2002) *A Statistical Approach to Non-Line-of-Sight BS Identification*, in Proceeding IEEE International Symposium on Wireless Personal Multimedia Communications (WPMC), Vol. **1**, pp. 296–300.
15. Gezici, S., Kobayashi, H., Poor, H.V. and Molisch, A.F. (2005) *Performance evaluation of impulse radio UWB systems with pulse-based polarity randomization. IEEE Transactions on Signal Processing*, **53**(7), 1–13.
16. Turin, G.L. (1960) *An introduction to matched filters. IRE Transactions on Information Theory*, **IT-6**(3), 311–29.
17. Chung, W.C. and Ha, D.S. (2003) *An Accurate Ultra Wideband (UWB) Ranging for Precision Asset Location*, in Proceeding IEEE Conference on Ultra Wideband Systems and Technologies, pp. 389–93.
18. Gezici, S., Sahinoglu, Z., Molisch, A.F. *et al.* (2008) *Two-step time of arrival estimation for pulse based ultra-wideband systems. EURASIP Journal on Advances in Signal Processing*, Article ID 529134, 12 pages, doi: 10.1155/2008/529134.
19. Basseville, M. and Nikiforov, I.V. (1993) *Detection of Abrupt Changes: Theory and Application*, Prentice-Hall.
20. Lee, J.-Y. and Scholtz, R.A. (2002) *Ranging in a dense multipath environment using an UWB radio link. IEEE Transactions on Selected Areas in Communications*, **20**(9), 1677–83.
21. Win, M.Z. and Scholtz, R.A. (1997) *Energy Capture versus correlator resources in Ultra-Wide Bandwidth Indoor Wireless Communications Channels*, in Proceeding IEEE Military Communications Conference (MILCOM), Vol. **3**, pp. 1277–81.
22. Yang, L. and Giannakis, G.B. (2003) *Low-Complexity Training for Rapid Timing Acquisition in Ultra-Wideband Communications*, in Proceeding IEEE Global Telecommunications Conference (GLOBECOM), Vol. **2**, pp. 769–73.
23. Yang, L. and Giannakis, G.B. (2004) *Blind UWB Timing with a Dirty Template*, in Proceeding IEEE International Conference on Acoustics, Speech and Signal Processing (ICASSP), Vol. **4**, pp. 509–12.
24. Yang, L. and Giannakis, G.B. (2004) *Ultra-wideband communications: an idea whose time has come. IEEE Signal Processing Magazine*, **21**(6), 26–54.
25. Yang, L., and Giannakis, G.B. (2005) *Timing ultra-wideband signals with dirty templates. IEEE Transactions on Communications*, **53**(11), 1952–63.
26. Homier, E.A. and Scholtz, R.A. (2002) *Rapid Acquisition of Ultra-Wideband Signals in the Dense Multipath Channel*, in Proceeding IEEE Conference on Ultra Wideband Systems and Technologies, pp. 105–9.
27. Maravic, I., Kusuma, J. and Vetterli, M. (2003) *Low-sampling rate UWB channel characterization and synchronization. Journal of Communications and Networks*, **5**(4), 319–27.
28. Mazzucco, C., Spagnolini, U. and Mulas, G. (2004) *A Ranging Technique for UWB Indoor Channel Based on Power Delay Profile Analysis*, in Proceeding IEEE Vehicular Technology Conference (VTC), Vol. **5**, pp. 2595–9.
29. Tian, Z., Yang, L. and Giannakis, G.B. (2002) *Symbol Timing Estimation in Ultra Wideband Communications*, in Proceeding IEEE Asilomar Conference on Signals, Systems, and Computers, Vol. **2**, pp. 1924–8.
30. Guvenc, I. and Arslan, H. (2004) *Design and Performance Analysis of TH Sequences for UWB Impulse Radio*, in Proceeding IEEE Wireless Communications and Networking Conference (WCNC), Vol. **2**, pp. 914–9.
31. Fishler, E. and Poor, H.V. (2005) *On the tradeoff between two types of processing gain. IEEE Transactions on Communications*, **53**(10), 1744–53.
32. Gezici, S., Kobayashi, H., Poor, H.V. and Molisch, A.F. (2004) *The Trade-Off Between Processing Gains of an Impulse Radio System in the Presence of Timing Jitter*, in Proceeding IEEE International Conference on Communications (ICC), Vol. **6**, pp. 3596–600.
33. McGuire, M., Plataniotis, K.N. and Venetsanopoulos, A.N. (2003) *Location of mobile terminals using time measurements and survey points. IEEE Transactions on Vehicular Technology*, **52**(4), 999–1011.
34. Gezici, S., Kobayashi, H. and Poor, H.V. (2003) *A New Approach to Mobile Position Tracking*, in Proceeding IEEE Sarnoff Symposium On Advances in Wired and Wireless Communications, pp. 204–7.
35. Wylie, M.P. and Holtzman, J. (1996) *The Non-Line of Sight Problem in Mobile Location Estimation*, in Proceeding IEEE International Conference on Universal Personal Communications, Vol. **2**, pp. 827–31.
36. Al-Jazzar, S., and Caffery, J. Jr. (2002) *ML and Bayesian TOA Location Estimators for NLOS Environments*, in Proceeding IEEE Vehicular Technology Conference (VTC), Vol. **2**, pp. 1178–81.

37. Al-Jazzar, S., Caffery, J. Jr. and You, H.-R. (2002) *A Scattering Model Based Approach to NLOS Mitigation in TOA Location Systems*, in Proceeding IEEE Vehicular Technology Conference (VTC), no. 2, pp. 861–5.
38. Le, B.L., Ahmed, K. and Tsuji, H. (2003) *Mobile Location Estimator with NLOS Mitigation Using Kalman Filtering*, in Proceeding IEEE Conference on Wireless Communications and Networking (WCNC), Vol. **3**, pp. 1969–73.
39. Shimizu, Y. and Sanada, Y. (2003) *Accuracy of Relative Distance Measurement with Ultra Wideband System*, in Proceeding IEEE Conference on Ultra Wideband Systems and Technologies, pp. 374–8.

11

UWB Transceiver for Indoor Localization

Gunter Fischer

ULP Group, Department of Circuit Design, Innovations for High Performance Microelectronics GmbH, Technologiepark 25, 15236 Frankfurt, Germany

11.1 General Consideration of Indoor Localization

Indoor localization systems with a positioning accuracy better than 1 m are the key enabler for providing position-specific information on or about mobile devices in buildings and halls. Such location-aware services could provide guidance in complex facilities, finding the location of particular persons or objects (for instance in case of emergency), and tracking and watching of active badges or tags (for instance in high-security areas, airports, storage halls). Usual technical options to estimate positions of mobiles are the distance estimation between the mobile and anchor knots at known positions by time measurement of propagation delays of the radio waves (time-of-arrival, TOA), by analyzing channel properties such as received signal strength or bit error rate (BER), and the angle estimation of transmitted signals of the mobile on anchor knots at known positions with known angle alignment (angle-of-arrival, AOA).

The impulse radio based UWB technique provides the best opportunity to realize an indoor localization system based on the TOA measurement principle. In particular, the short duration of the transmitted impulses (about 1 ns) is perfectly suited for high time accuracy measurement as well as allowing for multi-path resolution, which becomes important in indoor environment. The localization by propagation delay time measurements is based on trilateration. It measures the distance between a point of unknown position (mobile) and three points at known positions (anchor knots or base stations). The name 'mobile' doesn't necessarily mean moving; it describes all objects at unknown positions, which should be localized. In practice, the propagation time of an electromagnetic wave is measured and the distance is calculated by multiplication by the velocity of light (it takes about 3.3 ns for a radio wave to travel 1 m

Short-Range Wireless Communications Rolf Kraemer and Marcos D. Katz

through air). Assuming a feasible time measurement accuracy of about 100 ps, one could end up with a localization accuracy of about 3 cm per measurement.

Measuring propagation times requires the knowledge of the time of transmission (TOT) at the transmitter and the time of arrival (TOA) at the receiver. A real implementation would oblige their clocks to be synchronized with the expected accuracy, which is practically impossible for a wireless system. One work-around is to transmit precise information about the time of transmission (TOT), like the procedure in the global positioning system GPS. Depending on the application of the localization system, the mobile device could either be the transmitter, with the anchor nodes acting as receiver, or *vice versa*. Preferably, the calculation of the position will be performed, where the ranging data are fused. It is possible to omit the knowledge of the time of transmission by introducing an additional base station (anchor node) at a known position and measuring the time differences of arrival (TDOA). Note that the fourth base station has to be off the plane of the others. While the mobile device and the anchor nodes no longer need to be time aligned, the base stations still have to be tightly synchronized with each other. In today's applications this is ensured by fiber optic cables or high-bandwidth cable connections. Furthermore, to eliminate the time synchronization between the anchor nodes the more advanced scheme called 'Differential time difference of arrival' (DTDOA) might be used [1]. Here the tight time synchronization between the base stations is replaced by an additional signal transmission from the master station to provide a virtual synchronization for a limited time. Taking into account the known distances (or propagation delays) between the master station and the other base stations, this scheme allows the calculation of the unknown position from the measured time differences at each of the receivers. The clocks in the receivers can run independently without the need of an overall synchronization. Of course, clock mismatches will introduce additional ranging errors, but this seems to be tolerable in exchange for an expensive infrastructure for many applications.

A special kind of propagation delay measurement is the so-called two-way-ranging (TWR) [2], which requires a mobile device capable to receive and transmit (transceiver). This kind of distance measurement was historically used to determine the distance between satellites in orbit and the earth. The distance measurement works as follows: The base station starts an internal stopwatch and transmits a signal to the mobile at the same time. When the mobile receives the signal it immediately responds by transmitting a signal back to the base station. When the base station receives the response signal it stops the stopwatch. The measured time is twice the propagation delay of the radio waves plus an internal processing delay of the mobile device. Under the special condition of indoor usage the propagation delays are much smaller than the internal processing delay of the mobile device. Therefore, the mobile device waits a particular predefined time before responding on the request. This predefined delay should be as short as possible to keep the overall error small, but it has to be longer than the minimal internal processing time of the mobile device. In addition, it should be longer than the typical power decay profile of the channel. The advantage of this distance measurement is its autonomy with respect to other devices; only two devices are involved and need to communicate (wirelessly) to each other. This reduces the requirements on the infrastructure or even allows creating *ad hoc* networks for localization. On the other hand, the mobile device has to have the ability to precisely respond in time after receiving a request within the required tolerance. As already mentioned for the DTDOA scheme, clock drifts of the involved devices will increase localization errors in the TRW scheme as well. The symmetric double-sided two-way ranging [3] may help to reduce these errors at the expense of increased radio traffic.

The described localization methods demand the exchange of data between the mobile devices itself and/or the anchor nodes somehow. Therefore they need to establish communication links and provide data for identification. This could either happen in a centralized and coordinated fashion via a stationary infrastructure (star topology) or in uncoordinated fashion via an *ad hoc* multi-hop network (mesh topology). The communication link doesn't necessarily use the same hardware as the localization system, but for cost and power budget reasons this is preferred. On typical localization sides all mobile devices are located in the same space and therefore have to share the same communication medium. If the number of mobile nodes is large, a coordinated communication using time division multiple access (TDMA) schemes seems to be essential. For a limited number of mobile nodes uncoordinated communication is also feasible, introducing some multi-user interference [4]. A proper system design for a certain application has to spread the required features such as data communication, exchange of localization data, time measurement, position calculation from ranging results, and so on throughout the involved anchor nodes and mobile nodes. The decisions will be affected by system specifications such as localization accuracy, number of mobile devices and base stations, mobility of mobile devices, position update rates and channel conditions. This obviously leads to a large variety of hardware implementations, depending on the desired application.

11.2 Indoor Channel Properties

The main difference between outdoor and indoor channels is the larger number of reception paths due to reflections from walls and objects (multi-path propagation). The superposition of the different copies of the same transmitted signal at the receive antenna leads to so-called fading. Depending on the different propagation delays, either intersymbol or intrasymbol interference occurs. Inter-symbol interference may severely disturb the communication, the bit error rate (BER) will go up. Intra-symbol interference may not affect the data integrity, but degrades the signal-to-noise ratio (SNR) at the receiver. In usual indoor environments mainly intrasymbol interference will occur. In contrast to pure data communication, for localization the selection of the direct line-of-sight (LOS) propagation path is essential, since all NLOS paths indicate wrong (longer) distances and therefore lead to wrong positions. In addition, receivers tend to synchronize (lock) on the strongest received signal. In case of a blocked LOS path (object in between), one NLOS path appears stronger and relying on it would give a wrong distance. It is clear, that ignoring these facts will drastically reduce the achievable accuracy of the localization system [5]. Therefore, special attention has to be paid to the ability to resolve the first received path (LOS) from other possibly stronger paths (NLOS).

11.3 Impulse Based UWB Transceiver Architectures

The design of the overall architecture of transceivers suited for the TOA localization principle is usually driven by the required time measurement functionality. Compared with traditional transceivers with only data communication capability, this means some sort of high clock rate circuitry for TOA detection clocked in accordance to the demanded time resolution. One typical value is the chip rate (chip = time instance, which may contain one pulse), which is about 500 MHz in today's implementations [2]. The detection of just one impulse for localization is simply not enough due to uncertainties concerning the channel (propagation delay, multi-path) and multi-user interference. Instead, coded impulse sequences (for instance ternary M-sequences [2]) are transmitted. The usage of a cross correlation unit on the receiver side allows the exact determination of the arrival time of such a signal.

The desirable ideal cross correlator should permanently correlate the received signal with an expected signal (template) providing a time stamp when the signal appears. The sample rate of the correlator has to be larger or equal than the pulse chip rate to catch each received pulse securely. Realizations in the digital domain as well as in the analog domain are feasible. While an implementation in the digital domain requires a high-rate ADC plus proper DSP running on a high system clock, an analog implementation requires a correlation bank with proper template generation plus an integration and decision unit. Both approaches demand parallel processing in the correlation device since one never knows the exact reception time (when the received signal comes in) and since the template signal is usually longer than the sample period. A permanent cross correlation becomes possible only by providing N parallel taps (sequence duration divided by sample period), each shifted in time by one sample and operating them in a loop. This means, that N correlation results will appear at once in parallel per sequence duration.

In terms of power consumption a permanent cross correlation might not be desirable. To reduce the power consumption one can either disrupt the permanent correlation in time or one can reduce the number of taps (branches) of the correlator. Both options introduce the risk of missing the received signal entirely. Nevertheless, they become feasible if a preceding synchronization phase allows the prediction of the arrival time of the received signal with a certain accuracy. Then, an observation time window with the dimensions of the uncertainty of the prediction would be sufficient to catch the signal (temporary continuous cross correlation). The time window should be large enough to include the maximum allowable fluctuation of the channel between two sequences, otherwise the synchronization could be lost. The main decision points concerning the number of taps (branches) of the parallel correlation bank are: chip area (cost), power consumption, synchronization requirements and channel variability.

The cross correlation on the chip rate provides a medium localization accuracy, which is a good compromise between effort and performance. On the other hand, low-cost solutions demanding less hardware effort with finally weaker performance as well as highest performance solutions might be of particular interest. Low-complexity solutions and even software localization support can provide certain accuracy of about 1 m under LOS conditions. By averaging, the accuracy can be improved at the expense of lower update rates. In case of non-coherent energy peak detection, the detector cannot derive any information from the received signal, except stating the existence of received energy at a particular time. Therefore, there is no way to distinguish between different reception paths, multi-user interferences, and so on. In case of coherent impulse detection, even phase information of the impulses might be extracted, allowing for a maximum sampling rate up to the Nyquist rate of the center frequency of the impulses. This leads to high-performance solutions, which are required under harsh radio channel conditions (for instance industrial environment NLOS). Higher sample rates than the chip rate are used to achieve better time measurement resolution and better multi-path resolution. Of course, this provides a much better time resolution, but also requires a much higher postprocessing rate. Again, realizations in the digital as well as in the analog domain are feasible today.

For high-performance localization systems the coherent impulse detection seems mandatory to provide sufficient headroom for final localization accuracy much better than 1 m, especially under harsh conditions. The optimum coherent detector demands the generation of a template signal (right in time) including the channel response (channel estimation). This becomes a tough job if the channel variability is high (moving devices). However, employing rectangle or sinusoidal periodic templates for coherent detection still provides usable impulse phase information for good time measurement resolution and causes lower complexity.

In contrast to simple data communication systems the pulse generation becomes important in localization systems, if coherent impulse detection is employed on the receiver side. In such case the pulse shaping at the transmitter is important for the overall performance of the system. If the phase of the pulses carries any information, they have to be generated properly as well as predictably shaped after the propagation through the channel. Two pulse shapes are often used, the Gaussian shape and the root-raised cosine shape. For proper phase control during the impulse generation either direct digital generation, including analog filtering for bandwidth limitation (reconstruction), or direct up-conversion schemes (gated oscillator) seem to have the best ratio between performance and complexity.

References

1. Fischer, G., Dietrich, B. and Winkler, F. (2004) *Bluetooth Indoor Localization System*, in Proceeding Workshop on Positioning, Navigation and Communication, pp 147–56.
2. IEEE 802.15.4a (2007), http://www.ieee802.org/15/.
3. Hach, R. (2005) *Symmetric Double Sided Two-Way Ranging*, IEEE P802.15 Working group for WPAN, Doc. P. 802.15-05-0334-00-004a.
4. Luediger, H., Kull, B. and Perez-Guirao, M.D. (2007) *An Ultrawideband Approach Towards Autonomous Radio Control and Positioning Systems in Manufacturing and Logistics Processes*, in Proceeding Workshop on Positioning, Navigation and Communication, pp. 291–5.
5. Denis, B. (2003) *UWB Localization Techniques*, Presentation on UWB Summit in Bercy.

12

UWB Higher Layers

Maria-Gabriella Di Benedetto
INFOCOM Department, University of Rome La Sapienza, Via Eudossiana 18, 00184 Rome, Italy

12.1 State of the Art

Ultra Wide Band radio (UWB) is, in principle, a physical transmission technique suitable for all kinds of applications. Given the strong power emission constraints imposed by the regulatory bodies in the United States., and the further more restrictive rules recently out in Europe, UWB has emerged as a particularly appealing transmission technique for applications requiring either high bit rates over short ranges or low bit rates over medium-to-long ranges. The high-bit-rate/short-range case includes Wireless Personal Area Networks (WPANs) for multimedia traffic, cable replacement such as wireless USB and wearable devices, for example wireless Hi-Fi headphones. The low-bit-rate/medium-to-long-range case applies to long-range sensor networks such as indoor/outdoor distributed surveillance systems, nonreal-time data applications, for example e-mail and instant messaging, and in general all data transfers compatible with a transmission rate of the order of 1 Mb/s over several tens of meters. A recent release of an improved version of the IEEE 802.15.4 standard for low rate WPANs [1], together with the successful standardization process of a UWB-based amendment to the standard (IEEE 802.15.4a) led to increased attention to the low-bit-rate case.

The most appealing scenarios of application for low rate refer to networks that commonly adopt the self-organizing principle, that is, distributed networks. Examples of these networks are *ad hoc* and sensor networks, that is groups of wireless terminals located in a limited-size geographical area, communicating in an infrastructure-free fashion, and without any central coordinating unit or base-station. Communication routes may be formed by multiple hops to extend coverage. This paradigm can be viewed as different in nature from the cellular networking model where, typically, nodes communicate by establishing single-hop connections with a central coordinating unit serving as the interface between wireless nodes and the fixed wired infrastructure.

Short-Range Wireless Communications Rolf Kraemer and Marcos D. Katz

With respect to resource management and control as well as routing, which are functions that are typically demanded to the Medium Access Control (MAC) and Network layers, typical UWB features such as the need for operating at low power versus a rather accurate ranging capability have proved to have a significant impact on the design. The Impulse Radio (IR) principle, in particular, offers, because of the impulsive nature of the transmission, a boost to innovation in designing efficient algorithms; IR intrinsically partitions time in a peculiar way, because of the short pulse duration, that is the time interval where most energy in the received pulse is concentrated is rather limited and typically lies between about 70 ps and tens of nanoseconds, depending upon pulse shape and channel behavior. The spectrum of the IR signal is usually shaped by encoding data symbols using Time Hopping (TH) pseudorandom sequences that also serve as users' signatures, and ensure access to the medium by multiple users. This resource partitioning scheme called Time Hopping Multiple Access (THMA) falls at first glance in the CDMA category where different users adopt different codes. As in CDMA, each code modifies the transmitted signal in such a way that a reference receiver is capable of isolating the useful signal from other user signals that are perceived as interferers. The possibility of removing these unnecessary contributions mainly depends upon the characteristics of the codes used for separating data flows, and upon the degree of system-level synchronization. In realistic scenarios where devices do not achieve ideal synchronization, and codes lose orthogonality because of different propagation delays on different paths, the receiver may not be capable of removing the presence of the undesired signals completely, and as a consequence system performance is affected by Multi-User Interference (MUI). MUI features in continuous transmissions, such as CDMA, versus impulsive radio transmissions may differ in nature substantially, especially when a reduced number of pulses travels over the air interface as may occur in any of the following cases: (i) a reduced number of transmitters is present over a given geographical area, that is the density of users is low; (ii) transmitters are characterized by low data rates; (iii) the number of pulses per bit is low. Examples of these application scenarios are sensor networks that are typically characterized by low data rates, and sparse topologies. In the impulsive case, MUI can be better modeled following the concept of pulse collision rather than, as is common practice, in conventional systems, as continuous disturbance. Given the short pulse duration, pulse collision might result in a rather rare event. Moreover, since information bits are usually grouped into packets by the MAC before being sent over the air interface, it is on these packets that the effect of interference should be evaluated. MUI can be revisited under this perspective, by observing that interference is truly perceived as such in the event of both packet collision and pulse collision within the collided packet.

Based on this approach Di Benedetto *et al.* [2] proposed a UWB-IR tailored MAC design for asynchronous UWB networks in which asynchronous UWB users are allowed to transmit in an uncoordinated manner: the $(\text{UWB})^2$ approach, further refined in [3]. Results of network simulation showed that the probability of successful packet transmission is fairly high, for uncoordinated transmission of a reasonable number of users, leading to the conclusion that a UWB-tailored MAC for low-bit-rate applications may adopt a pure Aloha approach, the simplest decentralized access policy. The other side of the coin is an increased overhead required by the presence of a synchronization trailer in each transmitted packet. However, this drawback may be an acceptable price to pay for low-bit-rate applications, while an additional advantage offered by the proposed approach is the possibility of collecting distance information between transmitter and receiver during control packet exchange. This information can enable the introduction in the MAC of new functions, such as distributed positioning. The Aloha approach proposed in $(UWB)^2$ was presented in 2005 at a IEEE 802.15.4a TG and Aloha was eventually selected as the main access strategy in the standard that was recently completed.

12.2 Challenges

The design of a communication network has been traditionally subdivided into a hierarchy of levels or layers. At the top of the hierarchy, the application layer defines the high-level services to be supported by the network. These services are subdivided into classes of services at the network layer, each class having different requirements, for instance in terms of quality of service (QoS). Following a layered approach, communication networks have not been designed from scratch in a principled way to optimize global system properties. The traditional design of a communication network has rarely been able in fact to focus on the optimization of global system properties. By creating a new physical layer, UWB offers unique opportunities for the principled design of new classes of communications networks based on the optimization of complex trade-offs between use of resources, quality of service, costs and other relevant parameters.

Developing a general methodology for the principled design of UWB-based communication systems is definitely a future challenge. In the network layer the overall architecture and communication strategies of the system, such as routing, are defined, taking into account the constraints imposed by the application and the physical layer through the interface provided by the MAC layer. In this perspective, the *de novo* design of a rational communication network should first focus on the network layer, taking into account the goal of the network as well as the fundamental constraints originating from the physical layer. By focusing first on the network layer, the design is driven by the optimization of a properly tailored cost function in the network layer, that is flexible enough to subsume a range of possible applications and systems and to support *ad hoc* networking. A straightforward example of such an approach is multi-hop networks, where routing costs label different possible paths. A routing cost might interestingly depend upon a variety of parameters, including battery life, interference patterns, delay, and so on, where features may selectively become priority according to the nature of the network.

A major challenge consists in incorporating into the above cross-layer approach the concept of a cognitive radio, capable of adapting to the environment and of adjusting its principles of operation as a function of both external and internal unpredictable events. This concept translates into developing smart wireless devices able to sense the environment, whether this refers to channel or interference patterns, and modifying accordingly the spectral shape and other features of radiated signals while maintaining compatibility with regulations on emitted radiations. The final goal remains to form wireless networks that cooperatively coexist with other wireless networks and devices. Given their ultra wide bandwidth, Ultra Wide Band (UWB) radio signals must in principle coexist with other radio signals. The problem of possible interference from and onto other communication systems that must be contained within regulated values is thus intrinsic to the UWB radio principle, and is a mandatory aspect in conceiving the design.

The cognitive radio concept focuses on improving the utilization of the wireless resource, that is, the electromagnetic spectrum. As such, it mainly applies to the behavior of a single node regarding both its transmitter and receiver components, and as a direct consequence to the logic ruling communication over a single link.

The introduction of the cognitive principle in the logic of the UWB network as regards resource management and routing will be a major challenge. This operation will require extending the cognitive concept to rules of operation that take into account the presence of several nodes in the network as well as their instantaneous configuration. Cognitive principles must be integrated in the rules of interaction between nodes in the network, that is, the set of wireless nodes forms a social network that must be modeled and analyzed as one entity in order to optimize the design.

12.3 Vision

When cognitive principles affect rules of interaction, cognitive scientists refer to a phenomenon called consciousness. While consciousness appears as a unique feature of the human brain, it is interesting to map the concept onto our context. Consider a set of UWB nodes forming a self-organizing *ad hoc* network. Suppose some nodes are cognitive, that is, they are gifted with some sort of intelligence based on adaptive algorithms that take into account the environment and the network in which they operate. Introducing consciousness in the network is thus modeling nodes that operate in a cognitive way and that apply conscious mechanisms to communicate in the network, and adapt their behavior to current network topology and status.

One important concept that needs particular attention is the time scale over which operations such as adaptation take place. A common understanding assumes that all operations occur at clock intervals, or multiples thereof. When constant changes are desired, the clock interval duration is reduced and the system is forced to operate at higher sampling rates with the aim of pushing its behavior to being continuous.

The above model is limited in its nature. Consider the example of a node in an *ad hoc* network. The input-output dynamics might be well described by classical discrete systems formalism, while the phenomenon that should force the node to change its rules of operation might be asynchronous with respect of node dynamics. To complicate matters further, uncertainty may affect system behavior substantially. Noise as well as unpredictable events such as atmospheric changes or mobility random patterns, to cite a few, are examples of uncertainty that must be incorporated in the model. An accurate modeling of a node in a wireless network requires a mathematical foundation where continuous and discrete dynamics are appropriately defined.

Hybrid systems are dynamical systems where continuous and discrete dynamics are embedded together with propositional logic. Continuous and discrete variables interact and determine the hybrid system evolution. The hybrid state of a hybrid system is made of two components: the discrete state q_i belonging to a finite set Q and the continuous state x belonging to a linear space R^n. The evolution of the discrete state q_i is governed by an automaton, while the evolution of the continuous state x is given by a dynamical system controlled by a continuous input and subject to continuous disturbances. Whenever a discrete transition occurs, the continuous state is instantly reset to a new value. Even if the intuitive notion of hybrid system is simple, the combination of discrete and continuous dynamics and the mechanisms that govern discrete transitions create serious difficulties in defining its operation precisely. Other complexity stems from the continuous state reset that occurs when the system undergoes a discrete transition.

Hybrid systems are powerful abstractions for modeling complex systems and have been the subject of intense research in the past few years by both the control and the computer-science communities [4]. Particular emphasis has been placed on a unified representation of hybrid models rooted in rigorous mathematical foundations. Moreover, hybrid models have been used in a number of applications to understand the behavior of systems where digital controls are applied to continuous and discrete processes. The interaction between heterogeneous semantics has been difficult to understand without the rigorous framework offered by hybrid system formalization. A particular area of interest in this domain is represented by networks of control systems acting in a coordinated fashion where communication can be implemented in a variety of ways, wireless and wired, including optical fibers. The effects of nonideal communication channels are well captured by hybrid systems. In addition, when the resource is not only scarce, but must also be shared, and moreover is time-varying as in wireless channels, an interesting hybrid control problem appears: how to manage the resource so that the overall requirements on the communication network are satisfied.

Hybrid system formalism offers the framework for modeling the behavior of self-organizing networks. Thanks to this formalism we may succeed in characterizing self-organizing network dynamics as a discrete finite-state automaton where, for each state, state-specific rules of operation govern the evolution of the network itself. As already noted above, the ultra wide bandwidth of radiated signals, radio devices operating under UWB rules must coexist with severely interfered environments and must control their behavior in order to favor coexistence. In other words, these radios must be able to adapt to ever-changing operating conditions. We foresee that this can be achieved by introducing conscious mechanisms in the analysis process that is used by nodes for determining whether changes in the global network state are appropriate [5].

References

1. IEEE 802 Part 15.4 (2006) *Wireless Medium Access Control (MAC) and Physical Layer (PHY) Specifications for Low-Rate Wireless Personal Area Networks (WPANs)*, available at http://standards.ieee.org/getieee802/download/802.15.4-2006.pdf.
2. Di Benedetto, M.-G. and Giancola, G. (2004) *Understanding Ultra Wide Band Radio Fundamentals*. Upper Saddle River, NJ: Prentice-Hall.
3. Di Benedetto, M.-G., De Nardis, L., Junk, M. and Giancola, G. (2005) $(\mathrm{UWB})^2$: uncoordinated, wireless, baseborn, medium access control for UWB communication networks. *ACM/Springer Journal on Mobile Networks and Applications*, **10**(5), 663–74, Special Issue on WLAN Optimization at the MAC and Network Levels.
4. Antsaklis, P.J. (2000) Special issue on hybrid systems: theory and applications a brief introduction to the theory and applications of hybrid systems. *Proceedings of the IEEE*, **88**(7), 879–87.
5. Di Benedetto, M.-G., Giancola, G. and Di Benedetto, M.D. (2006) Introducing consciousness in UWB networks by hybrid modelling of admission control. *ACM/Springer Journal on Mobile Networks and Applications*, **11**(4), 521–34, Special Issue on Ultra Wide Band for Sensor Networks.

13

UWB Sensor Networks for Position Location and Imaging of Objects and Environments

Reiner Thomä, Ole Hirsch, Jürgen Sachs and Rudolf Zetik
Institute of Information Technology, Department of Electrical Engineering and Information Technology, Ilmenau University of Technology, Germany

13.1 Introduction

Ultra-wideband (UWB) radio sensor networks promise interesting perspectives for position location and object identification in short-range environments. Their fundamental advantage comes from the huge bandwidth which may be up to several GHz, depending on the national regulation rules, for example frequency ranges of 3.1–10.6 and 1.99–10.6 GHz are deregulated for communications and wall penetrating radar, respectively, in the United States. (FCC). Even 50 MHz to 18 GHz are envisaged for ground and wall-penetrating radar in Europe (which will include specific protection measures of sensitive sites). Consequently, UWB access network infrastructure allows unprecedented spatial resolution in geolocalization of active UWB devices without range ambiguity [1, 2] and UWB radar sensors [3] allow high resolution in detection and localization of passive objects at short-range distances. With the lower frequencies involved in the UWB spectrum, looking into or through nonmetallic materials and objects becomes feasible. This is of major importance for applications such as indoor navigation and surveillance, object recognition and imaging, through wall detection and tracking of persons, ground penetrating reconnaissance, wall structure analysis, and so on. UWB sensors preserve their advantages – high accuracy and robust operation – even in multipath rich propagation environments. Compared with optical sensors, UWB radar sensors maintain their accuracy in bad viewing conditions and still produce useful results in non-LOS situations by taking advantage of multipath.

Short-Range Wireless Communications Rolf Kraemer and Marcos D. Katz

Despite the excellent range resolution capabilities of UWB radar sensors, detection and localization performance can be significantly improved by cooperation between spatially distributed nodes of a sensor network. This allows robust sensor node localization, even in the case of partly obscured links. Moreover, distributed sensor nodes can acquire comprehensive knowledge on the structure of the unknown environment and construct an electromagnetic image which is related to the relative sensor-to-sensor coordinate system. Distributed observation allows robust detection and localization of passive objects and identification of certain features of objects such as shape, dynamic parameters and time variant behaviour.

This all makes UWB a promising basis for autonomous navigation of mobile sensor nodes, for example maneuvrable robots – in an unknown or even hostile environment that may arise as result of an emergency situation. In this case UWB can help to identify hazardous situations such as broken walls, locate persons buried alive, roughly check the integrity of building constructions, detect and track intruders, and so on.

In the following we briefly explain the different types of sensor nodes and the basic structure of UWB sensor networks for the application described above. We refer briefly to localization of sensor nodes within a network. Then we review the basic imaging principles that are applied to recognize the structure of the static propagation environment and to detect moving objects. Finally, we demonstrate time variant feature detection for identification of human beings behind a wall by their respiratory activity.

We point out that the flavour of this paper is mainly experimental. Basic sensor node activities shall be demonstrated by measured examples. The experiments were carried out by using a multiple-input multiple-output (MIMO) channel UWB sounder based on a proprietary SiGe chipset [4, 5]. The frequency bandwidth was about 700 MHz to 4.5 GHz, with the lower frequency limit determined by the antennas used. The multiple transmitters launch periodic pseudorandom binary signals that are allocated to different time slots. The receivers, which are working in parallel know the transmit signal and its time allocation and produce MIMO impulse responses by correlation processing. In this way, advantage is taken of impulse compression.

13.2 Structure of UWB Sensor Networks

The sensor nodes may be heterogeneous in terms of their sensing capabilities and mobility. 'Dumb' nodes may act just as illuminators of the environment. This requires only transmitting operation and no sensing and processing capability. Deployable nodes, placed at verified positions may act as reference or anchor nodes for the localization of roaming nodes. Other nodes are spread around as disposable or moving observers, which acquire information about the structure of the environment by recording backscattered waves. The most 'sensible' nodes are equipped with multiple antennas, for example one transmitting antenna (Tx) and one or two receiving (Rx) antennas. In case of two receiving antennas, such a sensor already constitutes a small-baseline bistatic radar, which resembles the sensing capability of a bat and allows estimation of directions of arrival. We will also refer to these nodes as 'scouts'. Thus, with regard to capabilities of sensor nodes involved and to their mutual cooperation we distinguish three basic structures of sensor networks:

- The multistatic approach assumes a number of widely distributed and synchronized cooperating sensor nodes. The position of those nodes is estimated and tracked with respect to the anchor nodes. By applying coherent data fusion methods moving nodes will create an image of the propagation environment.

- The ‘bat-type’ approach consists of the ‘scouts’, which are able to detect and recognize characteristic features of the propagation environment on their own. This allows building up partial maps of the environment, investigating unknown objects in more detail, identifying object features, and so on.
- The mixed approach is characterized by a cooperation of all types of sensors. For example, multistatic sensors will support the localization of the scout sensors and these will deliver additional reference information that relates the local sensor coordinate system to the structural details of the environment.

13.3 Sensor-to-sensor Node Localization

Since imaging of environments and objects in essence is a combination of sequential observations of one moving sensor or of a number of spatially distributed sensors, the knowledge and tracking of the precise location of each sensor node is a prerequisite. Whereas existing indoor navigation systems mostly rely on fixed reference beacons belonging to the infrastructure of some wireless access network, localization in an unknown environment has to be autonomous and self-contained. The nodes must at first establish their own local coordinate system by estimating their relative position. Then the structure of the unknown environment has to be recognized. This will be achieved by imaging and object recognition methods. When the structure of the environment is finally recognized, it will be related to the sensor network coordinate system. Since imaging includes coherent combining of multistatic signals, we should aim at centimetre-level location accuracy. With its unprecedented temporal resolution, time of arrival (ToA) based localization methods are the natural choice in case of UWB. This also avoids usage of expensive antenna arrays which we would need for angle of arrival methods [1]. ToA, however, requires temporal synchronization between the nodes. This can be achieved by the RTToA (Round Trip Time of Arrival) approach which means that every sensor involved must be able to retransmit received signals. However, to reduce overall interference created by the UWB nodes, it makes sense to keep the number of transmitting nodes small. Those nodes, being capable of transmitting and receiving, play the role of deployable anchor nodes. Anchor nodes will establish a common time reference. They should be placed at ‘strategic’ positions, meaning that they span a large volume and ensure a complete illumination of the environment. Then *Rx*-only nodes can calculate their own position relative to the anchor nodes by applying TDoA (time difference of arrival) or related pseudorange methods which requires at least one additional synchronized anchor node.

The major source of sensor location error is nonline-of-sight (NLOS) propagation [2, 6]. NLOS between anchor nodes can be avoided by proper choice of anchor’s positions and selective choice of related anchors. Further error mitigation schemes for moving sensor localization include redundant anchor reference and selective anchor choice, sensor cooperation, soft sensor localization decision including multiple hypotheses test and localization tracking. More detailed discussion of these methods is beyond the scope of this chapter.

13.4 Multistatic Imaging of Environments

Provided the scene is sufficiently illuminated by the signals transmitted from the anchor nodes, the backscattered waves carry valuable information on the structure of the propagation environment. A focused image can be built if enough information is collected. The quality of

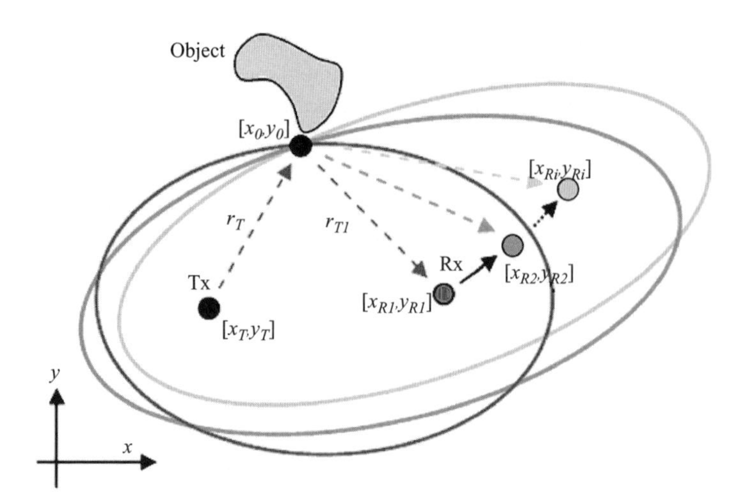

Figure 13.1 Imaging of the static environment by sequential Kirchhoff migration.

this image depends heavily on the number and the positions of the illuminating and observing nodes. If there are illuminators (e.g., the anchor nodes) and observers available simultaneously, then the available information is instantaneous. This allows real-time processing which will be used for localization of moving objects in Section 13.6. For imaging of the static propagation environment, one moving observer is enough. While moving around, the observer collects data and the image is built sequentially and enhances gradually.

The excellent resolution of UWB easily allows separation of the LOS component (which is used for mobile observer localization) and backscattered waves in the time domain. Moreover, time-of-flight information is unambiguous over a wide range. The respective time domain imaging methods are called Kirchhoff migration. They rely on Born's approximation, which presumes undisturbed ray-optical propagation [7]. The principle of migration imaging is depicted in Figure 13.1. The transmit signal is emitted at the fixed point $[x_\mathrm{T}, y_\mathrm{T}]$. At the variable position $[x_{R_i}, y_{R_i}]$, the receiver collects the total response $R_i(\tau)$. Assuming single bounce reflection and the propagation velocity v, the wave component reflected at the point $[x_0, y_0]$ can be found at the time delay τ_i.

$$\tau_i = \frac{1}{v}(r_\mathrm{T} + r_{Ri}) \tag{13.1}$$

at the recorded response which corresponds to an elliptical location in the image. The more the observer moves, the more energy 'migrates' to the pixel position $[x_0, y_0]$ in the focused image $o(x_0, y_0)$

$$o(x_0,\ y_0) = \frac{1}{N}\sum_{i=1}^{N} R_i(\tau_i) \tag{13.2}$$

where N is the number of available observations. Note that the data fusion described by Equation (13.2) may be coherent if $R_i(\tau)$ is complex or bipolar real valued (in case of baseband signal processing) or noncoherent if unipolar signal envelope is used.

From Figure 13.1, it can be deduced that signal energy does not only cumulate in the desired pixel. There also arise local maxima at every position where ellipses are crossing. In [8] it has been shown that those artefacts tend to be most severe if regular (e.g., rectangular) observer tracks are used. They are smeared for irregular random tracks. In the same conference paper [8] we have reported another possibility to enhance the image quality. These methods consist of cross-correlating the response of one or more auxiliary measurements. The spatial scheme for one additional reference receiver Rx_{ref} is indicated in Figure 13.2 and for two additional reference receivers in Equation (13.3). Integration is performed over the whole compressed transmit pulse length T.

$$o_{x\mathrm{corr}}(x_0, y_0) = \frac{1}{N}\sum_{i=1}^{N}\int_{-T/2}^{T/2} R_i\left(\frac{r_T + r_{Ri}}{v} + \xi\right) R_{\mathrm{ref1}}\left(\frac{r_T + r_{R\mathrm{ref1}}}{v} + \xi\right) \times R_{\mathrm{ref2}}\left(\frac{r_T + r_{R\mathrm{ref2}}}{v} + \xi\right) \mathrm{d}\xi \tag{13.3}$$

Figure 13.2 clearly shows that the spurious response is eliminated since the primary and the reference ellipse, in general, are not crossing at the same point. Since we have at least three anchor nodes available for x, y-localization, the optimum constellation may be to use one anchor node transceiver as illuminator and all other available anchor node receivers as cross-correlation reference. This procedure is repeated for every other anchor node taken as an illuminator (since the anchors are continuously transmitting anyway) and the results are fused. The influence on the image amplitude by the multiplication at the desired points can be corrected if necessary.

The example described below was calculated from measurements by standard Kirchhoff migration in an industrial environment, which is shown by the photograph in Figure 13.4.

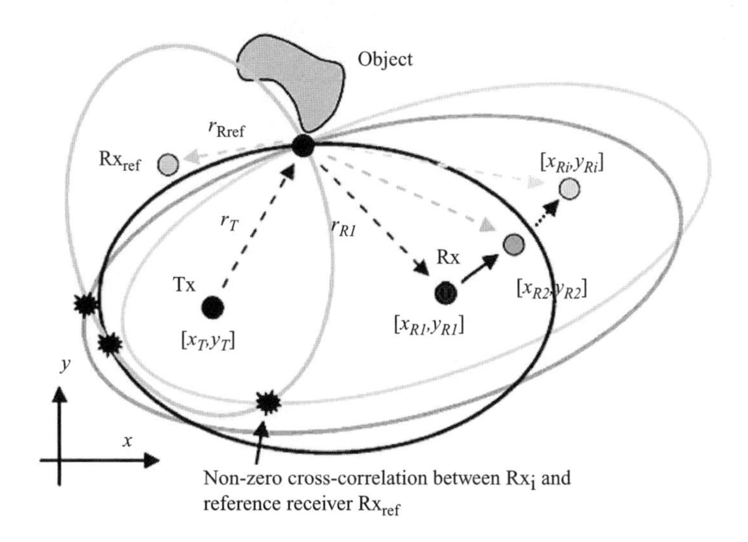

Figure 13.2 Cross correlated Kirchhoff migration.

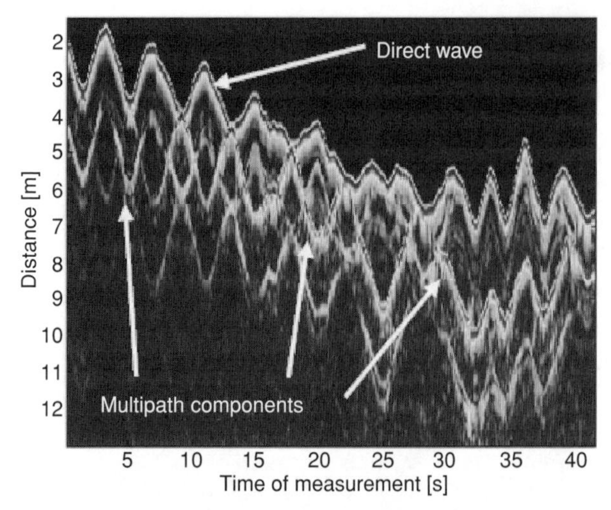

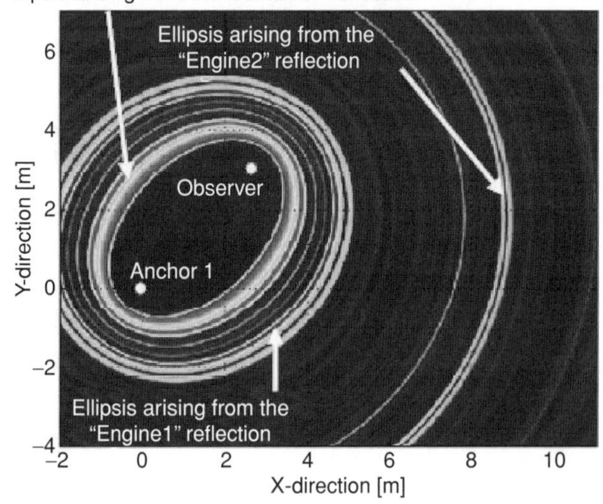

Figure 13.3 Measured sequence of impulse responses in the industrial environment (upper) and received nonfocused signal image at a single receiver position (lower).

The anchor nodes and the moving observer were synchronized by hardware. So in this case, only two anchor reference nodes were enough for the localization of the moving observer. Figure 13.3 (upper) shows the recorded impulse response sequence of the channel created by the anchor node 1 and the moving observer. The leading LOS component is used for localization of the observer and the remaining multipath components are used for image migration as described. A single contribution to the migrated image (at one observer position) is also depicted in Figure 13.3. The elliptic structure is clearly visible. Figure 13.4 shows the migrated image and the superimposed track of the moving observer. Here a random track was created by a roaming observer.

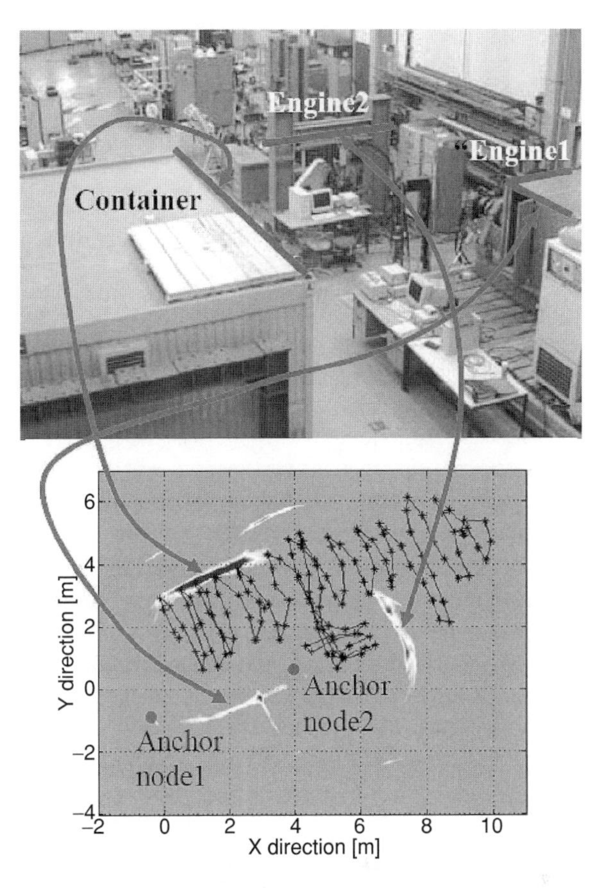

Figure 13.4 Industrial propagation environment (upper) and resulting migrated image (lower). The random track shows the moving observer and the points indicate the fixed anchor nodes.

13.5 'Bat-Type' Imaging

Whereas the multistatic imaging approach is most appropriate to submit an overview image of the global structure of the environment, the bat-type imaging approach to be described in this section is more suitable to investigate objects in more detail and to produce partial maps of the environment. Other than the multistatic approach it does not require coherent cooperation of widely distributed nodes. Since a 'bat' sensor node represents a-self contained bistatic radar platform, it can act autonomously as a 'scout'. Nevertheless, it can take advantage of cooperation with other nodes, for example for localization relative to anchor nodes and by noncoherent data fusion.

Being equipped with two receiving antennas, a bat sensor can submit a joint space-time radargram which is already an image as seen from the bat sensor. A sequence of observed data along a specifically chosen track can be fused for image enhancement. Figure 13.5 shows a sequence of space-time radargrams recorded by a scout on its track. The objects observed in this example are two point-like objects. Since the antenna baseline is short, the cross-range

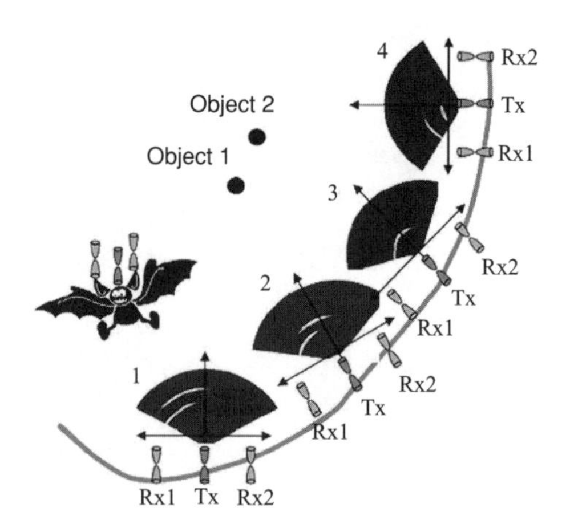

Figure 13.5 Scanning of an object by a bat-type sensor along an arbitrary track.

resolution (in terms of DoA) is low whereas the range resolution (in terms of delay) is high because of the huge bandwidth. Note that it easy to remove the direct coupling component because of the relatively fixed antenna arrangement.

A sharper image can be generated by fusing the images which are taken from different viewing angles. To this end, the space-time radargrams are appropriately shifted, rotated and stacked as indicated in Figure 13.6. To achieve this result we not only have to know the position of the scout, but also its attitude. The latter is not delivered by the localization

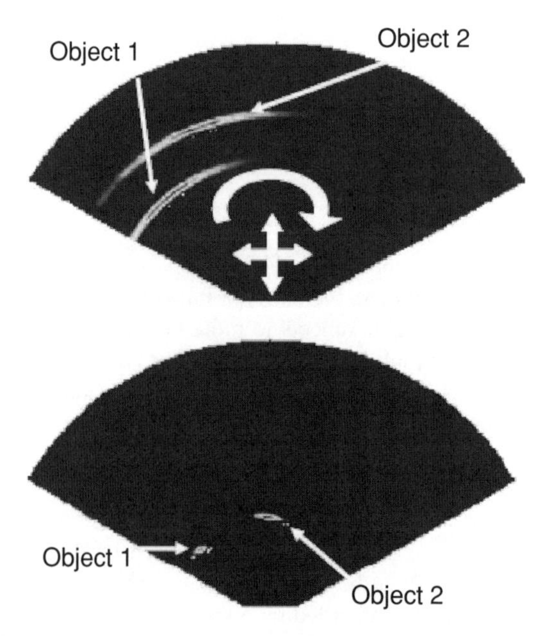

Figure 13.6 Image focusing by stacked averaging of correlated image snapshots. The upper figure demonstrates the shift and rotation operation to achieve the best match of image details.

approach described in Section 13.3. One possibility to get this additional information is to apply supplementary inertial sensors. Another possibility is based upon self-adjusting or auto-focusing as described in the following paragraph.

The basic idea of auto-focusing results from the observation that the stationary objects of a scenario usually dominate the backscattered signal [9]. Provided that those objects are appropriately illuminated, they show up in the image independent of the aspect angle and distance of the observer. Thus, if at least some dominant scatterers appear in an environment, these elements can be used as a landmark or anchor reference for auto-focusing. The individual images are stacked and averaged in a way that the reference objects experience the best focus. To this end, the individual images are shifted and rotated to achieve the best match at the chosen reference objects. If the object details are sharp and clear enough, explicit knowledge of the track is not required.

A specific problem of Kirchhoff migration is the appearance of artefacts, as already discussed in the previous section where we have proposed to use additional reference nodes for cross correlation (see Equation 13.3). The following example [10] demonstrates the resulting image enhancement. Figure 13.7 (upper part) shows a simulated result of simple Kirchhoff migration which is calculated from a sequence of signals, as described in Section 13.4. A scout sensor carrying only one transmitting and one receiving antenna is used in this case. Since the illuminator is always close to the observer this means that the object is illuminated as if the scout were using a flashlight. Here the scout moves on a rectangular track around the object, which is a vertical metallic pipe. The artefacts, which are most severe on those regular tracks, are clearly visible in the upper picture. The lower pictures show enhanced images which results from cross correlated migration as described in Equation (13.3) with one and two additional references, respectively. These reference nodes are placed separately from the scout node and may be fixed.

13.6 Detection and Localization of Time Variant and/or Moving Objects

In contrast to localization of active sensor nodes (that are acting as UWB transmitter or receiver), passive objects have to be detected and localized by backscattered radio waves. Thus, there are many analogies to imaging methods. However, the imaging methods discussed in Section 13.4 predominantly rely on synthetic aperture data principles, which restrict them to static environments. Reflection from moving objects will be smeared in the focused image and thus gradually disappear.

Therefore, detection and localization of moving objects requires real-time cooperation of sensor nodes. Moreover, detection of time-variant object features can help to identify a specific object. One example are human beings which will return characteristic radar patterns when walking. But even in a static position a human being returns time-variant signals because of its heart beat and respiratory activities. In this section we will summarize the basic methods of detection and location of moving or time-variant objects and we will give an example for through wall detection of human respiratory activity.

Moving and/or time-variant objects have to be detected in the presence of clutter from the static environment which in most cases will be much stronger. Therefore, the first processing step is background subtraction [17], which eliminates or at least reduces disturbing static signals. Background subtraction can considerably enhance the dynamic range for detection

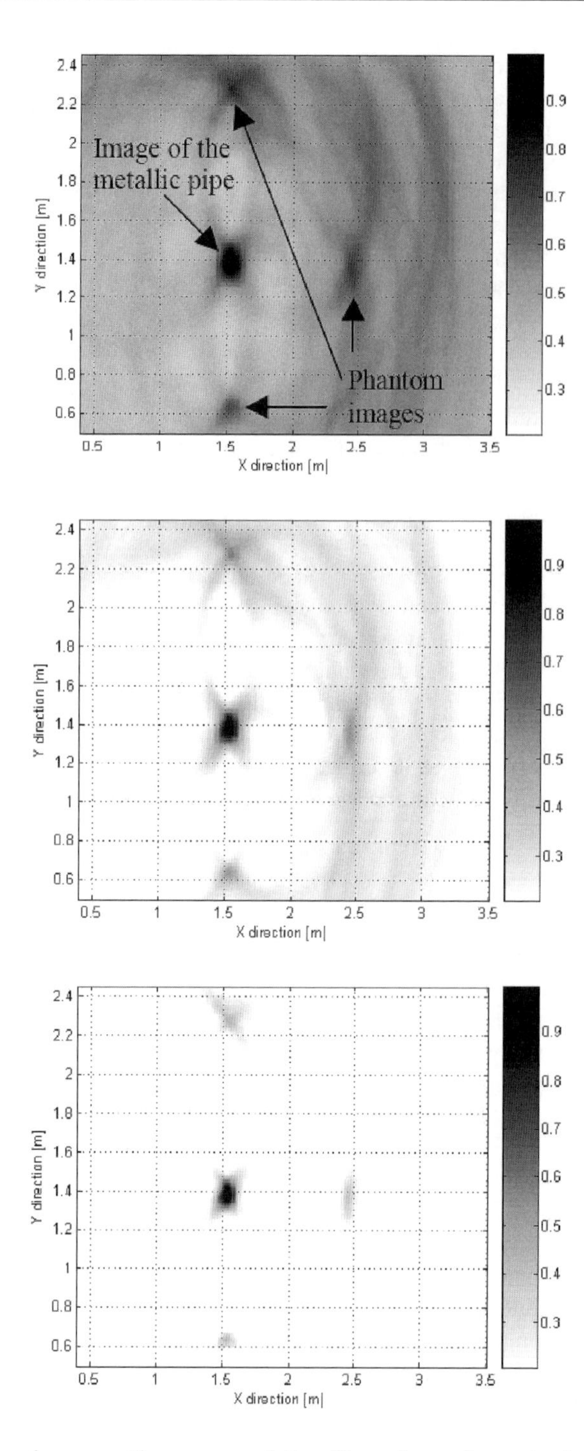

Figure 13.7 Image enhancement by cross correlation. Upper figure shows simple Kirchhoff migration according Equation (13.2), lower figures show cross correlated results according Equation (13.3) with one (middle) and two additional references (lower).

of weak time variant signal features. Static signal components result from direct *Tx-Rx* feed-through and from waves scattered at dominant static reflectors, for example walls, furniture or metallic devices.

A simple background subtraction approach starts with stacked averaging of the sequence of measured impulse responses. In this way, the static background scattering is estimated which in the next processing step is subtracted from the time-variant sequence of impulse responses.

The advantage of background subtraction is demonstrated by the following measurement example that is related to through wall detection of a human being. The person was sitting on one side of a wall and a bistatic radar node was arranged on the opposite side. The distance between the *Rx* and *Tx* antenna was about 20 cm. The distance between the radar and the person was about 2 m. The whole measurement scenario was static. The only time variations were caused by the vital activities (mainly due to the respiration) of the sitting person. The recorded signal is shown in Figure 13.8. The time variant reflection from the person is completely hidden by the static response, consisting of the direct *Tx-Rx* feed-through, the reflection from the walls, and scattering from other objects within the environment. The weak signal in Figure 13.8 (bottom) is the remainder of the background subtraction. It clearly reveals the time-variant component caused by the respiratory activity of the sitting person. The increase of the dynamic range for the detection of the time-variant feature is evident.

In a variety of realistic scenarios, estimation of the static background just by averaging is not enough, as illustrated by the following two problems:

- The background signal can be time-variant. This time variance is caused, for example by undesired antenna movement, or by movement of objects that are not of interest.
- The object of interest can change its state of motion. This means, after having been moving, it can also be stationary over some time interval and 'misinterpeted' by the algorithm as part of the undesired background clutter.
- The object can be both, time-variant and moving. In this case it may be necessary to separate between both variations.

To solve these problems, we need more elaborate two-dimensional filtering procedures which may also include object tracking. Similar problems are known from video surveillance. The respective solutions reported in [11] also describe promising approaches for our goals.

In the following we at first continue discussing the problem of through wall detection of a human being. Provided the background subtraction was successful, we can further analyse the time-variance of the object. In this case it is the respiratory activity; heart beat can also be detected. Promising applications are related to localization of persons buried alive [12], public safety, surveillance, health care, driver monitoring, and so on. The sequence of measured impulse responses is processed by the simple background subtraction algorithm, as described above. The result is shown in Figure 13.9 as a 2D t, τ- sequence where τ corresponds to range distance. The 2D profile provides the following information on the target: it is 2 m away from the radar, it is not moving since the distance does not change with time, its state changes periodically firstly with the period of 4 s, then with the period of 1.1 s and then with about 5 s. The amplitude variation is stronger at the beginning.

The mere localization of moving objects is different from the example of identification of object's time variance. Again, provided the background subtraction was successful, algorithms similar to those described in Section 13.4 can be applied for detection and location estimation of the moving objects. The requirements are different from those for the imaging methods.

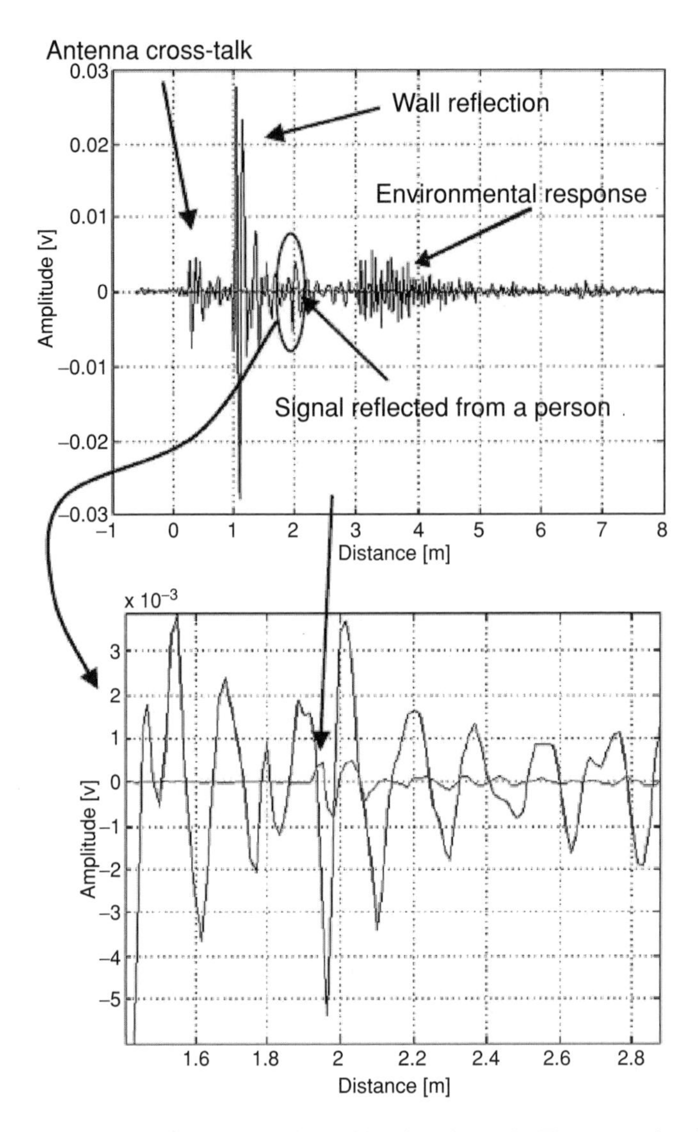

Figure 13.8 Signal measured through wall by a bistatic radar node. The strong signal is the averaged static response, which is mainly due to direct antenna feed-through and reflections from the environment whereas the weak signal describes the instantaneous time-variant response due to the vital activities of the human body.

Since sequential synthetic aperture methods are no longer possible because of real-time requirements, only a small number of observations can be fused; this is limited by the number of available sensors. So we cannot get a detailed image of the object. On the contrary, we need a reliable indicator that describes the presence and the location of the object independent of its shape and attitude. In this sense we would like to interpret the result of the algorithm as a score function that describes the location probability map of the object which would reveal its

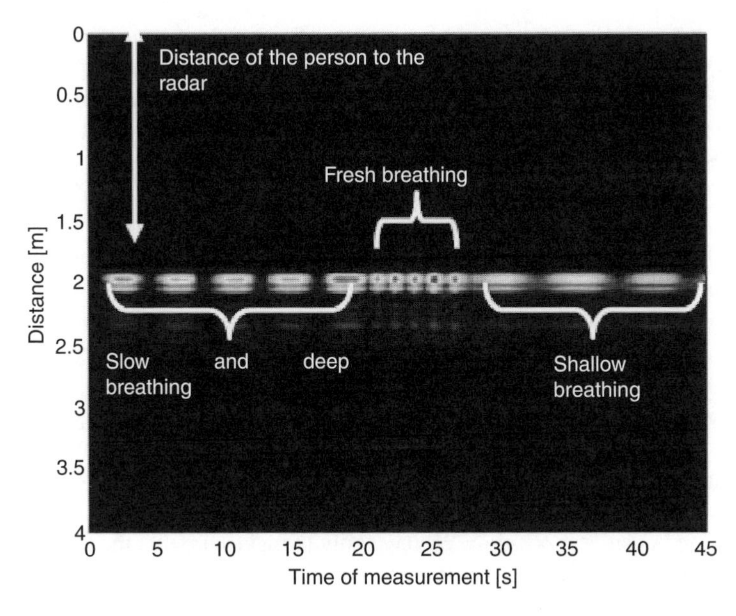

Figure 13.9 Human respiratory activity from radar returns with background subtraction.

maximum likelihood location estimate. However, depending on the object's shape and aspect angle (related to both Tx and Rx since illuminator and observer are in general not collocated), the received signal will undergo rapid fluctuations which are known as fading or scintillation. In conventional radar detection this is considered a nuisance effect since these RCS (radar cross section) fluctuations degrade detector performance. On the other hand, from wireless communications it is well known that a receiver can significantly enhance its detection performance if it receives the same information on different links, provided that the signals undergo independent fading. This effect is known as diversity gain. Among other possibilities, it can be achieved by sending the same information from different transmitters that are well separated in space so that the respective propagation paths undergo independent interactions with the environment (independent fading). Spatial diversity is especially effective in multipath-rich environment. Only recently it has been proposed to employ the same advantage in radar signal processing. This principle is known as statistical MIMO radar [13, 14]. Since we have widely distributed sensors in the sensor network anyway, this benefit comes almost without cost in our application. Two variations of this approach will be explained now.

From the viewpoint of our discussion in this section and in Section 13.4, the most natural choice for location of passive objects seems to be fusion of elliptical location lines. Because of the real-time requirement and the necessity for background subtraction discussed above, we would need at least three static synchronous Tx/Rx-nodes (which could be the anchor nodes) to achieve an unambiguous location estimate in 2D. However, since this gives only the minimum number of observations, we have no diversity gain and we will be plagued by the image artefacts which cause local maxima in the object's location probability map. Increasing the number of widely distributed anchor nodes in space increases the number of independent observations. With M transmitting and N receiving nodes we have NM independent radar observations. In case of N nodes acting simultaneously as receiver and transmitter (e.g., anchor nodes) we still have $N!/2(N-1)!$ independent observations. The reduction comes from the reciprocity of the Tx/Rx, link provided that the same antennas and antenna positions are used at both sides. Every radar link submits independently fading amplitudes at the object's ToA. Proper statistical combining of those data would stabilize the location probability estimate. So we shift our focus from coherent to noncoherent combining at the receiver.

However, there are further facets of statistical MIMO radar detection in distributed sensor networks. Remember, the imaging method described in Section 13.4 in essence was a ToA based method that required precise knowledge of the position and synchronous multistatic cooperation of the any pair of nodes involved. A modified version of location estimation is based on time difference of arrival (TDoA) measurement. In this case we need pairwise synchronous receiver nodes. The two received signals are cross correlated which results in a hyperbolic location base line for any object. Unambiguous location can again be achieved by three observations which require at least three pairwise synchronous receivers. These can again be the anchor nodes. Interestingly enough, it is easy to see that in this case the synchronization requirements related to the transmitter are widely relaxed. Strict (phase-related) synchronization is no longer required. We do not even have to know the transmit sequence and the transmitter position. We have only to make sure that the transmitter signal is fitted to the transmit time slot regime described in Section 13.1. This is necessary in order to have selective access to the individual transmitter signals.

Strict coherency, on the other hand, is required at the receiver antenna pair in order to achieve direction of arrival resolution by TDoA. This requires not only synchronous receiver operation, but also the two received signals to be mutually coherent. Therefore, the two receive antennas have to be closely spaced (as is the case for the bat-type sensor) to receive the signal which is reflected from the same object detail.

On the transmitter side, noncoherent operation offers far-reaching advantages. Since precise knowledge of the transmitter position is not necessary, we can locate illuminators at any position. Even multiple illuminators can be placed and also their reflection from the environment contributes to illuminating the object, as indicated in Figure 13.10. The situation is similar to optical perception where we experience that visibility of an object is better in diffuse light than under a spotlight. Note that the UWB correlation receiver benefits from the diversity gain

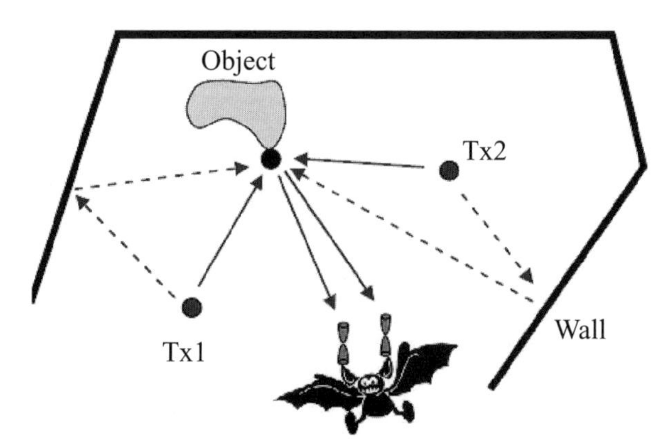

Figure 13.10 TDoA object location by pairwise synchronous receivers and multiple noncoherent transmitters.

because of the huge bandwidth of the signal which (from a technical perspective) can easily extend beyond 10 GHz, corresponding to a coherency distance of less than 3 cm.

13.7 Conclusions

UWB sensor networks are capable of autonomous navigation in unknown environments. Multistatic coherent cooperation of widely distributed sensors and noncoherent fusion of data from independent bistatic nodes can produce overview images of the environment and partial images of environmental details, respectively. This information is used to build a map of the environment which can be related to the relative sensor-to-senor coordinate system. Real-time processing and background subtraction are necessary to detect, identify and track unknown moving and/or time-variant objects.

The interplay between coherent and noncoherent data fusion in a widely distributed MIMO sensor network seems to be a very interesting topic, which not only allows control of the synchronization requirements of the distributed nodes. It also gives the possibility of a considerable diversity gain, which enhances the robustness of object detection by sacrificing the information on the object's shape. This tends to make the detection probability independent of the object's instantaneous attitude.

Simultaneous localization and mapping (SLAM) has been a recent research topic in robot navigation [15, 16]. Since mainly optical and millimetre-wave radar sensors have been discussed in this context, we believe that UWB sensors with their different frequency range and bandwidth and their unprecedented capability of range resolution and penetration through materials can be an interesting alternative to those sensors or at least submit valuable additional information which may interesting for robot navigation. On the other hand, the signal processing and data fusion methods used in SLAM can obviously be adopted for localization and navigation in UWB sensor networks. So, compared with the examples given in this chapter, further performance enhancement can be achieved by advanced multiple sensor data fusion methods, soft detection based on probability distribution instead of hard location decision, fusion of local and distributed information and iterative enhancement for computational requirement reduction, particle filter based object tracking and multiple hypothesis tests for track initiation, and so on.

In addition to an object's location, its attitude and dynamic parameters such as speed and direction of movement are also of interest. Advanced signal processing methods such as space time and time frequency methods are used to recognize and identify additional object features such as shape, orientation and other morphological and time-variant object features.

Object detection must be robust against various error sources and missing (incomplete) information which includes missing links and obstructed LOS [17]. Knowledge-based approaches, such as utilization of *a priori* information and adaptation of network resources may help to optimize the available resources and enhance cooperative detection performance. Moreover, since sensor cooperation may require considerable information exchange between nodes, optimization of local computational effort versus message distribution payload is necessary.

13.8 Acknowledgements

The research reported in this article was partially supported by the projects PULSERS and UKoLos. PULSERS (Pervasive Ultra-wideband Low Spectral Energy Radio Systems) is an integrated project within the sixth Framework of the Commission of European communities.

PULSERS aims to investigate the challenges and possibilities associated with the UWB technology and its application. More details can be found at www.pulsers.eu. UKoLos (UWB Radio Technologies for Communication, Localization and Sensor Technology) is a German project cluster supported by 'Deutsche Forschungsgemeinschaft DFG'. More details can be found at http://www-emt.tuilmenau.de/ukolos/.

References

1. Gezici, S., Tian, Z., Giannakis, G.B. *et al.* (2005) Localisation via ultra-wideband radios: a look at positioning aspects for future sensor networks. *IEEE Signal Processing Magazine*, **22**(4), 70–84.
2. Pahlavan, K. *et al.* (2006) Indoor geolocation in the absence of direct path. *IEEE Transactions on Wireless Communications*, **13**(6), 50–8.
3. Zetik, R., Sachs, J. and Thomä, R.S. (2007) UWB short-range radar sensing. *IEEE Instrumentation and Measurement Magazine*, **10**(2), 39–45.
4. Sachs, J., Kmec, M., Herrmann, R. *et al.* (2006) *An Ultra-Wideband Pseudo-Noise Radar Family Integrated in SiGe:C*, in Proceeding International Radar Symposium (IRS), pp. 1–4.
5. Zetik, R., Sachs, J. and Thomä, R.S. (2006) *Distributed UWB MIMO Sounding for Evaluation of Cooperative Localization Principles in Sensor Networks*, in Proceeding European Signal Processing Conference (EUSIPCO).
6. Sayed, A.H., Tarighat, A. and Khajehnouri, N. (2005) Network-based wireless location: challenges faced in developing techniques for accurate wireless location information. *IEEE Signal Processing Magazine*, **22**(4), 22–40.
7. Daniels, D.J. (2004) *Ground Penetrating Radar*. IEEE, pp. 278–332.
8. Zetik, R., Sachs, J. and Thoma, R. (2005) *Modified Cross-Correlation Back Projection for UWB Imaging: Numerical Examples*, in Proceeding IEEE International Conference on Ultra-Wideband, pp. 650–54.
9. Sachs, J., Zetik, R., Friedrich, J. and Peyerl, P. (2005) *Autonomous Orientation by Ultra Wideband Sounding*, in Proceeding International Conference on Electromagnetics in Advanced Applications (ICEAA), pp. 1–9.
10. Zetik, R., Sachs, J. and Thomä, R. (2005) *Modified Cross-Correlation Back Projection for UWB Imaging: Measurement Examples*, in Proceeding International Scientific Conference DSP-MCOM, pp. 25–8.
11. Piccardi, M. (2004) *Background Subtraction Techniques: A Review*, in Proceeding IEEE International Conference on Systems, Man and Cybernetics, Vol. **4**, pp. 3099–3104.
12. Zetik, R., Crabbe, S., Krajnak, J. *et al.* (2006) *Detection and Localization of Persons Behind Obstacles Using M-Sequence Through-the-Wall Radar*, in Proceeding SPIE Defense and Security Symposium, Vol. **6201**, pp. 823–38.
13. Fishler, E., Haimovich, A., Blum, R.S. *et al.* (2006) Spatial diversity in radars – models and detection performance. *IEEE Transactions on Signal Processing*, **54**(3), 823–38.
14. Fishler, E., Haimovich, A., Blum, R.S. *et al.* (2004) *MIMO Radar: an Idea Whose Time has Come*, in Proceeding IEEE Radar Conference, pp. 71–8.
15. Lima, P. (2007) A Bayesian approach to sensor fusion in autonomous sensor and robot networks. *IEEE Instrumentation and Measurement Magazine*, **10**(3), 22–7.
16. Wijesoma, M.S., Perera, L.D.L. and Adams, M.D. (2006) Toward multidimensional assignment data association in robot localization and mapping. *IEEE Transactions on Robotics*, **22**(2), 350–64.
17. Algeier, V., Demissie, B., Koch, W. and Thomä, R.S. (2008) State space initiation for blind mobile terminal position tracking. *EURASIP Journal on Advances in Signal Processing*, Special issue on Track-Before-Detect Algorithms, pp. 1–14.

14

Low-power UWB Hardware

Robert W. Brodersen and Ian O'Donnell

The Department of Electrical Engineering and Computer Sciences, University of California at Berkeley, Berkeley, CA 94720-1770, USA

Ultra-Wideband signaling offers tremendous communication capacity, even at the low, mandated level of transmit power and including interference from pre-existing channel users. While this low transmit power limits high data rate transmission to shorter distances, UWB presents an attractive opportunity for both power-efficient and low-power communication in addition to offering the ability to perform localization. To date, development has generally focused on OFDM and pulse-based architectures, but as the physical signaling is not explicitly regulated the possible design space is enormous.

Using well-established sinusoidal techniques, OFDM UWB extends radio communication to very wide bandwidths placing stricter performance criteria upon building block circuits, such as the analog to digital (A/D) and digital to analog (D/A) converters, gain stages and backend digital complexity. For applications with a fixed power budget, for example battery powered devices, these constraints must be met with comparable, or lower, power consumption than existing narrowband solutions. This need to scale operation to faster speeds, higher bandwidths and increased computation has spurred research interest at the block level. Low cost, and hence high levels of integration, is also a goal, further directing the effort towards CMOS implementations. Recent A/D results have re-examined the choice of architecture with pipeline and successive approximation approaches posting performance comparable to the most efficient, traditional flash-based architectures. Likewise, wideband filters have been codesigned with the front-end gain stages to provide improved flatness and matching over the transmission bandwidth. Also, digital area and power continue to improve as observed in 'Moore's Law', accommodating the increased computational load. Power consumption efficiency, in mW/Mbps, is achieved through a vastly increased throughput without additional power expenditure. In this way, the inherent capacity of UWB signaling contributes to more power efficient communication.

Short-Range Wireless Communications Rolf Kraemer and Marcos D. Katz

Motivated by the promise of a straightforward implementation, pulse-based UWB offers a competing approach to OFDM for power efficient, high-rate, short-distance communication. Additionally, a pulse-based radio can achieve low-power operation by slowing the pulse rate and duty-cycling the radio operation between received pulses. Several low-rate, pulse-based UWB radios in CMOS with power consumption of the order of a milliwatts have been published. Based on a correlating receiver, these implementations range from simple analog architectures to a more flexible, digital architecture based on a 1-bit A/D conversion. The energy efficiency of these low-power, pulse-based radios prompts a question about the interplay between the physical signaling and an efficient, if not more optimal, implementation. Indeed, research using signaling methods other than pulses and sinusoids, such as chirps or chaotic waveforms, have also shown desirable characteristics. This large design space encourages investigation at the building block level and the system architecture level as well as in the interaction between the two.

In addition to increasing the demands on a circuit block by scaling performance to wider bandwidths, faster speeds and lower power, the challenge of efficient UWB implementation extends to the system level and the trade-offs that can be made throughout the design. For example, one important issue that has yet to see much circuit realization is that of interference cancellation. To ease requirements on the A/D converter and gain stages, it would be advantageous to adaptively estimate and cancel interference as soon after the antenna as possible. However, this implies an analog-based approach that may not take advantage of the benefit of digital processing. The impact of this decision upon system performance is complicated and difficult to predict without closer scrutiny. In particular, the ideal location of the analog and digital boundary within the system is open to debate for a given architecture. Another example that shares a complex relationship between the system and circuit blocks is that of channel estimation and the cost associated with the speed and accuracy needed. Circuit block specifications made to lower power consumption (e.g., a very low resolution A/D or very noisy gain stages) may disproportionately impair or slow the estimation algorithm.

With many different proposed architectures UWB radios inspire the creation of new circuit blocks and combinations of blocks, producing novel timing and waveform generation circuits or merging functionality to lower power consumption. The wealth of applications for short-distance, high-rate and low-power, low-rate radios also stimulates development. In addition to communication, there is strong interest in ranging, locationing and imaging. The chief requirement for operation is 500 MHz of occupied bandwidth at low power spectral density. Approaching the theoretical performance will require innovation at all levels within the system.

From a power consumption perspective, results have already been published showing efficiency on the order of 1 Mbps/mw at 3 m. Low-rate radios capable of communication and ranging have also shown power consumption on the order of 1 mW. In an UWB radio, the power consumption tends to be dominated by the receiver and within the receiver, the wideband, front-end gain stages tend to dominate. Depending upon the architecture, the A/D converter may also be a big concern. Sampling timing circuitry may be implemented using digital delays referenced to accurate or calibrated sources for power cost on the order of a milliwatt. With the continued improvement in process technology, we may expect the digital components to scale in size, power and frequency. (Although issues of leakage and reduced reliability will need to be addressed.) The improvements in digital capability may also be utilized to compensate and correct for less accurate analog front-ends. High-speed, low-resolution A/D converters have reported figures of merit of better than 0.2 pJ/conversion/level.

At 4-bits and 1 GHz this implies a 3-mW power consumption. Advances in architecture and circuit designs are expected to further reduce this value by up to an order of magnitude. LNA power consumption is limited by the low input impedance and noise figure requirements. Co-designing to the antenna may relax the impedance constraint, delivering savings around 50%. Creative current reuse techniques may reclaim another 50–75%. Duty-cycling the gain stages will provide added power savings proportional to the pulse rate. The net effect of all of these improvements suggests that more than an order of magnitude in power savings is still possible for UWB radios as currently designed. These savings may be realized directly or used to improve performance by adding complexity to channel and interference processing. Allowing for the unpredictability of innovation, even lower power, more efficient architectures may be on the horizon as well.

15

Analog-to-digital Converters for UWB

Takao Waho

Department of Information and Communication Sciences, Sophia University, 7-1 Kioicho, Chiyoda-ku, Tokyo 102-8554, Japan

15.1 State of the Art

The performance of analog-to-digital converters (ADC) has been improved with the evolution of semiconductor technology. This extends the ADC application fields from special-purpose data acquisition setups to various consumer electronics products. High-performance, cost-effective ADCs have also been developed for the IF signal processing in wireless receivers. For conventional narrow-band signals, oversampling delta-sigma ADCs are quite effective, in particular those using bandpass-type delta-sigma modulators, where the quantization noise can be suppressed (shaped) for a particular narrow range of frequency [1]. The sampling frequency, f_s, can be selected independently on the center frequency of the input signal, f_0, although f_s should be much larger than the signal bandwidth, f_B. (For Nyquist rate ADCs, $f_s > 2f_0$.)

On the other hand, in UWB the signal bandwidth is > 500 MHz. This excludes the use of delta-sigma ADCs. Even if an oversampling ratio of 32 is assumed, a sampling frequency of 16 GHz is required, which is not impossible, but impractical. Therefore, we have to rely on the Nyquist rate ADCs, in particular flash ADCs with a sampling frequency of a few GHz. There is a trade-off relationship between the sampling frequency and the bit-resolution in ADCs. For $f_s > f_B > 500$ MHz, a practical bit-resolution is limited at 6–8 bits. This is usually too low for narrow-band applications, but hopefully acceptable in UBW. It should be noted that such flash ADCs have been developed for commercial use in the HDD read channel as well as digital oscilloscopes. Therefore, the ADC for UWB is not a futuristic dream, but a real challenge.

Short-Range Wireless Communications Rolf Kraemer and Marcos D. Katz

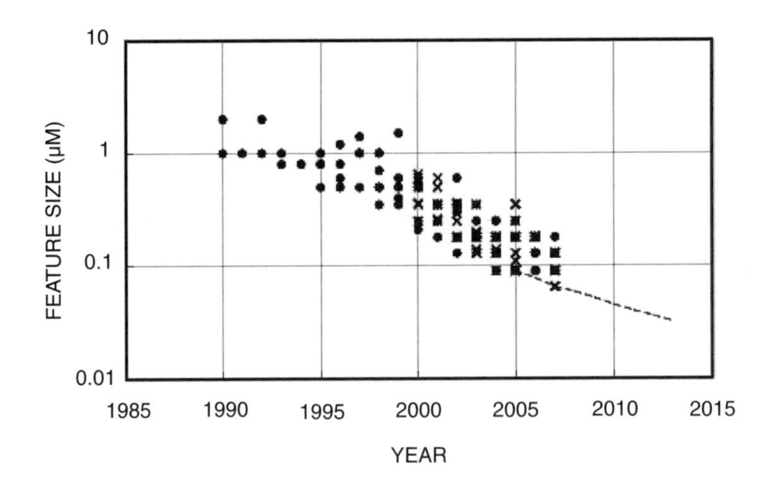

Figure 15.1 Feature size used in ADCs published at the IEEE International Solid-State Circuits Conferences (ISSCC). Circles and crosses represent Nyquist-rate and oversampling delta-sigma ADCs, respectively. The dashed line shows the International Technology Roadmap of Semiconductor (ITRS) prediction for the metal half pitch in CPU/ASIC.

15.2 Technology

Like CPUs and other VLSIs, ADCs are implemented by using the most advanced mass-production technology available at the moment. Figure 15.1 shows the feature size used in ADC fabrications. According to the prediction of the International Technology Roadmap of Semiconductors (ITRS) (http://public.itrs.net/), the decrease in feature size will continue for the next 10 years or so. This indicates in general that the ADC performance will also be improved in the near future.

15.3 Performance

Figure 15.2 shows the SNR, which means the bit-resolution in the ADC case, as a function of the sampling frequency of the Nyquist-rate ADCs, such as pipelined and flash ADCs. (The SNR value divided by 6 corresponds approximately to the bit-resolution of ADCs.) For oversampling delta-sigma ADCs the horizontal axis represents the bandwidth. These are the most widely used ADC architectures. As shown in this figure, each type of ADC covers its own application area. For UWB applications, the flash ADC is the unique selection because of the width. To obtain such a high operation speed, not only Si, but also compound semiconductor devices have been used to implement flash ADCs [2–5].

It should be noted that the SNR is limited by the clock jitter and the thermal noise. In particular, the jitter should be suppressed for the flash ADC to improve the bit-resolution. According to the theoretical analysis, the SNR decreases by 20 dB (about 3 bits), if the jitter increases by a factor of 10. Since the jitter is >1 ps in practical systems, the bit-resolution is <6. This is almost independent of the scaling-down of MOSFET, and the UWB system will be designed based on this value even in the future.

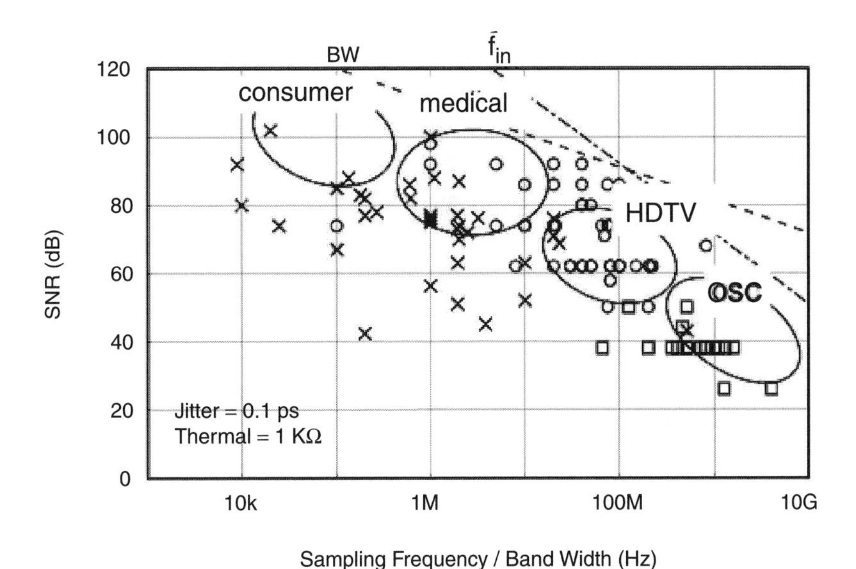

Figure 15.2 ADC performance published at ISSCC. Squares, circles and crosses represent flash, pipelined, and delta-sigma ADCs, respectively. Several applications and predicted limits are shown. The dashed and dash-dotted lines represent the theoretical limit due to the thermal noise and sampling jitter, respectively.

15.4 Figure of Merit (FOM)

The MOSFET scaling merit is mainly in the power efficiency. This is usually represented as a figure of merit (FOM) shown in Figure 15.3. The FOM, which is defined as the energy required in converting the analog input into the digital output, decreases by a factor of 5 every 5 years. The ADC power in 2015 will be less than one-tenth of that needed for today's ADC, which will result in low-power operation of the system.

15.5 Challenges and Visions

Flash ADC challenges are:

- High-speed sample-and-hold (S/H) circuit and time-interleave configuration.
- Jitter suppression to increase the bit-resolution.
- Digital correction technique to alleviate conditions required for imperfect analog circuits.
- Adaptive structure to accommodate the change in environment and standards.
- Low-power design while keeping the high-speed operation.

To develop hardware for new systems such as UWB needs interdisciplinary research. Filling the gap between system, circuit, device, and even materials is the best way to reach the optimum ADC realization. This is beyond individual capability, so constructive collaboration as well as information exchange between different disciplines play an essential role.

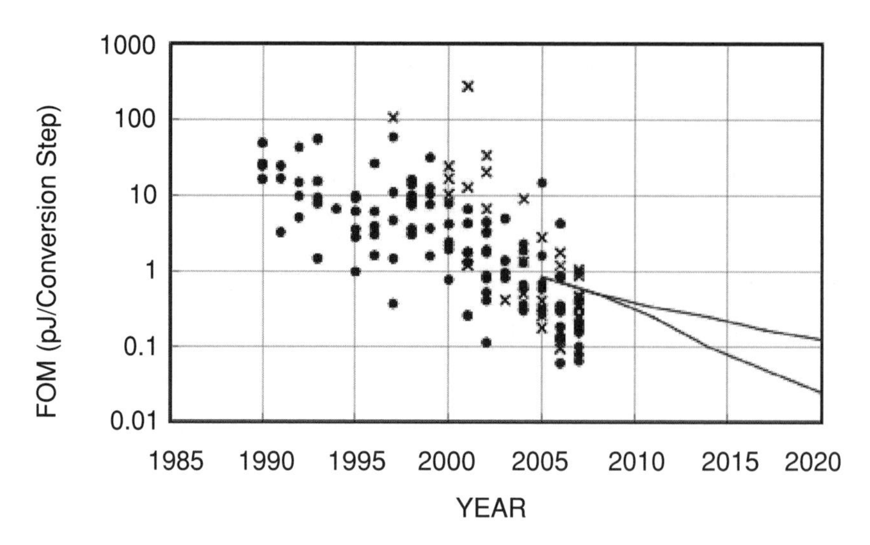

Figure 15.3 Figure of merit of ADCs published at ISSCC. Circles and crosses represent Nyquist-rate and oversampling delta-sigma ADCs, respectively. The lines show the region of the International Technology Roadmap of Semiconductor (ITRS) prediction.

Recently, heterogeneous integration has attracted increasing attention. This allows us to combine new technologies, such as compound semiconductor and quantum devices [6, 7], with Si CMOS platform, which results in a novel architecture of ADCs that is suitable to UBW applications. For instance, such high-speed devices are used at the analog front end as a high-accuracy, high-speed S/H circuits, which are followed by parallel Si circuit blocks.

References

1. Shreier, R. and Temes, G. (2005) *Understanding Delta-Sigma Data Converters*, John Wiley & Sons.
2. Cheng, W. *et al.* (2004) *A 3b 40GS/s ADC-DAC in 0.12-μm SiGe*, in Proceeding IEEE International Solid State Circuits Conference (ISSCC), Digest Technical Papers, Vol. **1**, pp. 262–3.
3. Nosaka, H. *et al.* (2004) *A 24-Gsps 3-bit Nyquist ADC using InP HBTs for Electronic Dispersion Compensation*, in Proceeding IEEE MTT-S International Microwave Symposium, Vol. **1**, pp. 101–4.
4. Schvan, P., Pollex, D., Wang, S. *et al.* (2006) *A 22GS/s 5b ADC in 013-mm SiGe BiCMOS*, in Proceeding IEEE International Solid State Circuits Conference (ISSCC), Digest Technical Papers, pp. 2340–9.
5. Mohan, A., Zayegh, A. and Stojcevski, A. (2007) A high speed analog to digital converter for ultra wide band applications, in *Emerging Directions in Embedded and Ubiquitous Computing* (eds M. Denko *et al.*), Springer-Verlag, pp. 169–80.
6. Frazier, G., Taddiken, A., Seabaugh, A. and Randall, J. (1993) *Nanoelectronic Circuits Using Resonant Tunneling Transistors and Diodes*, in Proceeding IEEE International Solid State Circuits Conference (ISSCC), Digest Technical Papers, pp. 174–5.
7. Waho, T., Itoh, T., Maezawa, K. and Yamamoto, M. (1998) *Multi-GHz A/D Converter Using Resonant-Tunneling Multiple-Valued Logic Circuits*, in Proceeding IEEE International Solid State Circuits Conference (ISSCC), Digest Technical Papers, pp. 258–9.

16

UWB Coexistence Scenarios

Sven Zeisberg
GWT-TUD GmbH, Chemnitzer Strasse 48b, D-01187 Dresden, Germany

16.1 State of the Art

Traditionally, existing legacy radio communications systems are designed following the assumption that the received signal is an attenuated replica of the transmitted signal, mainly disturbed by the radio channel characteristics and superseded by thermal noise in the receiver. This holds if the number of radio services and transmitting devices is limited, which was true for a relatively long period of time during the last century. Technical realization of system separation was achieved using nonadaptive traditional analog filters with their inherently low degree of flexibility. Consequently the frequency spectrum regulation policy tended to grant exclusive frequency use rights for dedicated radio services, valid for longperiods of time. Following this historical development, the current radio frequency regulation paradigm is still based on frequency separation, which is defined for long periods of time (tens of years), as described in [1].

With the introduction of the mass market for bidirectional cellular radio systems the situation in system design has changed significantly. It has become necessary for the system design paradigm is take account of additional signals and interference. Interference from adjacent channels allocated by nearby users (due to nonideal filters) and nearby transmitters (due to frequency reuse in cellular systems) is considered in general to be inevitable and is taken into account in system design and rollout. This so called intrasystem interference is accepted by the radio service provider, called the *operator*, mainly due to the fact that it is possible to strictly control this interference by means of network planning, intelligent resource allocation and traffic load control. Therefore it is relatively easy for the operator to guarantee a certain link quality and thus subsequently a certain Quality of Service (QoS) for the customer. The major aspect of this procedure is the consideration of appropriate link margin during the network planning.

Short-Range Wireless Communications Rolf Kraemer and Marcos D. Katz

In the analog radio age analog frequency filters provided an easy means to separate wanted and unwanted received signals. Based on this technology the radio systems were supposed to operate in separate frequency bands. Consequently the multiple access technique as well as the duplex technique applied in those systems, frequency division multiple access (FDMA) and frequency division duplex (FDD), respectively, were based on the well-known analog frequency filtering technique. This traditional approach to frequency regulation is still followed in principle for the separation of radio systems and services, with a few very recent exceptions (namely UWB [2] and nano-FM transmitters [3, 4]). However, with the beginning of the digitalization of radio communications, several innovative options became feasible. The digitalization of radio communications became economically and technically feasible, enabled by the revolutionary advances in the semiconductor industries since the 1980s allowing the implementation of complex digital signal processing algorithms at reasonable cost. The feasibility of time and code division multiple access (TDMA and CDMA), time division duplex (TDMA) and the combination of diversity techniques with forward error correction coding (FEC) enabled in principle more efficient and innovative frequency resource sharing methods. In general one can state that radio systems became more robust, and it has become feasible to accept interference from other radio systems without harmful degradation of the system performance. Furthermore, flexible air interfaces adapting to the current radio environment on a 'call by call' basis are feasible now and are currently being investigated and optimized in several European research projects.

To provide an illustration of the fundamental change of possibilities one can imagine two commonly used cases of service provision. First, a traditional analog service, such as terrestrial analog audio and video broadcast reception is considered. These analog services are based on frequency separation and can hardly accept interference from other radio systems, as any interference will directly result in a degradation of the perceived quality of the audio or video. Even allowed small out-of-band emissions of other radio systems (such as the spurious emissions of some cellular phones) result in clearly noticeable noise in anaolg audio receivers (frame structure spectral line). This can be explained by the earlier analog system design, which was of course state-of-the-art when these services were introduced. Therefore, exclusive frequency use rights appear as the only feasible approach to resource sharing for such systems. Due to the fact that the equipment costs for the customer as well as for the broadcast service provider are significant, the method of assigning the frequencies for longer terms appears to be justified.

In contrast, today's modern digital radio services, such as digital audio and video broadcasting or wireless local area networks can accept certain (higher) level of interference without degrading the perceived system performance noticeably. Advanced error correction coding techniques either blank out or restore disturbed received signals by means of complex digital signal processing. ISM band operation in the 2.4 GHz range is a good example of radio systems designed to operate in an interference scenario-however, quality of service is still not guaranteed in such 'resource sharing'.

An interesting early technical approach to air interface flexibility practiced by cellular systems operators today is so-called network roaming – where a radio communications service can be provided temporarily by using an alien network, if the native network, to which the user is subscribed, is not available for whatever reason (bad radio channel, high traffic load, no native network coverage). This approach works because the user radio terminals are able to connect to the network infrastructure of the alien network service provider by means of switching to other frequency bands. Even if it is basically not allowed for primary radio services to share radio spectrum (except of ISM bands), each radio system operator has to accept by law that there is unintentional out-of-band emission from the other radio systems in operation and has to cope with such interference. What is the current legal and actual practical situation for interference to be accepted by legacy radio systems from other radio systems? From a legal point of view, the allowed out-of-band emission for radio systems is defined in Europe in the appropriate ETSI standards and CEPT recommendations [5, 6]. These standards are implemented or revised under the R&TTE directive released by the Radio Spectrum Committee (RSCOM) of the European Commission in 1999 [7].

16.2 Challenges

With the start of the UWB regulation process in Europe [8–10], in Asia and at ITU level it became obvious that there is a number of challenges associated with the regulation of such a kind of modern example of spectrum sharing technology. Modern radio technology is able and required to operate based on coexistence and cognitive radio principles. Besides technical challenges there are also administrative ones, mainly convincing the stakeholders in the (still clustered) European regulation process that this approach will provide a more efficient spectrum use from an overall economic point of view and will open a new market for several applications enabled by modern 'intelligent' radio technologies, such as UWB.

The technical developments are also challenging several areas and need a considerable amount of research and development effort in various basic disciplines, still considered as precompetitive R&D activities. Just to mention some, without claiming completeness:

- design of radio communications systems based on the 'cognitive radio' principle and having no exclusive spectrum usage rights while still ensuring a certain quality of service for the application;
- providing extremely low protection limits to passive radio astronomy services;
- providing sufficient low protection to live critical services such as aviation radar;
- providing sufficient protection to indoor fixed wireless access terminals operating close to the sensitivity limits without introducing harmful interference;
- protecting strategic services such as military radar and military sensing and communications without disclosing details of their system specification;
- coordinating colocated radio systems (several radio platforms within a single user device, such as PDA, PC or mobile phone);
- a major challenge for the UWB systems is to cope with many potential victim services addresses due to the inherently huge spectrum band used, and these are still not completely specified.

From the current point of view these challenges listed above can be dealt with by applying certain techniques, such as mitigation techniques and attempts at creating common signaling carriers which are still essentiallyundeveloped. First steps have been taken to verify the effectiveness of these techniques and intensive research work is ongoing to further develop these visions and turn them into reality [11–14].

Since February 2007 an initial legally binding European harmonized UWB radio regulation has been in place [2] which is expected be complemented subsequently by means of amendments. The first update of the initial EC regulation complementing the initial release concerning mitigation techniques and adjusting the associated protection level, is expected to be released by the end of 2008, following the reception of the CEPT ECC decision update [15] and an ECC answer to RSCOM, in response to the third UWB mandate of the RS-COM to the CEPT ECC.

There will be very likely no worldwide harmonized approach to frequency regulation covering the whole UWB band from 3.1 to 10.6 GHz. However, for subparts of it there will be similar regulation rules in place in the USA, in Europe and in Asia. In order to allow the creation of the UWB enabled application market in Europe it has been decided that first-generation devices will be allowed to operate for a limited time under a 'phased approach' in a sub-band from 4.2 to 4.8 GHz until a certain cut-off date, 31 December 2010 [2].

For the medium- and long-term future of UWB applications there are two major directions of development ongoing and expected to be implemented. The first is the continuous development the '*cognitive radio*' concept allowing a highly sophisticated data-aided spectrum sharing between several different heterogeneous radio systems including UWB (e.g., by defining a standardized cognitive signaling channel as one possibility for universal recognition of a radio terminal and its radio isolation to the current receiver) [11–13]. The second is to further research and develop the various kinds of non-data-aided mitigation techniques such as activity factor limitation (like the initial LDC approach [16, 17]), enhancement of sensing techniques such as listen before talk (LBT) and detect and avoid (DAA), as investigated in CEPT [18] and ETSI [14]. One further possibility, which sounds exotic at the moment, but is practical to implement in many cases, would be the introduction of spectrum avoidance algorithms based on absolute geographic location, the so-called location- based recognition of exclusion zones.

It is expected that, for the majority of UWB devices in the distant future the possibility to operate in the 'upper band' (6–9 GHz) will be included due to advances in semiconductor technology and due to the relaxed requirements of certain applications, such as Bluetooth, concerning distance [19]. Further it is expected that for a certain number of applications the 'lower band' (for which the target is 3.1–4.8 GHz) will still be available as there are certain applications requiring operation in the lower band. Further current developments in R&D and standardization [20, 21] show that future UWB systems will cover the bands below 10 GHz as well as the bands around 60 GHz. The 60 GHz range allows higher transmission power, enabling even increased data rates (3–5 Gbps, e.g., allowing uncompressed HD video), but this requires higher energy consumption, so it may be mainly applied in stationary devices, which have fixed power plug connections instead of portable/mobile devices, which may be operated with UWB below 10 GHz only, which is much more power efficient.

The robustness of all UWB devices will be increased and the UWB market will follow basically two major trends – the very high data rate applications (approx. 1–5 Gbit/s physical layer burst data rate) and the low-energy consumption sensor and location tracking applications, enabling lowest energy consumption wireless connection of sensors and/or real-time accurate positioning of the terminal equipment. Further it is expected that UWB will be used for some niche applications enabled by the imaging and radar features of UWB, such as building material analysis, ground penetration radar, vital function monitoring and other sensor applications related to home entertainment [22].

Acknowledgements

The author would like to acknowledge the support of DG INFOSYS, Unit D1, of the Commission of the European Communities, which is partly funding the R&D project *EUWB – Coexisting Short Range Radio by Advanced Ultra-Wideband Radio Technology* under the umbrella of the 7th Framework Programme of the European Community for research, technological development and demonstration activities [23].

References

1. URL of *the International Telecommunication Union, Radio-communication Sector (ITU-R) History*, http://www.itu.int/net/about/history.aspx (accessed 23 May 2008).
2. 2007/131/EC: *Allowing the use of the radio spectrum for equipment using ultra-wideband technology in a harmonised manner in the community*. URL of the CEPT ERO mirroring the official EC document, http://www.erodocdb.dk/doks/filedownload.aspx?fileid=3398&fileurl=http://www.erodocdb.dk/Docs/doc98/official/pdf/2007131EC.PDF (accessed 23 May 2008).
3. German national administration Bundesnetzagentur Vfg 7/2006: *Allgemeinzuteilung von Frequenzen in den Frequenzbereichen 87.5-108 MHz, 863-865 MHz und 1795-1800 MHz für drahtlose Audio-Funkanwendungen*, Germany.
4. Ofcom: http://www.ofcom.org.uk/consult/condocs/exemption/statement/.
5. CEPT/ERC/Recommendation 74-01E (Siófok 98, Nice 99, Sesimbra 02, Hradec Kralove 05): *Unwanted Emissions in the Spurious Domain*. URL of the CEPT ERO, http://www.erodocdb.dk/doks/filedownload.aspx?fileid=1695&fileurl=http://www.erodocdb.dk/Docs/doc98/official/pdf/REC7401E.PDF (accessed 23 May 2008).
6. ECC/REC/(02)05, Revised ECC Recommendation, 21/10/2002: *Unwanted Emissions*. URL of CEPT ERO, http://www.erodocdb.dk/doks/filedownload.aspx?fileid=2658&fileurl=http://www.erodocdb.dk/Docs/doc98/official/pdf/REC0205.PDF (accessed 23 May 2008).
7. Directive 1999/5/EC of the European Parliament and of the Council of 9 March 1999 on *radio equipment and telecommunications terminal equipment and the mutual recognition of their conformity*.
8. RSCOM 04-08: *Mandate to CEPT to harmonise radio spectrum use for ultra-wideband systems in the European union*, 2004-02-18. URL of RSCOM, http://ec.europa.eu/informationsociety/policy/radiospectrum/docs/current/mandates/rscom0408mandateuwb.pdf (accessed 23 May 2008).
9. RSCOM: *Second mandate to CEPT to identify the conditions necessary for harmonising radio spectrum use for ultra-wideband systems in the European union*, 2005-06-06. URL of the EC RSCOM, http://ec.europa.eu/informationsociety/policy/radiospectrum/docs/bytopics/final2nduwbmandate.pdf (accessed 23 May 2008).
10. RSCOM: *Mandate to CEPT to identify the conditions relating to the harmonised introduction in the European union of radio applications based on ultrawideband (UWB) technology* (MANDATE 3), 2006-07-15. URL of RSCOM, http://ec.europa.eu/informationsociety/policy/radiospectrum/docs/current/mandates/3ectoceptuwb0606.pdf (accessed 23 May 2008).
11. URL of ETSI RRS, http://portal.etsi.org/rrs/rrstor.asp (accessed 19 May 2008).
12. URL of Radio Access and Spectrum cluster of the EC DG INFSO and MEDIA of the European Commission, http://www.newcom-project.eu:8080/Plone/ras (accessed 19 May 2008).
13. URL of IEEE SCC 41, http://www.scc41.org/ (accessed 19 May 2008).
14. URL of ETSI STF 350 on *DAA parameters and test procedures for UWB communications applications* (ETSI STF 350), http://portal.etsi.org/STFs/STFHomePages/STF350/STF350.asp (accessed 19 May 2008).
15. ECC/DEC/(06)04: *UWB technology in bands below 10.6 GHz*. http://www.erodocdb.dk/doks/filedownload.aspx?fileid=3189&fileurl=http://www.erodocdb.dk/Docs/doc98/official/pdf/ECCDEC0604.PDF.
16. ECC Report 094: *Technical requirements for UWB LDC devices to ensure the protection of FWA systems*. URL of CEPT ERO, http://www.erodocdb.dk/doks/filedownload.aspx?fileid=3299&fileurl=http://www.erodocdb.dk/Docs/doc98/official/pdf/ECCREP094.PDF (accessed 23 May 2008).
17. ECC/DEC/(06)12: *Low duty cycle UWB in the band 3.4-4.8 GHz*. URL of CEPT ERO, http://www.erodocdb.dk/doks/filedownload.aspx?fileid=3287&fileurl=http://www.erodocdb.dk/Docs/doc98/official/pdf/ECCDEC0612.PDF (accessed 23 May 2008).

18. TG3#23_03R0: Draft ECC Report 120 on DAA, *Technical requirements for UWB DAA (Detect And Avoid) devices to ensure the protection of radiolocation in the bands 3.1-3.4 GHz and 8.5-9 GHz and BWA terminals in the band 3.4-4.2 GHz* (under public consultation until 19 May 2008). URL of CEPT ERO, http://www.ero.dk/C1C484C5-26C4-490F-B70A-391DE79F1DB1?frames=no& (accessed 23 May 2008).
19. *Announcement of bluetooth v3.0 PHY to include ultra-wideband (UWB)*, URL of Bluetooth SIG, http://www.bluetooth.com/Bluetooth/Press/News/HighSpeedemBluetootheme.htm (accessed 23 May 2008), ref BT 3.0 SIG PHY discussion.
20. ETSI TG31c, ETSI TR 102 495-7: System Reference Document: Part 7: *Location tracking and sensor applications for automotive and transportation environments operating in the frequency bands 3.1-4.8 GHz and 6-9 GHz.*
21. URL of ECMA TC48: *High rate short range wireless communications.* URL of ECMA, http://www.ecma-international.org/memento/TC48-M.htm (accessed 23 May 2008).
22. Zeisberg, S. and Schreiber, V. (2008) *EUWB – Coexisting Short Range Radio by Advanced Ultra-Wideband Radio Technology*, ICT Mobile and Wireless Communications Summit. URL of EUWB consortium, http://www.euwb.eu (accessed 23 May 2008).
23. ICT – Information and Communication Technologies, A Theme for research and development under the specific programme '*Cooperation*' implementing the Seventh Framework Programme (2007–2013) of the European Community for research, technological development and demonstration activities, Work Programme 2007-08.

17

UWB Regulation and Standardization

Isabelle Bucaille

Thales Communications, Land and Joint Systems, Secured Wireless Products Department, 146 Boulevard de Valmy, 92704 Colombes Cedex, France

17.1 Introduction

This chapter deals with worldwide UWB regulation and standardization status as well as recommendations for future regulation in Europe.

17.2 Worldwide UWB Regulation Status

17.2.1 UWB Spectrum Regulation Status in USA

In the United States the FCC approved the UWB emission mask in 2002 [1]. Figure 17.1 gives the emission mask for communications devices in an indoor environment. For generic UWB devices, a maximum mean EIRP of −41.3 dBm/MHz is authorized between 3.1 and 10.6 GHz.

17.2.2 UWB Spectrum Regulation Status in Europe

The European Commission Radio Spectrum Committee (RSC) has mandated CEPT to undertake all relevant work for harmonized introduction of UWB based applications in the EU.

In March 2006, CEPT ECC TG3 has given a first approval for generic UWB devices allowing a mean EIRP of −41.3 dBm/MHz in the band 6–8.5 GHz (ECC/DEC/(06)04) [2, 3].

The EC decision on the harmonization of the radio spectrum for equipment using ultra-wideband technology was published in February 2007 [4]. The maximum EIRP densities without any mitigation techniques are defined in Figure 17.2.

Short-Range Wireless Communications Rolf Kraemer and Marcos D. Katz

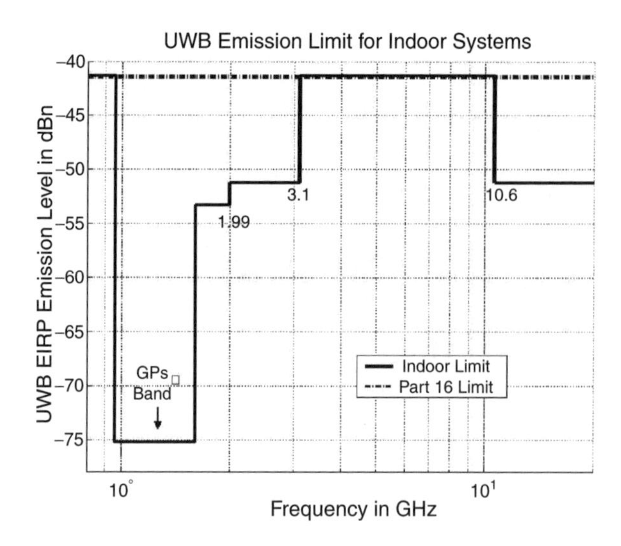

Figure 17.1 US Emission mask for communications devices in an indoor environment.

Frequency range (GHz)	Maximum mean e.i.r.p. density (dBm/MHz)	Maximum peak e.i.r.p. density (dBm/50 MHz)
Below 1.6	-90.0	-50.0
1.6 to 3.4	-85.0	-45.0
3.4 to 3.8	-85.0	-45.0
3.8 to 4.2	-70.0	-30.0
4.2 to 4.8	-41.3 (until December 31st 2010) -70.0 (beyond December 31st 2010)	0.0 (until December 31st 2010) -30.0 (beyond December 31st 2010)
4.8 to 6.0	-70.0	-30.0
6.0 to 8.5	-41.3	0.0
8.5 to 10.6	-65.0	-25.0
Above 10.6	-85.0	-45.0

Figure 17.2 EC decision (February 2007).

In the EC decision, a maximum mean EIRP density of −41.3 dBm/MHz is allowed in the 3.4–4.8 GHz band provided that a Low Duty Cycle (LDC) restriction is applied. This mitigation technique specifies that the sum of all transmitted signals per device must be less than 5% per second and less than 0.5% per hour, provided that each single transmitted signal does not exceed 5 ms. With this restriction UWB devices applying this LDC mitigation technique are not allowed to transmit more than 18s par hour. LDC parameters have been defined in CEPT/ECC TG3 working group, and ECC approved in December 2006 the decision ECC/DEC/(06)12 [5] detailing these parameters. The introduction of LDC in the lower band will enable a class of so called location tracking applications based on UWB in Europe.

As stated in Figure 17.2, this EC decision also includes the so-called 'phased approach' which authorizes placing UWB devices with a maximum mean EIRP of −41.3 dBm/MHz between 4.2 and 4.8 GHz on the market before 31 December 2010.

In the EC decision [4], it is also stressed that equipment using ultrawideband technology may be allowed to use the radio spectrum with EIRP limits other than those set out in the previous table (Figure 17.2), provided that appropriate mitigation techniques other than LDC set out in the previous section are applied, with the result that the equipment achieves at least an equivalent level of protection to that provided by the limits in the table.

In the near future, this EC decision may be revised on the following basis:

- For the changes performed in July 2007 in ECC/DEC/(06)04 authorizing the use of UWB devices in road and rail vehicles between 6 and 8.5 GHz and between 3.4 and 4.8 GHz. This operation is subject to the implementation of Transmit Power Control (TPC) with a range of 12 dB with respect to the maximum permitted radiated power. In the lower band, this means that UWB devices not implementing LDC mitigation technique for vehicle applications must apply TPC.
- For the work performed in ECC TG3 concerning the flexible Detect And Avoid (DAA) approach for the protection of Broadband services (WiMAX) between 3.1 and 4.8 GHz, and radars between 3.1 and 3.4 GHz and between 8.5 and 9 GHz.

17.2.3 UWB Spectrum Regulation Status in Japan

In Japan the regulation authorities have approved a phased approach solution for the lower band (4.2–4.8 GHz), where it would be allowed to get a product certification until the end of 2008. The products could then be sold and used as long as considered useful. The out-of-band emission is defined as −70 dBm/MHz.

In Japan, the upper band starts at 7.25 GHz and goes up to 10 GHz. This causes difficulties for the WiMedia alliance to define common band groups for the United States, Europe and Japan, because this would need at least 1.5 GHz of common radio band. But from 7.25 GHz (current Japanese lower limit) and 8.5 GHz (current European upper limit) there is only 1.25 GHz available. The work ongoing for the compatibility studies between 8.5 and 9 GHz are then very important for worldwide use of High Data Rate UWB devices.

Figure 17.3 gives a summary of US, European and Japanese UWB regulation.

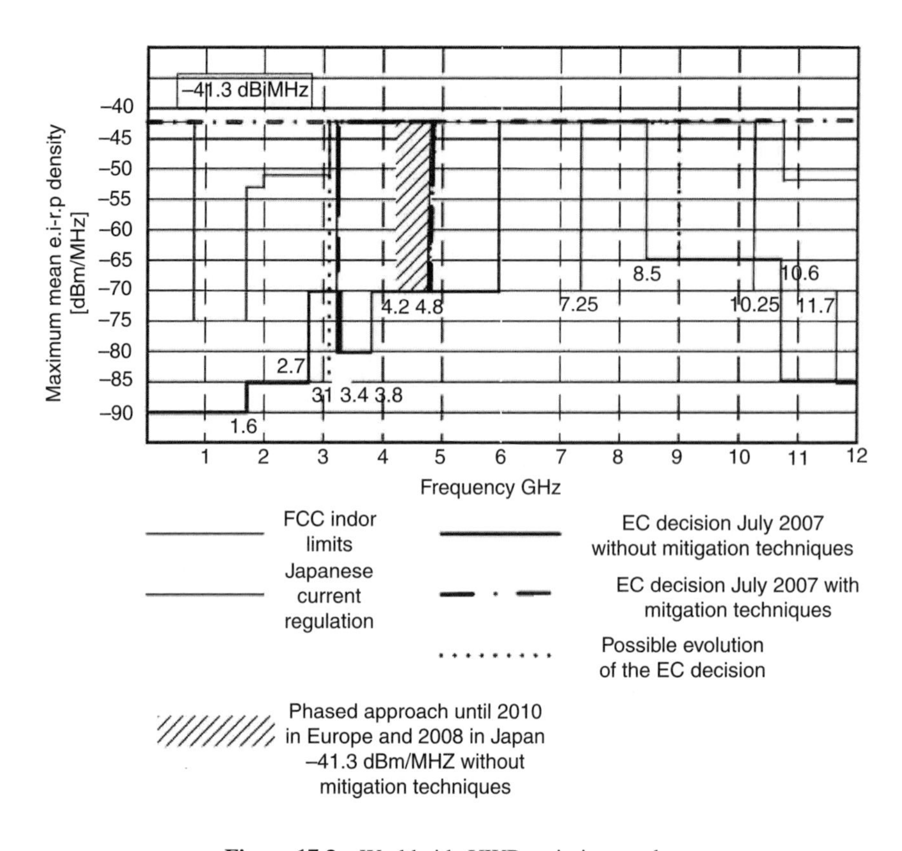

Figure 17.3 Worldwide UWB emission mask.

17.3 Worldwide UWB Standardization Status

UWB applications are classified in two main categories:

- High Data Rate (HDR) applications dedicated to very short range (less than 10 m) with data rate up to 480 Mb/s;
- Low Data Rate (LDR) applications for large networks with data rate in the order of 1 Mb/s with low power consumption and localization capabilities.

17.3.1 UWB Standardization in USA

The IEEE 802.15.3a group that was launched in 2003 in order to elaborate a standard for HDR applications has been dissolved. Indeed, the two main approaches, DS-SS based pulse radio technique and the OFDM approach led by the WiMedia Alliance were competing without any possible consensus.

In 2004, a new group, IEEE 802.15.4a, dedicated to LDR systems was launched by the IEEE. The standard was approved in March 2007. Two physical layers are described in this standard: the first based on impulse radio technology and the second based on chirp waveform.

In the IEEE, two other groups are developing UWB standards. The first, IEEE 802.15.3c is working on a standard for HDR UWB devices at 60 GHz. A first draft has been published. The second group, IEEE 802.15.6, is working on a standard for LDR devices dedicated to body area network applications. Several proposals will be made by this group, based on UWB.

17.3.2 UWB Standardization in Europe

In December 2005, the ECMA-368 standard dedicated to HDR UWB applications was published. This standard is based on the Multi Band OFDM solution pushed by the WiMedia industrial consortium; it defined the physical layer as well as the MAC layer for UWB devices having a radio data rate up to 480 Mb/s and a range of few meters. Today, this standard is the reference for all HDR UWB devices.

In Europe, UWB standardization is achieved through the European Telecommunications Standards Institute (ETSI) with its Technical Committee ERM (Electromagnetic compatibility and Radio Spectrum Matters) which aims to define a harmonized European standard. Three groups are defined:

- *ETSI TG31A,* dedicated to UWB communications devices and Ground Probing Radar and Wall Probing Radar applications (GPR/WPR). In this group two standards have been approved and published in February 2008: ETSI EN 302 065 covering essential requirements for UWB communications devices and ETSI EN 302 066 for GPR/WPR applications.
- *ETSI TG31B,* in charge of short-range radar applications in the 24 and 77 GHz frequency bands.
- *ETSI TG31C,* in charge of UWB devices dedicated mainly to location and tracking applications. Draft harmonized standards for Building Materials Analysis (EN 302 435 1 and 2) and for Object Discrimination and Characteristics (EN 302 498) have been published in 2006. A new draft version of Location tracking Type 1 harmonized standard (EN 302 500 for devices between 6 and 8.5 GHz with outdoor usage) has been available since May 2007. Location tracking Type 2 standard, location tracking and sensor applications for automotive and transportation environments standard as well as Location tracking Application for Emergency Services (LAES) standard and Object Identification Surveillance standard are ongoing in this group.

17.4 Visionary Approach to Improve Regulation

Today spectrum sharing in Europe is based on the traditional regulation approach, but in the near future this will probably change. Indeed, it now seems necessary to move from traditional radio service protection towards promotion of viable coexistence scenarios, avoiding harmful interference based on cognitive principles. This new regulation approach should allow a friendly sharing environment.

UWB regulation in Europe is a good example of the need to build new regulation rules allowing innovative services and technologies to be developed. The regulation and standardization innovative rules could be the following:

- Definition of realistic coexistence scenarios, taking into account real usage (indoor/outdoor usage, estimated number of devices, likelihood of worst case scenarios, etc.)

- Definition of new radio architectures in order to support coexistence mechanisms. This includes flexible air interfaces enabled to work at different frequencies. The DAA mechanisms which are under investigation for UWB in the lower band are good examples of the need to have new mechanisms for the air interface. The coexistence mechanisms can also be based on activity factors, as this is the case for LDC mechanisms defined for UWB LDR systems.
- Introduction of new mechanisms at MAC and higher layers. This could offer the possibility of determining what kind of radios are in the vicinity of the user and then to put in place mechanisms in order to avoid time slots or frequencies used by the 'victim services' (at physical and MAC layers).

These new rules necessary for the evolution of the regulatory system are likely to have a great impact on the new wireless standards definition.

Acknowledgements

This work has been partly founded by the European Commission within the PULSERS Phase II project, IST-FP6-027142. PULSERS (Pervasive Ultra-wideband Low Spectral Energy Radio Systems) is an integrated project within the 6th Framework of the Commission of European communities and aims to investigate the challenges and possibilities associated with UWB technology and its application (http://www.pulsers.eu).

References

1. FCC 02-48: *Revision of Part 15 of the Commission's Rules Regarding Ultra-Wideband Transmission Systems –* First Report and Order.
2. ECC/DEC/(06)04: ECC Decision of 24 March 2006 on *the harmonised conditions for devices using Ultra-Wideband (UWB) technology in bands below 10.6 GHz.*
3. ECC/DEC(06)04: ECC Decision of 24 March 2006 amended 6 July 2007 at Constanta on *the harmonised conditions for devices using Ultra Wideband (UWB) technology in bands below 10.6 GHz.*
4. 2007/131/EC, 23 February 2007 allowing the use of the radio spectrum for equipment using ultra-wideband technology in a harmonised manner in the Community.
5. ECC/DEC/(06)12, ECC Decision of 1 December 2006 on *the harmonised conditions for devices using Ultra-Wideband (UWB) technology with Low Duty Cycle (LDC) in the frequency band 3.4–4.8 GHz.*

Part III

60 GHz Communication Systems: Concepts and Implementation Aspects

18

An Introduction to 60 GHz Communication Systems: Regulation and Services, Channel Propagation and Advanced Baseband Algorithms

Isabelle Siaud[1], Anne-Marie Ulmer-Moll[2], Nadine Malhouroux-Gaffet[3] and Guillet Valery[3]

[1]*France Telecom R&D-Orange Labs, WIN Department, Cesson Sevigné-France*
[2]*France Telecom R&D-Orange Labs, WASA Department, Cesson Sevigné-France*
[3]*France Telecom R&D-Orange Labs, NET department, Belfort-France*

18.1 Introduction

Very high data rates for radio communications require a very large transmission bandwidth in accordance with the Shannon capacity theorem. Depending on the allocated spectrum, it may involve interference with other WLAN and WPAN systems operating simultaneously in the same band. One alternative to reduce interference and successfully allow WPAN/WLAN coexistence is to limit the radiated transmitted power density per frequency unit combined with ultra wide transmission bandwidth sizes. In February 2002, FCC regulation parts permitted Ultra WideBand (UWB) radio communication systems to operate as license-exempt short-range systems in the {3.1–10.6} GHz [1] band. This was the starting point of WPAN UWB system interests for very high data rate radio-Communications with high QoS. The very low radiated power significantly limits radio coverage. On the other side it ensures WLANs deployment without interference. On the other side, WPANs UWB systems are affected by WLAN interferers.

Short-Range Wireless Communications Rolf Kraemer and Marcos D. Katz

Another alternative is to explore millimeter wave frequency bands for WPAN applications, especially the 60 GHz band. In July 2003, a new IEEE802.15.3 SIG was created to deal with 60 GHz radio communications [2]. The first investigations on 60 GHz regulations in Europe prove that a minimum of 7 GHz bandwidth should be addressed in the {59–66} GHz band to license-exempt short-range applications. In 2003, a first IEEE802.15.3c SIG was created to define initial technical requirements for Multi-Gigabit Wireless Systems (MGWS). The objective of this chapter is to position recent technical studies carried out on UWB-OFDM and millimeter-wave short-range radio transmissions, especially within the framework of the MAGNET Beyond project and internal studies lead by France Telecom R&D.

The technical background oriented towards regulation and standardization issues is given in draft form in IEEE802.15.3c TG. This group deals with the 60 GHz short-range MGWS PHY/MAC layer. Section 3 is dedicated to 60 GHz propagation characteristics issued by France Telecom R&D from measurements carried out within the framework of the IST-MAGNET project (http://www.ist-magnet.org). This section presents a novel propagation model able to simulate and characterize UWB propagation channel in different bands. The final section presents the UWB-OFDM system defined by France Telecom R&D to successfully deliver Multi-Gigabit data rates in {3.1–10.6} GHz and {57–64} GHz bands. The conclusion presents future France Telecom R&D studies that will be carried out through the IST-FP7 OMEGA project.

18.2 Technical Background to the WPAN Concept

18.2.1 Regulation and Standardization Issues

In February 2002, FCC regulation parts permitted Ultra WideBand (UWB) radio communication systems to operate as license-exempt short range systems in the {3.1–10.6} GHz band [1]. This was the starting point of standardization activities concerning WPAN communications for telecommunication applications. In Europe, the CEPT opened the {6–8.5} GHz band for Very High Data Rate (VHDR) UWB systems [3]. Following the ECC meeting in July 2006, it was decided to enforce the First Generation of UWB, (UWB (1G)) development on the market in a 'phased approach' [4] where the first generation of UWB would be allowed to operate in the {4.2–4.8} GHz band with a −41.3 dBm/MHz PSD level. After 2010, (1G) UWB would evolve towards (2G) associated with mitigation techniques in this band.

The IEEE802.15.3a standardization group designed two WPAN systems in the {3.1–10.6} GHz band intended for bit rates up to 1 Gbp/s. Facing a deadlock between these two IEE802.15.3a standards, the Wi-Media alliance [5] used the fast track procedure of the ISO/IEC ECMA organization to submit, in December 2005, a UWB OFDM based proposal to the ECMA Technical Committee TC32-TG20 [6]. The proposal specifies the PHY/MAC layer of UWB-OFDM system as a slightly improved version of the UWB-OFDM system from the Wi-Media organization [5]. UWB-WPAN systems require mitigation techniques to coexist with other WLAN and WMAN systems operating in similar bands. Despite some efficient detect and avoid techniques proposed by the group IEEE802.15.3a, it appears relevant to explore millimeter-wave bands around the 60 GHz frequency band.

In July 2003, a new IEEE802.15.3 SIG was created to deal with 60 GHz radio communications [2]. The first investigations on 60 GHz regulations in Europe prove that a minimum of 7 GHz bandwidth should be addressed in the {59–66} GHz band to license-exempt short-range applications. In 2007, following the SE_24 meeting held in September 2007, it was decided

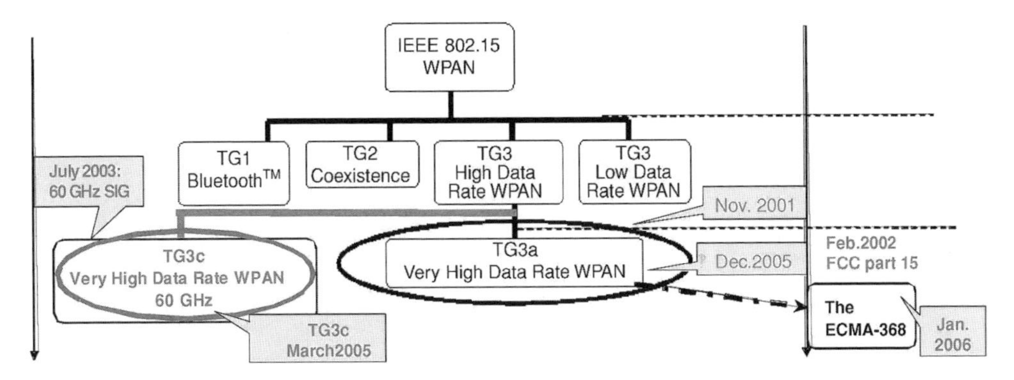

Figure 18.1 IEEE802.15.3c TG in the scope of WPAN activities.

to translate the 7 GHz exempted license band into {57–64} GHz in considering the use of {57–59} GHz band. This decision has been motivated by the high oxygen absorption in the {57–64} GHz b nad which is foreseen as an advantage in the frequency reuse of resource allocation between adjacent WPAN cells. A call for intent was edited in November 2005 to seek partners on PHY/MAC contributions and to finalize a proposal at the end of 2008. Targeted data rates were up to 480 Mbps for UWB-OFDM system (Wi-Media) and are now up to 5 Gbps at 60 GHz (Figure 18.1).

In 2003, Newlans introduced Gi-Fi and Giga Ethernet To The Desktop (GTTD) applications to illustrate the relevance of 60, 70 and 90 GHz frequency band use for P2P high data rate transmissions. Motorola and Oki initiated the SIG group activity over several millimeter-wave WPAN concepts. In November 2005, IHP proposed the first RF front end for 60 GHz WPAN applications [7]. Thanks to exchanges within the Wireless World Research Forum, in May 2007, IHP and France Telecom R&D initiated a common PHY/MAC proposal for MultiGigabit Wireless short range Systems (MGWS) resorted from enhanced UWB-OFDM transmission modes [8, 9]. Actually, the wireless HD consortium, ComPA group, IMEC, Intel, France Telecom and other companies have produced two main PHY/MAC layer proposals [10, 11] combining single carrier and multi-carrier transmissions with common OFDM parameter sets and different FEC coding schemes.

18.2.2 European Consortium: Overview

In parallel, since 1996, several European consortiums have been formed to define 60 GHz WPAN/WLAN systems.

Within the IST MAGNET project [2002–2005] (http://www.ist-magnet.org), milimeter-wavelength UWB-OFDM systems have been designed and evaluated by France Telecom R&D to achieve bit rates up to 1 Gbps over 528 MHz subchannels (http://www.ist-magnet.org) [8]. Multi-carrier Spread Spectrum techniques (MC-SS) have also been investigated to aggregate flexible data rates per user under a TDMA process. France Telecom R&D designed an advanced multi-user access MC-SS scheme involving 200 MHz channels [9] and UWB channels [12] for 60 GHz transmissions.

Following these first studies, France Telecom R&D defined a common UWB baseband PHY layer proposal intended to operate either within the {3.1–10.6} or {57–64} GHz bands [8, 13] validated thanks to UWB propagation analysis in these two bands [14]. Millimeter-wave UWB-OFDM systems exhibit similar radio coverage for UWB-OFDM

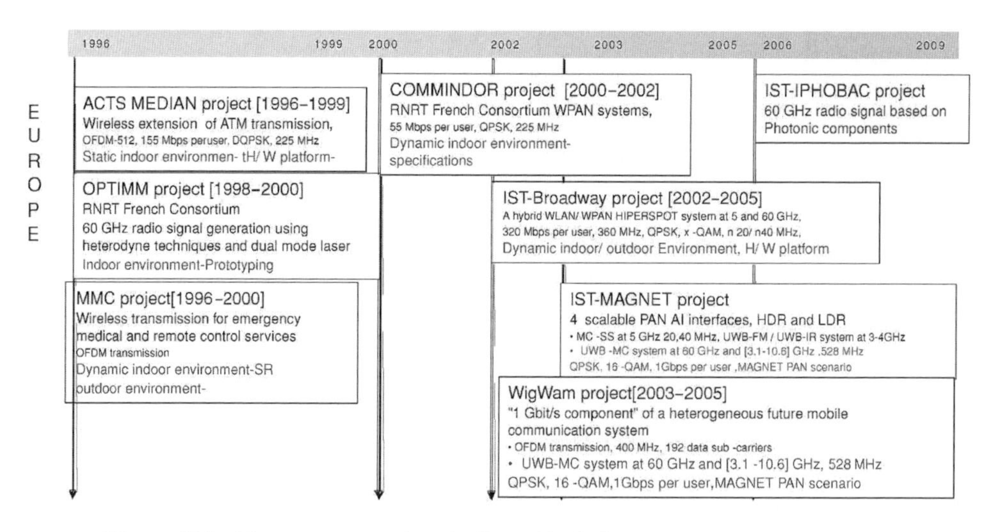

Figure 18.2 European consortium dealing with 60 GHz wireless communications.

systems defined in {3.1–10.6} GHz band thanks to higher radiated transmitted power despite considerable path-loss attenuation due to the RF frequency effect. More recently, IHP and France Telecom R&D have designed a dual-band UWB-OFDM prototype operating at 60 GHz and in the {3.1–10.6} GHz band for downlink and uplink transmissions respectively [15]. The system introduces RF band scalability for bidirectional transmissions and proves H/W feasibility for short-range radio communications and promising data rates up to 1 Gbps. In a first step, this H/W platform exploits Wi-Media/ECMA baseband processing.

Within the scope of the IST IPHOBAC project [16], launched in June 2006, Radio over Fibre (RoF) topologies [17] have been investigated to extend the coverage of 60 GHz MGWS. The FTR&D system described in the Section 18.5 and emerging IEEE802.15.3c systems are envisioned for IPHOBAC RoF topologies. This architecture does not completely remove WPANs/WLANs interference within a cell, but extends the global radio coverage between separate cells using RoF transmissions. It strengthens the Personal Network concept promoted within the IST-MAGNET/MAGNET Beyond project (http://www.ist-magnet.org). Figure 18.2 summarizes European Consortium activity focused on 60 GHz radio-communications for multi-media applications.

Complementary German consortiums are presented by IHP in dedicated Chapter 20 of this book.

18.3 Millimeter-Wave Applications and Services

Millmeter-wave application and services are directed towards WPAN applications covered by the IST-MAGNET project [18], the recent IEEE802.15.3c usage models [19] and RoF architectures defined in the IPHOBAC project [16].

With the IST-MAGNET project, a Personal Network (PN) concept network has been introduced to establish communications between users. A PN is a federation of several PAN

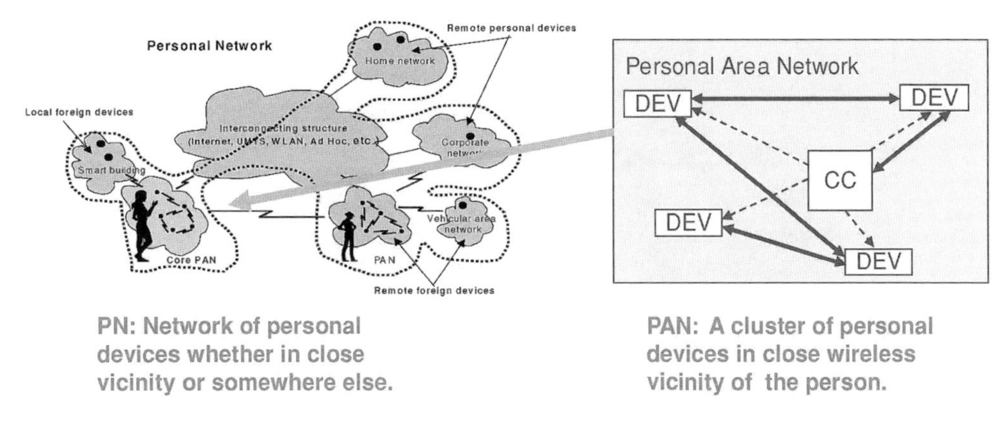

Figure 18.3 The Personal Network (PN) concept of the IST-MAGNET project.

networks based on peer-to-peer architectures, flexible air interfaces (AI) to cover a large bit rate range and common convergence layer to select the most appropriate AI in compliance with the expected QoS and radio coverage. A core PAN is centered on each user and is connected to several devices each bearing a dedicated service or several services into a single radio link (Private PAN). Several core PANs communicate together in a multi-PANs manner, using either common devices or dedicated FEDNET (Federation of Network) architectures [20]. The Personal Network concept is illustrated in Figure 18.3.

Three MAGNET PAN scenarios are defined in Figure 18.4 while Figure 18.5 shows typical LDR services:

- A Private PAN (P-PAN) is core PAN associated to a user enable to establish communications with dynamic/static collection of personal devices around a person (core PAN). The P-PAN is designed as 'personal wireless bubble' where the BAN is a particular case of P-PAN.

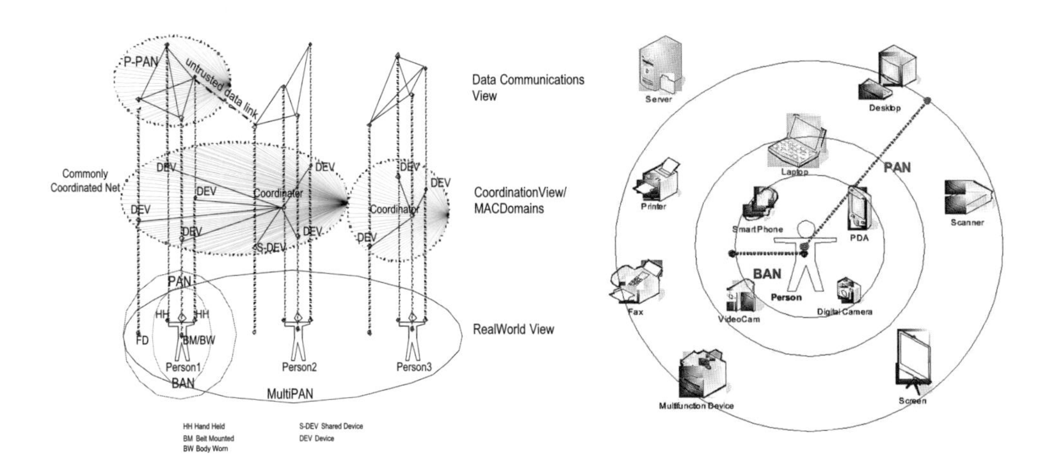

Figure 18.4 PAN scenarios in the IST Magnet project.

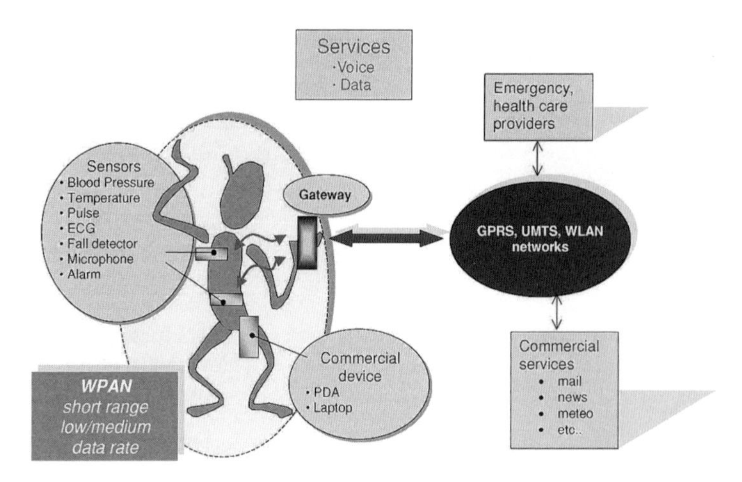

Figure 18.5 Typical LDR services connected to the IST-FP6 MAGNET project [22].

- A PAN is a P-PAN enable to communicate with remote foreigners linked to other P-PANs through common devices.
- A multi-PAN is a communication system where several PAN communicate together using the FEDNET concept.

For MAGNET Beyond air interface radio coverage assessments, France Telecom R&D assigned different path-loss models to PAN and Multi-PAN scenarios, leading to accurate link budget assessments [21].

In the IST-MAGNET project [18], dedicated WPAN services have been identified. Medical and remote control services have been associated with the low data rate UWB-FM air interface and data rates ranged from 10 to 100 Kbps. UWB-FM implements frequency hopped FM modulated pulses transmitted through sub-bands in the $\{3.1–4.5\}$ GHz band.

High data rate services are connected to video services and data file sharing. An identification of services has been carried out within the IEEE802.15.3a Task Group through usage models.

Other complementary wireless applications are defined in the IEEE802.15.3c TG, in compliance with multi-Gigabit data rates as proposed in [16, 19]. Applications are split into five different usage models, as illustrated in Table 18.1.

Another IEE802.15.3c contribution [23] oriented towards usage models, suggests adaptive antenna gain for each usage model. The idea is to adapt and select antenna gain in accordance with typical UMD distance range, data rates and deployment scenario. Mandatory and optional usage models according to IEE802.15.3c are depicted in Figure 18.6 and Figure 18.7, respectively.

Table 18.1 Usage models issued from the IEEE802.15.3c Task Group [19]

Usage model number	UM1	UM2	UM3	UM4	UM5
Description	P2P Video	P2MP Video	Wireless desktop	Conference	P2P kiosk
Total score	116	72	72	73	114
Average score	2.64	1.64	1.64	1.66	2.59
Ranking score	1	4	4	3	2
Modes	Mandatory	Optional	Optional	Optional	Mandatory

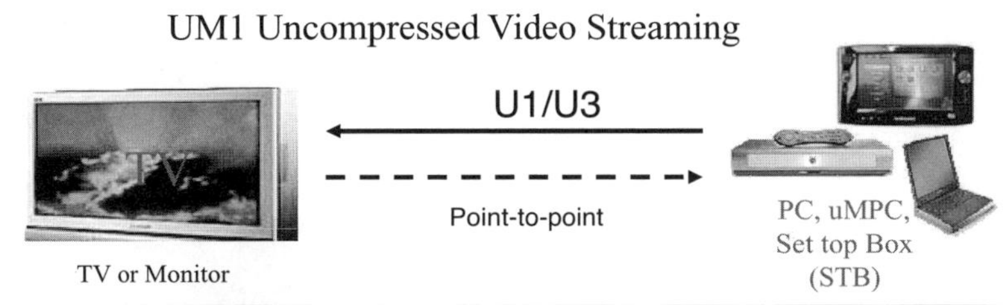

Environment	Throughput MAC SAP	BER/PiER	Distance	Note
NLOS, LOS Residential (STB-TV)	1.78 Gbps 1.49, W/O Blk Stream, Up to 1080i, 24, 60	10^{-6} BER for PHY Simulations	5	• No data retransmission required • Unidirectional data transmission noted by Solid line • Low bitratereverse link • Target of 10^{-9} TMDS CER for HDMI • Pixel is RGB, 24 bits
LOS, NLOS Residential (STB-TV)	3.56 Gbps 2.98, W/O Blk Stream, Up to 1080p, 24, 60	10^{-6} BER for PHY Simulations	10	

Submission Slide 2 Ali Sadri, Intel Corporation

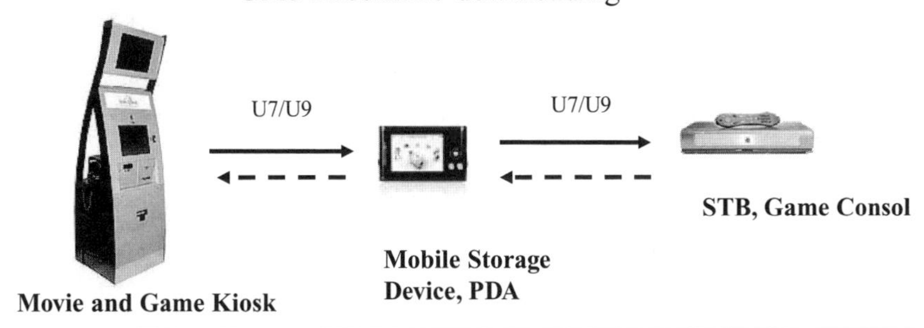

Environment	Throughput MAC SAP	BER/PER	Distance	Note
LOS-office (Server-PDA or PDA-STB)	**1.50 Gbps burst (Server-PDA or PDA-STB)**	8% PER before retransmission 2K Byte	**1 m**	**• Asymmetric download/Upload • Low data rate reverse link**
LOS-office (Server-PDA or PDA STB)	**2.25 Gbps burst (Server-PDA or PDA-STB)**	8% PER before retransmission 2K Byte	**1 m**	

Source: IEEE802.15.3c UMD (Doc: IEEE 15-06-0055-22-003c)

Figure 18.6 Mandatory usage models issued from the IEEE802.15.3c TG [19].

18.4 Frequency Regulation and Standardization Issues

A significant body of sharing and compatibility studies has to be conducted to support harmonization of regulations for license-exempt operations in the 60 GHz range between Europe, Asia, the United States, Canada and Australia.

In Europe, the allocation of the {57–66} GHz band is referenced in CEPT Recommendation TR 22-03 [24] the ERC Report 25 [25] and the ETSI DTR/ERM document [26] for potential license-band wireless applications. The ECC is currently adopting decisions through ECC

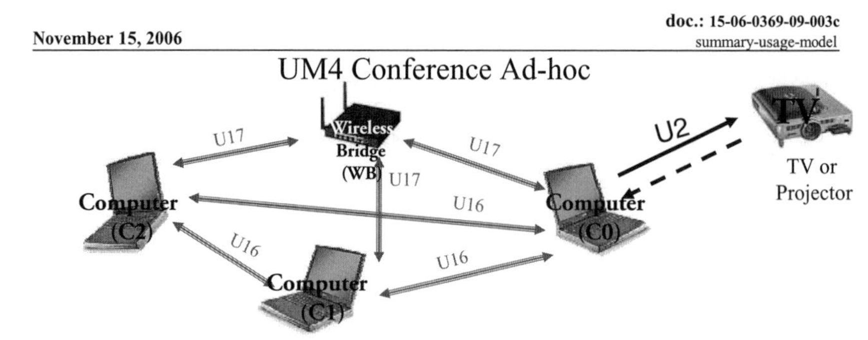

Environment	Throughput MAC SAP	BER/PER	Distance	Note
LOS, office. (C0-TV)	1.75 Gbps 1.49, W/O Blk Stream 1080i, 24, 60	10^{-6} BER for PHY Simulations	5 m	•No data retransmission required for TV1 •Unidirectional data transmission noted by solid line for U1 •Low bitrate reverse link for unidirectional link •device are not co-located •Target of 10^{-9} TMDS CER for HDMI •Pixel is RGB, 24 bits •One simulation for this Usage Model •Total Asyncthroughput of 1 Gbps
LOS, Desktop (C0-C1-C2)	0.0416 Gbps, , average async. Each direction	8% PER before retransmission 2K Byte	1 m	
LOS, office (C0,C1,C2)-WB	0.125 Gbps, , average async. Each direction for WB	8% PER before retransmission 2K Byte	3 m	

Figure 18.7 Optional UM4 usage models issued from the IEEE802.15.3c TG [19].

report 114 [27] on exempt-license bands use for MGWS as reported below. It may be safely assumed that MGWS WLAN & WPAN applications would be deployed pre-dominantly indoors leading to overall low risk of interference. Therefore it would appear that WLAN & WPAN applications might be allowed to be deployed across entire frequency range 57–66 GHz on the license exempt provisions with emission limitations considered in this study, based on +40 dBm EIRP (Emitted Isotropic Radiated Power). Possible technical measures to ensure indoor usage and give additional degree of interference protection could include obligations for integral antennas.

It was also noted that some kind of Dynamic Frequency Selection (DFS)/Detect-And-Avoid (DAA) mechanism may be introduced to ensure intra-system co-existence between WLAN/WPAN installations, which would also provide additional mitigation of inter-service interference, but practical implementation and feasibility of this measure was not further considered in this report as this was felt being outside the mandate of this study. Data rates are intended to scale up 10 Gbps and receiver sensitivities should be ranged from −68 dBm to −48 dBm for 1 Gbps −10 Gbps designs for operation at up to 10 m. Devices typically range from 500 to 2500 MHz of occupied bandwidth with a maximum transmitter radiated power set to 10 dBm. Figure 18.8 illustrates the gain vs. coverage angle of a flexible antenna.

Regarding the CEPT recommendation TR 22-03, the {54.25–66} GHz band is split into several parts dedicated to separate services. The {54.25–58.2} GHz band potentially aims for WPANs on condition that it does not cause harmful interference with intersatellite communications. The {58.2–59} and {64–65} GHz bands are allocated to military applications. The {59–64} {65–66} GHz bands addressed for mobile and fixed links should be

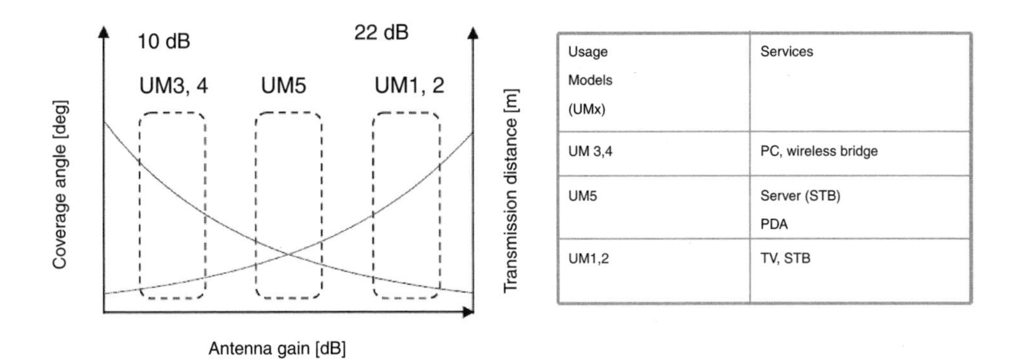

Usage Models (UMx)	Services
UM 3,4	PC, wireless bridge
UM5	Server (STB) PDA
UM1,2	TV, STB

Figure 18.8 Flexible antenna gain in connection with usage models [23].

used for WPAN applications. The {61–61.5} GHz band is designated for industrial, scientific and medical (ISM) applications. The use of this frequency band will require special authorization. The ETSI DTR/ERM document specifies different bands for military applications (Europe_2 in Figure 18.9). The {57–64} GHz band seems to be a contiguous band that should be addressed for licenseexempt radio applications.

In USA and Canada, regulations are harmonized in accordance with the FCC Part 15.255 requirements in response to the NPRM (Notice of Proposed Rulemaking) in ET Docket 94-124, addressing the {57.05–64} GHz band to unlicensed wireless applications with a 7 dBm/MHz spectrum mask.

In Japan and Korea [28, 29], the {59–66} GHz band is dedicated to unlicensed low-power radio stations with a maximum antenna gain set to 47 dBi and an occupied bandwidth up to 2.5 GHz. In Korea, the Millimeter Wave band Frequency Study Group (MWFSG) allocated the

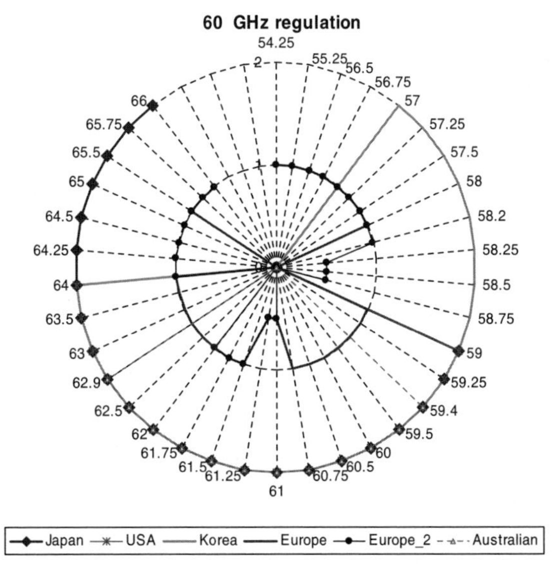

Figure 18.9 60 GHz regulation status for wireless transmissions.

Table 18.2 Common sub-band allocation in the {57–66} GHz band

Country	Common sub-band allocation (GHz)	Bandwidth size (GHz)
USA, Europe, Japan, Korea, Australia	59.4–61.0	1.6
	61.5–62.9	1.4
USA, Europe, Japan, Korea	59.0–61.0	2.0
	61.5–64.0	3.5
USA, Europe_2, Japan, Korea, Australia	61.5–62.0	0.5
USA, Europe_2, Japan, Korea	61.5–62	0.5
USA, Europe_2	61.5–62	0.5

{57–64} GHz band to indoor WPANs and point-to-point outdoor radio links with a maximum transmitter power set to 10 dBm.

In Australia, the Australian Communications and Media Authority (ACMA) authorize unlicensed frequency bands for Low Interference Potential Devices (LIPDs) in the {59.4–62.9} GHz band with a maximum peak transmitter power set to 10 dBm.

An international regulation harmonization is required between these regions. The most restrictive band allocation is due to ACMA specifications. The next figure stacks up potential license-exempt bands in the range 54.25–66 GHz for the five countries cited. For each sub-band, we attribute three different values. The value '2' is linked to the case where the band is considered exempt-license, the value '1' corresponds to authorized sub-bands without license-exempt band certifications. The value '0.5' is dedicated to the ISM band {61–61.5} GHz and other bands requiring special authorizations and the value '0' is associated with forbidden bands usually attributed to very restrictive cases. In regards with the ETSI DTR/ERM [26], we make the assumption that military applications in some special bands involve forbidden bands. The associated spectrum allocation is denoted Europe_2 (Figure 18.9). The common 60 GHz license-exempt bands are listed in Table 18.2.

18.5 Channel Propagation Characterization and Modeling

18.5.1 60 GHz Propagation Measurements

The 60 GHz multipath propagation channel has been investigated by France Telecom R&D through the RNRT COMMINDOR project, the IST-MAGNET project [30, 31] and IEEE802.15.3c TG [32–34]. France Telecom has carried out many measurements to characterize the 60 GHz multipath channel in the time and spatial/angular domains [30]. Siemens carried out 60 GHz dedicated measurements published through IEEE802.15.3c TG [35] that lead to similar 60 GHz propagation characteristics. Additionally, material measurements have been assessed to provide transmission and reflection coefficients used in a ray-tracing model. The NICTA elaborated a modified Saleh-Valenzuela (SV) model with extended spatial parameters for 60 GHz indoor applications [36, 37]. Taiwan University defined a single-cluster SV model studies and promoted it through the group [38]. Other research laboratories investigated the 60 GHz channel through the IST Broadway project, MMIC consortiums and advanced research laboratories. Tables 18.3, 18.4 and 18.5 present measurement conditions, selectivity parameters and RMS delay spread measured in 60 GHz channels.

Table 18.3 France Telecom R&D measurement conditions

	Scenarios			
	Residential 1	Residential 2	Office 1	Office 2
Parameter	Sectoral antenna	Directive antenna	Sectoral antenna	Directive antenna
Tx antenna beam width	60°	60°	72°	10°
Rx antenna beam width	72°	10°	60°	60°
Tx antenna gain	13 dBi	13 dBi	8 dBi	24.6 dBi
Rx antenna gain	8 dBi	24.6 dBi	13 dBi	13 dBi°
Length of measurement run	150 mm (30 λ)	150 mm (30 λ)	90 mm (18 λ)	90 mm (18 λ)
Number of CIR per measurement run	76 (0.4 λ step)	76 (0.4 λ step)	60 (0.3 λ step)	60 (0.3 λ step)
Frequency bandwidth	1024 MHz	1024 MHz	512 MHz	1024 MHz
Maximum CIR delay	250 ns	250 ns	500 ns	500 ns
Relative delay resolution	0.97 ns	0.97 ns	1.95 ns	0.97 ns

Table 18.4 60 GHz wideband selectivity parameters in vertical polarization and transmission band $B = 1024$ MHz

			Wideband parameters					
Scenario	Link	Percentile	σ_{DS} (ns)	$W_{90\%}$ (ns)	$I_{6\,dB}$ (ns)	$I_{12\,dB}$ (ns)	$Bc_{-50\%}$ (MHz)	$Bc_{-90\%}$ (MHz)
Residential 1 (sectoral antenna)	LOS	Q10	1.1	1.9	1.9	2.7	291.5	77
		Q50	2.0	2.1	3.1	2.9	317.5	106
		Q90	3.1	3.2	2.2	3.2	327	122.5
	NLOS	Q10	3.3	2.8	2.1	3	30.5	6.5
		Q50	7.1	17.2	3	12.2	101.5	12.5
		Q90	11.8	34.4	14.8	33	297	84.5
Residential 2 (directive antenna)	LOS	Q10	0.5	1.9	2.1	2.9	318.5	118.5
		Q50	0.7	1.9	2.2	2.9	322	126
		Q90	1.5	2.1	2.2	2.9	323	126.5
	NLOS	Q10	1.8	2.1	2.1	2.9	35	5.5
		Q50	6.3	11.3	2.1	11.2	216	19
		Q90	13.4	37	11.8	34	319	117
Office 1 (sectoral antenna)	LOS/NLOS	Q10	2.5	3.5	3.5	4.5	74.5	12.5
		Q50	45	6.5	3.5	6.5	169	37.5
		Q90	8.5	19.5	7.0	15.5	206	62
Office 2(directive antenna)	LOS/NLOS	Q10	0.9	1.85	1.9	2.6	119	4.7
		Q50	7.9	2.3	2.1	2.9	315.5	88
		Q90	17.7	58.3	2.3	16.6	327	127

Table 18.5 RMS delay spread in residential environments

Residential FTR&D environment	LOS/NLOS	RMS delay spread σ_{DS} (ns)	Excess delay (ns)	Distance range (m)
Tx-72°/Rx-60°	LOS	2.2	50–70	<7
Tx-72°/Rx-60°	NLOS	6.9	80–120	$4 < d < 12$
Tx-72°/Rx-10°	LOS	0.6	40–50	<5 m
Tx-72°/Rx-10°	NLOS	5.4	80–100	$6 < 12 < 6$ m

18.5.1.1 France Telecom Contribution to 60 GHz Propagation Channel Characterization

France Telecom carried out measurements to specify path-loss attenuation, multipath channel and spatial variations of the millimeter-wave channel in indoor environments [30–34]. The Selected channel sounding technique is based on a frequency sweep mode with a total bandwidth set to 1024 MHz using a VNA. Residential and office environments have been considered. Vertical polarization has been mainly studied. Some experimental conditions issued from [30] are summarized below.

Wideband selectivity parameters of the channel have been evaluated upon the Average Power Delay Profile (APDP) of the channel $P(\tau)$: the RMS delay spread σ_{DS}, the delay window $Wx\%$ associated with the centered window containing $x\%$ energy of the APDP of the channel, the delay interval Iy corresponding to the delay duration associated to a dynamic range set to y dB on the APDP and the coherence bandwidth $Bc\text{-}x$ of the channel. The coherence bandwidth is simply deduced from the FFT transform applied on the APDP under Wide Sense Stationnary Uncorrelated Scaterrers (WSSUS) assumptions. $Bc\text{-}x$ is the frequency spacing associated with power level x of the normalized correlation coefficient. Qx corresponds to the cumulative distribution function associated with Prob $\{X < x\}$.

Angular and spatial parameters associated with the and AoA distribution indoor environment are detailed in the MAGNET deliverable [30]. In connection with the scenarios defined in the Table 18.6, the CDF of angular spread in azimuth and vertical planes has been assessed, showing a higher angular spread in the azimuthplane. Figure 18.10 shows angular spreads in azimuth and elevation for scenarios A, B, C and D. Table 18.7 displays the selectivity parameters of the CEPD input files.

Table 18.6 Propagation scenario for angular characterization

	Scenarios			
Parameter	A	B	C	D
AP position in the room	Middle	Angle	Angle	Angle
Location of the sensor array (for DoA analysis)	T	T	AP	T
Array orientation	XY	XZ	XZ	XY
AP antenna	Dipole	Horn	Horn	Horn
T antenna	Dipole	Horn	Horn	Dipole

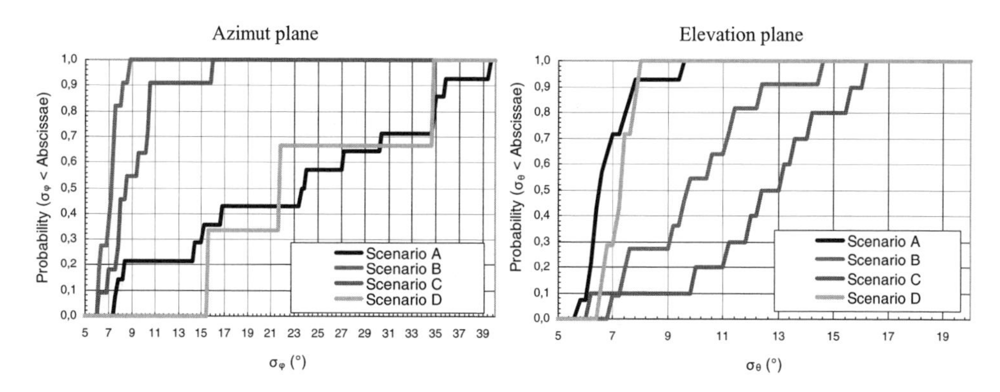

Figure 18.10 Angular spread in azimuth and elevation plane.

18.5.1.2 Siemens Contribution to the 60 GHz Propagation Channel Characterization

Siemens published similar propagation characteristics [35] where the impact of antenna was analyzed (Figure 18.11). The frequency selectivity of the channel is more marked than for FTR&D measurements, probably due to a smaller channel sounding bandwidth size, as observed in [31].

Many other companies and research laboratories have contributed to the 60 GHz propagation channel. Among them, Smulders and Correia [39] significantly contributed the 60 GHz propagation channel characterization.

18.5.2 Multipath Propagation Modeling

18.5.2.1 France Telecom Propagation Channel Models

Digital Signal Processing for the Generic Model: The UWB/WB propagation channel model developed by France Telecom differs from the classical tapped delay line model approach referring to a non-uniform sampling of the scattering function of the channel Ps(ν,τ) with a limited number of taps and an identical U-shape Doppler for each tap. The proposed model resorts from a two-dimensional multirate filter applied on selected measured Channel

Table 18.7 Selectivity parameters of the CEPD Input files (Rx = 60° and 10°)

	60° *Rx* Antenna				10° *Rx* Antenna	
	LOS		NLOS		LOS	
Type	Typical	Atypical	Typical	Atypical	Typical	Atypical
Input file: CEPD_xx	71	72	73	75	73	75
D(*Tx*-*Rx*) (m)	7.55	3.37	8.33	12.75	8.33	12.75
Transmitter location	A1	A2	A2	A1	A2	A1
Delay spread (ns)	1.73	2.28	6.92	12.40	6.92	12.40
I6/I12 dB (ns)	2/3	2/4	5/14	23/36	5/14	23/36
W 90% (ns)	2	5	16	35	16	35
B_c_0.9 (MHz)	110	59.3	17.6	5.7	17.6	5.7

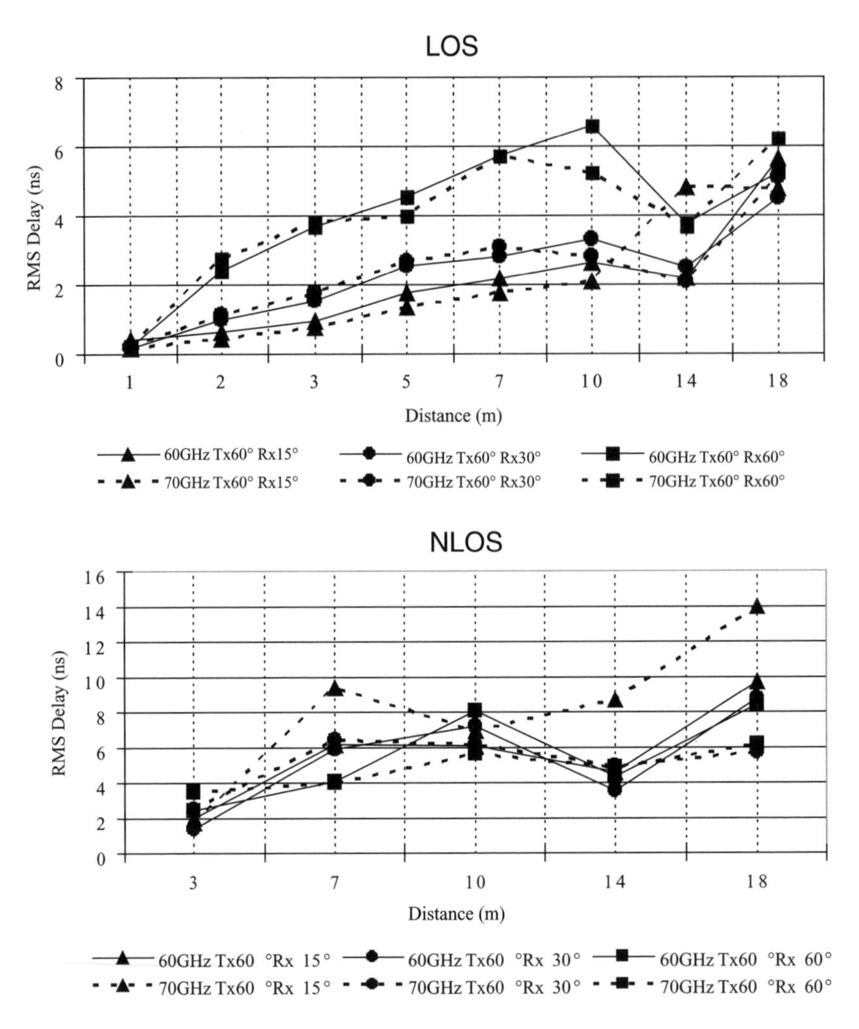

Figure 18.11 Hirose's RMS delay spread.

Impulse Responses (CIRs) of the channel issued from a statistical analysis of the selectivity parameters of the channel. The model generates a time-variant filter fit with the simulated system using the system bandwidth to filter the channel and carries out a conversion rate using an optimized interpolator-decimator filter. This modeling approach has been primarily introduced for WLANs at 5 GHz [40] and extended to many other applications [8, 30, 31, 41]. Actually, we use this concept to model the UWB channel. The main advantage results in a realistic representation of the channel in connection with measurements in the face of stochastic models.

Thanks to WSSUS assumptions, variables t and (τ) of the impulse response $h(t, \tau)$ of the channel shall be separated, hence the filtering processing is processed in a 2×1D assessment, reducing the complexity. First, the instantaneous transversal filter $h_F(t, \tau)$ describing the channel at the time t is generated thanks to an interpolator-decimator filter with a conversion

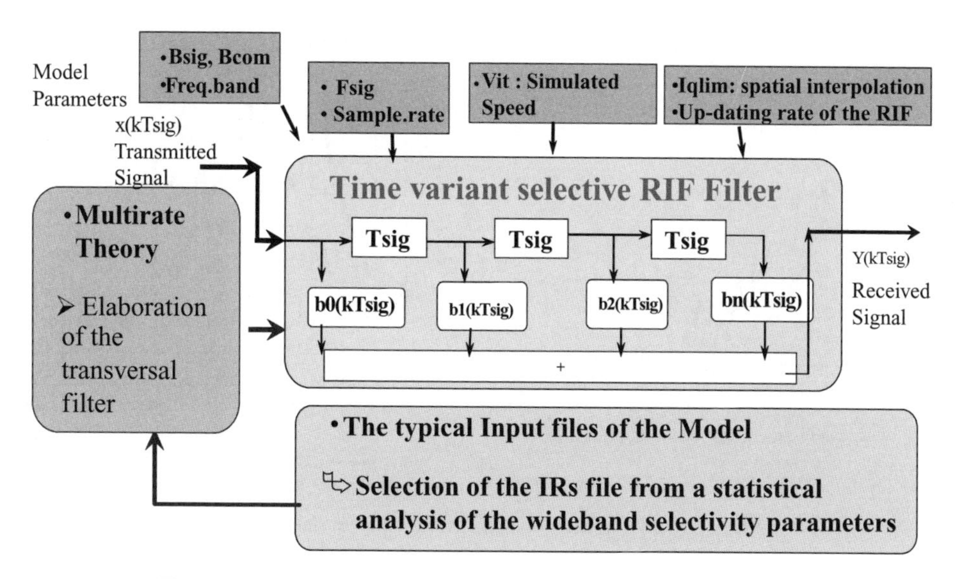

Figure 18.12 The CEPD propagation channel model principle [40].

rate $R = T_{\mathrm{sig}}/T_{\mathrm{e}} = p_{\mathrm{o}}/q_{\mathrm{o}}$ expressed as a ratio of two integers. T_{sig} is the sampling period of the communication system and T_{e} is the sampling rate of propagation measurements. In a second step, the renewal rate of every transversal filter $h_F(t, \tau)$ is processed in performing a conversion rate of each transversal filter tap. The renewal rate is chosen as a multiple of the largest Doppler value $\upsilon_{\max} = \frac{Vit}{\lambda}$ (λ is the wavelength associated to the RF frequency) and fitted to the data framing rate and time variations of the channel. The CEPD propagation channel model principle is shown in Figure 18.12. The mathematical expression of the model is given by

$$hF(t,\ \tau) = \sum_k \delta(t - kT') \sum_i c(k,\ i) \cdot \delta(\tau - iTsig) \tag{18.1}$$

And the coefficients $c(k, i)$ result from a an expansion rate $F = q_0 \times \mathrm{Fe}$, followed by a filtering processing and a decimation rate with a ratio p_0. $c(k, i)$ is expressed as follows:

$$c(k, i) = \sum_n b(k,\ i \cdot p_0 - n \cdot q_0) \cdot g'_F(n) \tag{18.2}$$

$b(k, n)$ are the coefficients of the baseband equivalent measured impulse response of the channel sampled to Fe converted to T' acquisition sampling, g'_F is the filter impulse response used in the multirate processing and T' is a time acquisition period. Figure 18.13 shows typical average power delay spread profiles of a WPAN working in 60 GHz.

The antenna impact and transmission bandwidth B_w are involved into the base-band filter realizations through measurements and input files of the model. The sub-carrier spacing and Spectrum Efficiency (SE) dependency is shown in Figure 18.14.

The input files of the model are issued from selected measured CIR files whose selectivity parameter values are included in a specific interval. Assuming the selectivity parameter

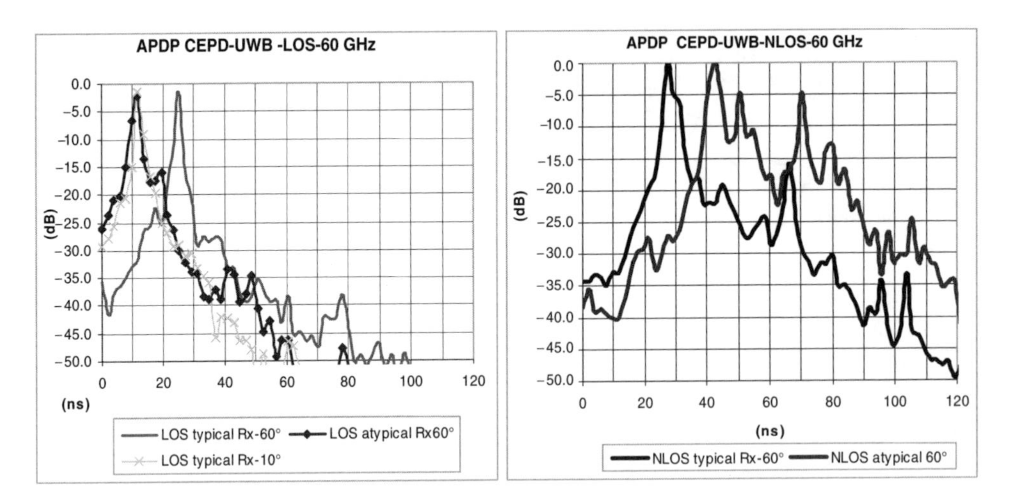

Figure 18.13 WPAN 60 GHz average power delay profile.

first-order statistics follow a Gaussian distribution, the typical and atypical input files of the model are CIR files verifying the equation:

$$\text{Prob}\left\{ I = \frac{\left| m_i^j - X^j \right|}{\sigma^j} < \alpha \right\} = \beta,\ \alpha = \{0.1 - 0.15\} \tag{18.3}$$

m_i^j is the average of the selectivity parameter j for the ith measurement position, $j = \{$delay spread, coherence bandwidth, delay interval to 6 and 15 dB, Power standard deviation$\}$. X^j is either the first-order statistic of the m_i^j variable for the *typical input files* or the Cumulative density function set to 0.8 for the *atypical input files* deduced from $i = 1, N$ measurement points.

Input Files of the CEPD Model in a Residential Environment: Input files of the CEPD model are a measurement point comprising Nri instantaneous impulse responses of the channel along a linear path of 30λ approximately. Typical and Atypical cases have been extracted from a

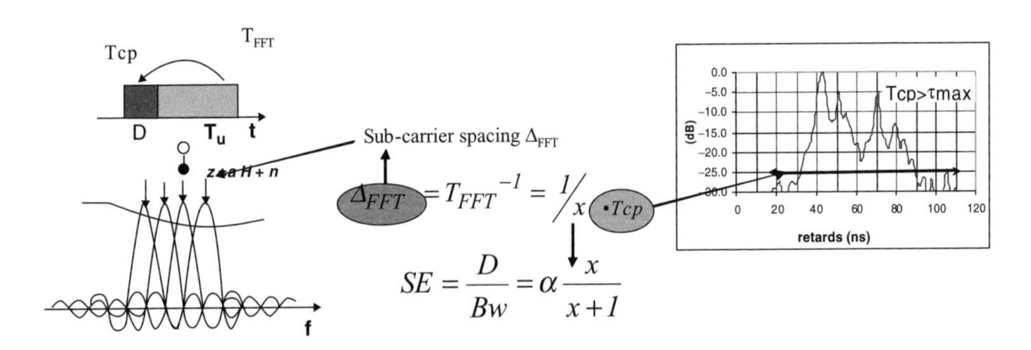

Figure 18.14 Sub-carrier spacing and Spectrum Efficiency (SE) dependency.

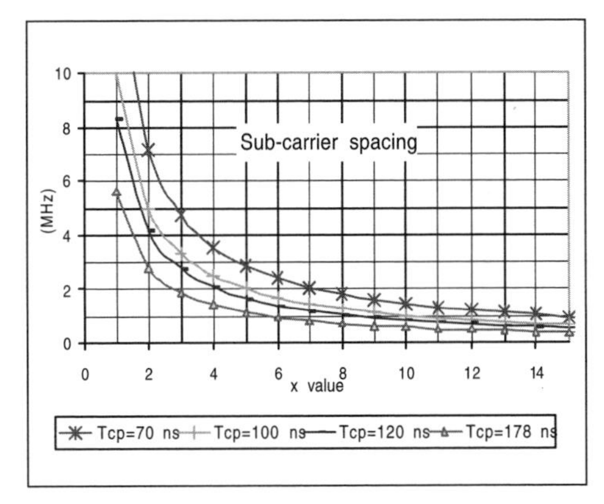

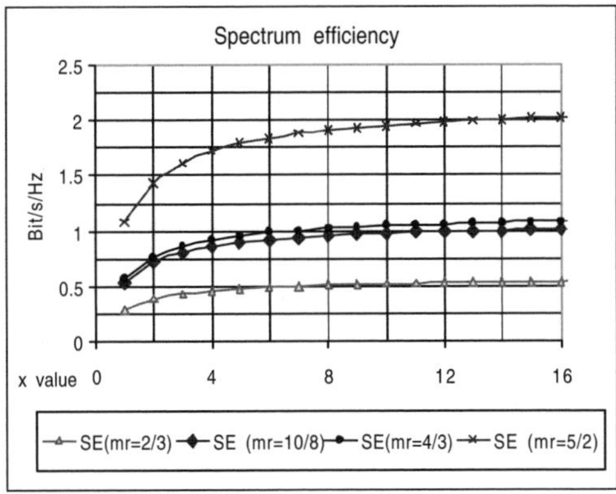

Figure 18.15 Cyclic prefix duration T_{cp} and subcarrier spacing combined with SE.

given scenario. Input files are denoted CEPD_xx. Figures 18.15 and 18.16 depict the cyclic prefix duration and subcarrier spacing combined with SE, as well as typical phase noise values, respectively.

Office environment is modeled in a similar way with typical and atypical input files of the CEPD model. Associated average power delay profiles are represented below for a sampling rate set to 528 MHz.

18.6 Physical Layer Baseband Algorithms and Architectures

18.6.1 Technical Background

Siemens proposed OFDM to deliver bit rates ranged from 34 to 134 Mbps using QPSK and 16-QAM modulated subcarriers over 100 MHz subchannels [42]. Access points are located at the ceiling. The CRL laboratory promoted single carrier transmission implementing FSK modulation over a 1.552 GHz bandwidth and a discriminator detection to supply bit rates up to 2 GHz [43]. The ACTS MEDIAN project (http://www.imst.de/mobile/median/median.html) developed OFDM based PHY/MAC layer and a H/W platform delivering bit rates up to 154 Mbps over 170 MHz subchannels. Adaptive TDMA access protocols were designed.

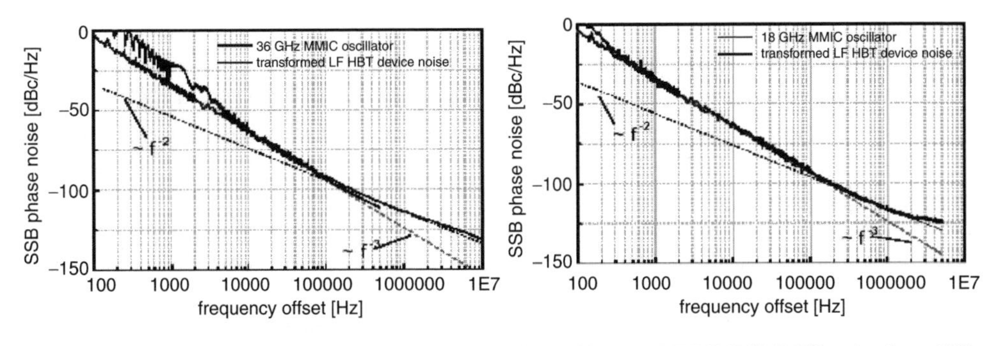

Figure 18.16 Typical gabarit phase noise based on GaInP/GaAs-HBT MMICMMIC technology [48].

In 2002, France Telecom R&D contributed to a French consortium to define a 60 GHz WPAN system integrating a moderate mobility of the environment and targeted bit rates near 200 Mbps. This consortium, denoted RNRT COMMINDOR (http://www.rnrt.commindor), carried out a comparison between several PHY layers based on multi-code transmission, single carrier with a turbo-equalizer and OFDM. OFDM has been selected due to the simple zero-forcing equalizer structure, independent of the targeted bit rate and BER performance. Turbo-coding and antenna diagram impact have been assessed, proving that a 60° receiver antenna was preferred to a 10° receiver antenna to collect multipath energy [44]. This project concludes that adaptive beam-forming or space diversity techniques should be relevant to cope with obstructed radio link transmissions.

The IST-Broadway project [2002–2005] (http://www.ist-broadway.org) led by Motorola, developed a hybrid HIPERSPOT system and associated H/W platform able to operate at 5 and 60 GHz to ensure seamless high QoS and throughput for WLAN/WPAN services [45, 46]. The system aggregates several elementary 20/40 MHz subchannels up to 360 MHz subchannel size and OFDM transmission. A complete H/W platform has been developed; able to operate either at 5 or 60 GHz with a common baseband processing. RF front end has been investigated in a three-stage approach.

In the IST-MAGNET project, France Telecom R&D selected multi-carrier transmissions and CDMA techniques for 60 GHz WPAN systems and extended the system to an unified system, able to operate either in the {3.1–10.6} GHz UWB spectrum mask or at 60 GHz [8]. The PHY layer is derived from the ECMA-368 proposal without MB processing [6]. UWB-OFDM refers to a transmission bandwidth higher than 500 MHz in accordance with part 15 FCC rules. A common subchannel size is considered for low bands and millimeter-wave frequency bands. Additionally, the system has been conceived to carry out the multi-carrier spread spectrum technique in order to increase the user's capacity and introduce frequency diversity [12, 47].

18.6.2 FTR&D UWB-OFDM Physical Layer Proposal

France Telecom introduced a common baseband OFDM PHY layer proposal intended to extend the Multiband OFDM Alliance system in a unique baseband system able to operate either at 60 GHz or in the UWB spectrum mask {3.1–10.6} GHz. This approach has been introduced in [8] and detailed in [13, 14]. Optimizations are directed towards spectrum efficiency maximization, considering high OFDM boosted modes and advanced baseband processing, mainly directed towards interleaving processes.

18.6.2.1 OFDM Boosted Modes

Sub-carrier spacing and cyclic prefix from the MBOA proposal [5] have been refined to strengthen the spectrum efficiency and extend the system radio coverage. A trade-off has been achieved to cope with RF phase noise and high spectrum efficiency, as illustrated in the following figures and explained in [8, 14].

The subcarrier spacing is adjusted to the excess delay of the channel to ensure high spectrum efficiency and a high number of data subcarriers, ensuring efficient interleaving at the binary and subcarrier levels.

The subcarrier sizes (Table 18.9) are compatible with a three-stage RF front end using novel GaInP/GaAs-HBT MMIC [48] or CMOS device oscillators. Considering single side band

Table 18.8 Cyclic prefix size of the system

	$x = T_{FFT}/T_{cp}$	T_{cp} max (ns)
Mode 1	8.17	178
Mode 2	5.44	178
Mode 3	4	242.5
Mode 4	4	178

phase noise levels at 100 and 700 kHz frequency offsets respectively, the phase noise PSD is inferior to −91 and −100 dBc/Hz.

To remove ISI, a cyclic prefix Tcp is preferred to a zero-padded suffix in connection with time domain power to noise issues and synchronization aspects. Tcp is enlarged to 178 ns instead of 70 ns as proposed in Wi-Media to envision higher radio coverage compliant with enhanced baseband processing.

Nevertheless, other cyclic prefix durations are proposed, as described in Table 18.8 to perform scalable data rates and radio coverage. Sub-channel size is a multiple of ECMA with an elementary size set to 528 MHz.

The OFDM parameter sets are given in the Table 18.9. The OFDM symbol duration T_{FFT} (FFT period), the total OFDM symbol duration T_{SYM}, including cyclic prefix duration, the number of data subcarrier Npm, the FFT size, the subcarrier spacing (Δ_F) and the efficient bandwidth B_w are given below. Interleaving depth is set to 6 OFDM symbols with respect with the ECMA proposal.

Table 18.9 FTR&D UWB-OFDM parameter sets

Parameters	FFT size	F_{sig} (MHz)	B_w (MHz)	N_{pm}:	Δ_{FFT}: (MHz)	T_{FFT} (ns)	x	T_{SYM} (ns)	Interleaving depth (λ/**q**, μ/ms)
				Sub-channel 528 MHz					
Mode 1	768	528	506.7	736	0.69	1454	8.17	1632	λ/510/9.79
Mode 2	512	528	512.5	496	1.03	969.6	5.44	1148	λ/726/6.89
Mode 3								1213	λ/687/7.278
Mode 4	384	528	506.7	368	1.375	727.3	4	909.1	λ/916/5.454
				Subchannel 2 × 528 MHz					
Mode 1	1536	1056	1015.7	1472	0.69	1454	8.17	1632	λ/510/9.79
Mode 2	1024	1056	1025	992	1.03	969.6	5.44	1148	λ/726/6.89
Mode 3								1213	λ/687/7.278
Mode 4	768	528	506.7	736	1.375	727.3	4	909.1	λ/916/5.454
				Subchannel 3 × 528 MHz					
Mode 1	2208	1584	1523.5	2208	0.69	1454	8.17	1632	λ/510/9.79
Mode 2	1536	1584	1537.5	1488	1.03	969.6	5.44	1148	λ/726/6.89
Mode 3								1213	λ/687/7.278
Mode 4	1152	1584	1523.5	1104	1.375	727.3	4	909.1	λ/916/5.454

18.6.2.2 Advanced Interleaving Processing [13, 49]

The interleaving algorithm $L(k)$ is set aside for fulfilling proper technical requirements.

First, the maximization of the *interleaving spreading* $\Delta L(s)$ *between* samples separated by $s - 1$ samples is an important criterion to ensure reduced correlations at the input decoder.

$\Delta L(s)$ is the minimum distance between samples separated by $s - 1$ samples in the output sequence with:

$$L(k) = I_{p,q}^{(j)}(k)$$
$$\Delta L(s) = \Delta I_{p,q}^{(j)}(s) = \underset{k}{Min}\left\{\left|I_{p,q}^{(j)}(k+s) - I_{p,q}^{(j)}(k)\right|\right\} \tag{18.4}$$

For that purpose, several s values are selected in a *multi-level interleaving spreading maximization* criterion to select interleaving patterns providing high spreading in a multi-level scale.

Secondly, interleaving scrambles data and breaks special data multiplexing used to optimize system performance. We focus the interleaving design on a dedicated algorithm which preserves special data multiplexing, denoted *interleaving partitioning preservation*. The size of the interleaving partitioning size is given by one parameter of the interleaver, the parameter p. This property is translated by the equation:

$$[L(k) - k]_p = 0 \tag{18.5}$$

This property is illustrated in Figure 18.17.

The interleaving spreading is predicted using algebraic formula for the interleaving algorithm description.

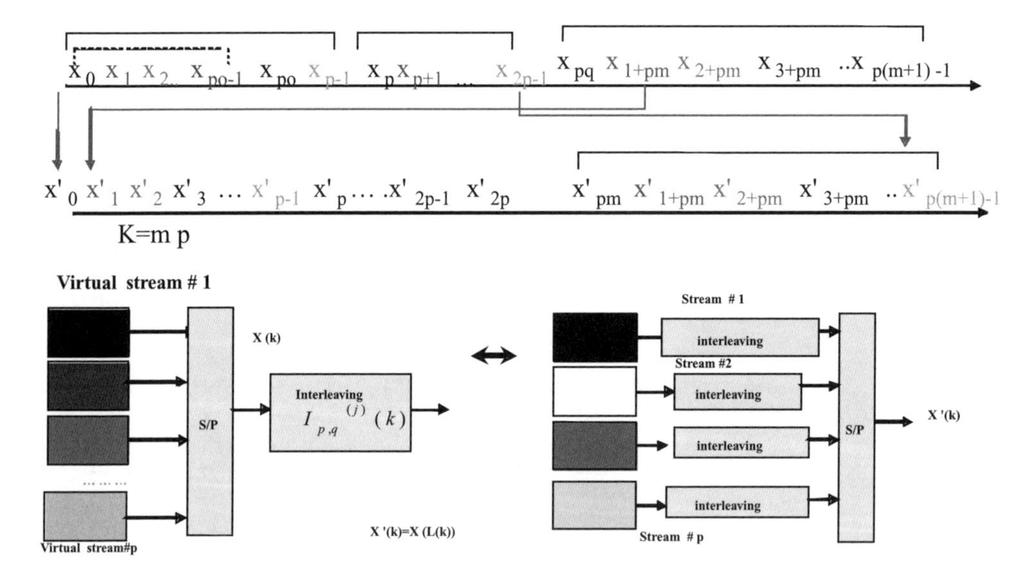

Figure 18.17 Preservation of an interleaving partitioning of size p.

The principle consists in selecting the parameters of the interleaver which provides the largest interleaving spreading $\Delta L(s)$ considering several criteria of maximization. These criteria are defined by the choice of several s values to calculate the interleaving $\Delta L(s)$.

To satisfy these two conditions, the algorithm has a turbo-based structure, as illustrated in Figure.

The interleaver structure $L(k)$ depends on four parameters:

- The interleaving block size K;
- The integer parameter p which provides the interleaving partitioning size and is submultiple of K;
- The parameters q and j.

This structure reduces complexity of the algorithm set-up in using a unique interleaving unit I integrated in an iterative structure to generate different interleaving pattern in a scalable way.

$L(k)$ is built on a serial combination of two algebraic functions L_0, p and $L_{1,p,q}$, characterized by two inputs and one output implementing modulo operations (Figure 18.18). Inputs of the function $L_{0,p}$ are fed with the input index position k (position index of samples $0, 1, \ldots, K^{-1}$) and the feedback of the previous iteration of the algorithm, while function $L_{1,p,q}$ is fed with the input index position k and the output of the function $L_{0,p}$ associated with the current iteration.

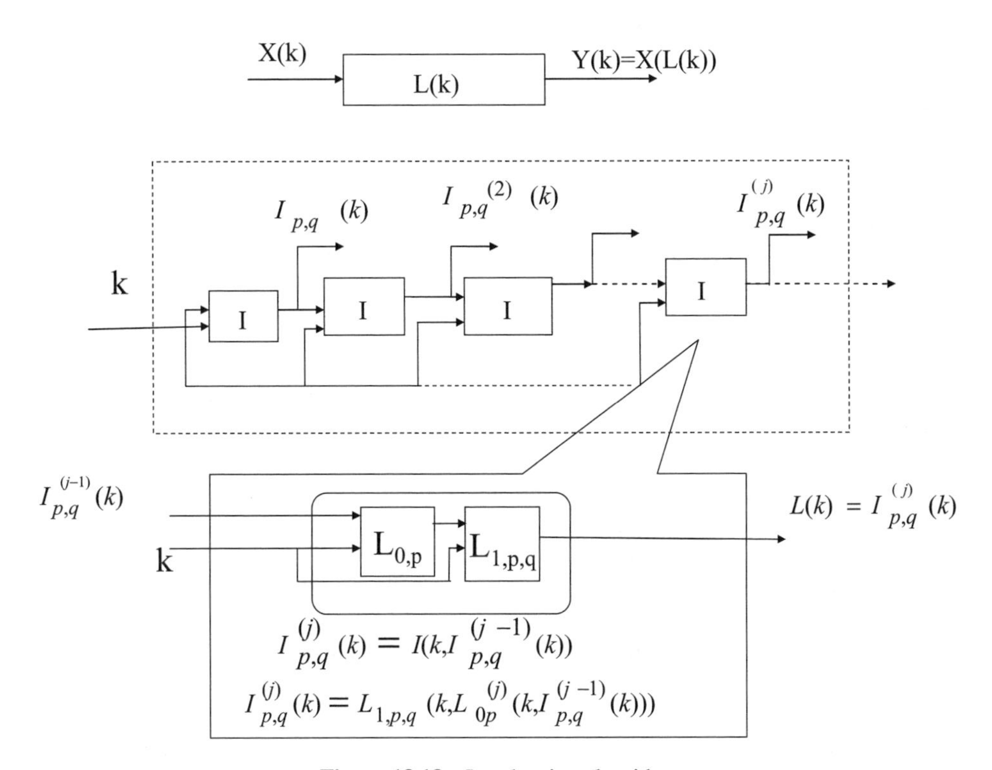

Figure 18.18 Interleaving algorithm.

$L_{0,p}$ and $L_{1,p,q}$ functions are expressed, for $j = 1$, as follows:

$$L_{0,p}(k, k_1) = [-k - k_1 \cdot p]_K\,, \quad k = \{0, \ldots, K-1\}$$

$$I_{p,q}(k) = L_{1,p,q}(k, k_1 = L_{0,p}(k,k)) = [K - p + k + q \cdot p \cdot k_1]_K \qquad (18.6)$$

$$I_{p,q}(k) = \left[K - p + k + q \cdot p \cdot [-k - k \cdot p]_K\right]_K$$

The extension to the *j*th iteration is given by:

$$L_{0,p}^{(j)}\left(k, I_{p,q}^{(j-1)}(k)\right) = \left[-k - p \cdot I_{p,q}^{(j-1)}(k)\right]_k$$

$$I_{p,q}^{(j)}(k) = L_{1,p,q}^{(j)}\left(k, L_{0,p}^{(j)}(k)\right) = \left[K - p + k + q \cdot p \cdot L_{0,p}^{(j)}(k)\right]_k \qquad (18.7)$$

$$I_{p,q}^{(j)}(k) = \left[K - p + k + q \cdot p \cdot \left[-k - p \cdot I_{p,q}^{(j-1)}(k)\right]_k\right]_k$$

Binary Interleaving Process: At the binary level, interleaving is carried out upon encoded bits. The *interleaving spreading maximization* is performed within every data symbol and between data symbols as follows:

- The interleaving spreading between adjacent bits within every data symbol consists in selecting $\{p, q, j\}$ values giving the largest $\Delta L(s)$ for s values ranged from 1 to $N_{\mathrm{BPSC}} - 1$. N_{BPSC} is the number of encoded bits per data symbol (data subcarrier).
- The maximization of the binary interleaving spreading between adjacent symbols consists in selecting $\{p, q, j\}$ values giving the largest $\Delta L(s)$ for s values multiple of N_{BPSC} ($[s]_{\mathrm{NBPSC}}=0$), that is between equally weighted bits associated with adjacent data symbols.

Common interleaving parameters $\{p, q, j\}$ fulfilling these two technical requirements are candidate interleaving parameters to generate interleaving pattern $\{L(k), k = 0, \ldots, K^{-1}\}$selected for the binary interleaving process. The process is illustrated in Figure 18.19.

Multi-Carrier Subcarrier Mapping Allocation: Subcarrier mapping allocation results from two types of processing. A multi-dimensional interleaving spreading maximization is performed for adjacent subcarrier mapping allocation and OFDMA subcarrier distribution through subchannels.

Subcarrier mapping allocation is dynamically implemented, as shown in Figure 18.20. Different permutation rules are successively applied to blocks of data subcarriers, usually associated with an OFDM symbol. The dynamic implementation introduces a higher time variation of the propagation channel which strengthens the binary interleaving role. Furthermore, it may whiten colored noise in the presence of narrow-band interferers and selective propagation channel thanks to interleaving diversity techniques. Dynamic interleaving may involve interference diversity in considering narrow band interferers. Spreading coding techniques are usually used to whiten noise which involves multi-code interference. The dynamic interleaving process is not impaired by this interference. Furthermore, the additional time variations of the channel introduced by dynamic interleaving process is not carried forward

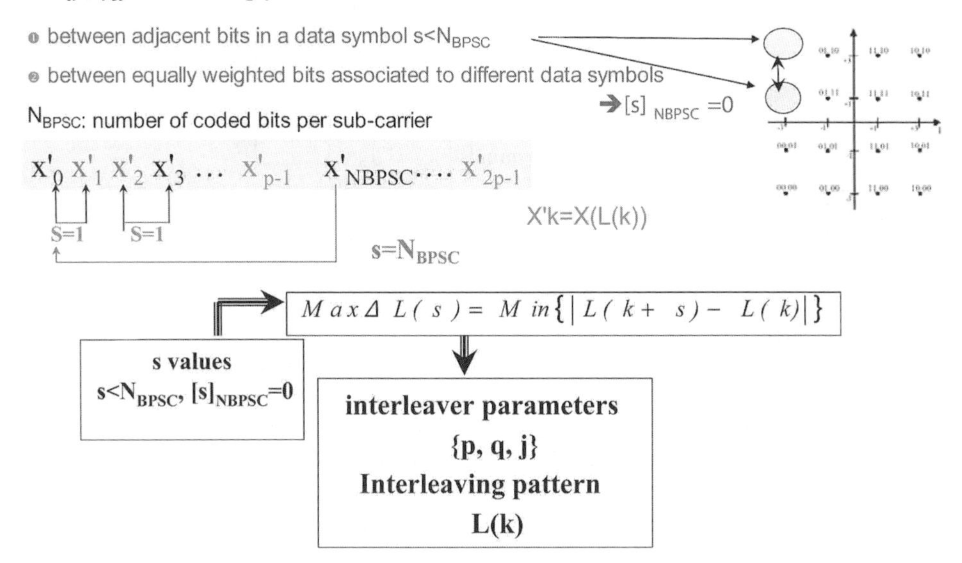

Figure 18.19 Binary Interleaving spreading maximization procedure.

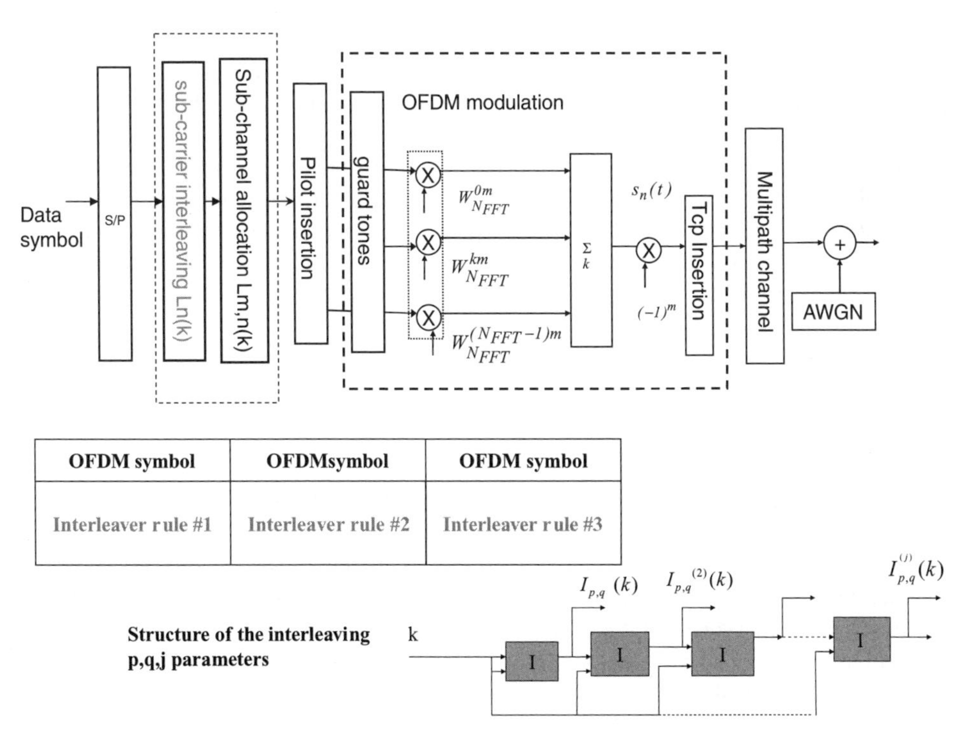

Figure 18.20 Dynamic subcarrier mapping allocation combined with OFDMA process.

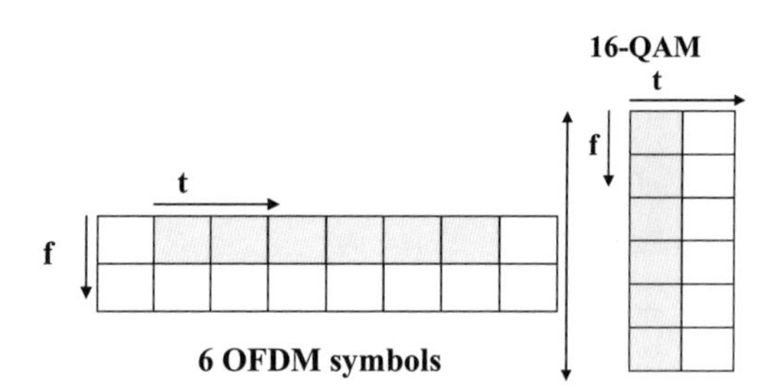

Figure 18.21 OFDMA subchannel allocation (left) and OFDM subchannel allocation (right).

as the subcarrier orthogonality loss usually observed in the presence of a highly time-variant propagation channel.

In this context, we consider data subcarrier interleaving in the frequency domain. The interleaving depth K' corresponds to the number of data subcarriers per OFDM symbol. The interleaving spreading ΔL′(s) is maximized within a subchannel of size Km for s values inferior to Km-1 and between subchannel for s submultiples of Km. Common interleaving parameter sets $\{p', q', j'\}$ providing the highest interleaving spreading are selected to generate interleaving patterns. The subcarrier interleaver size is set to $K' = N_m \times K_m$.

The interleaving process is also optimized to generate interleaving subpattern diversity techniques. The interleaving $L_n(k)$ is selected to present different *interleaving subpatterns* $L_{n,m}(k)$ when they are distributed in the time domain to the same subset of subcarriers with a size K_m $(K' = N_m K_m)$.

The interleaving subpattern diversity consists in selecting the interleaving pattern $L_n(k)$ with a size K' in connection with the subchannel size (K_m) to generate different subpatterns $L_{n,m}(k)$ upon adjacent subchannels.

In the case of OFDM process, the Nm subchannels are mapped in the frequency domain, as illustrated in Figure 18.21. In the case of OFDMA process, the interleaving is performed upon K' subcarriers and subchannels are allocated in the time domain using a single frequency subset group of subcarriers composed of K_m subcarriers. If successive subpatterns present different patterns from one subchannel to another , we benefit from a diversity subchannel mapping allocation in the time and frequency domains.

Figure 18.22 illustrates the concept in considering $K' = 736$, $K_m = 23$ and $N_m = 32$.

18.6.2.3 PHY Layer Description

Information data are first encoded and punctured with a convolutional code $\{g_0 = 133_o$, $g_1 = 145_o$ and $g_2 = 171_o\}$. Turbo-coding schemes may also be implemented with significant gains, doubling the radio coverage in the face of simple convolutional code [44]. Six punctured code rates {1/3, 1/2, 5/8, 2/3, 3/4, 5/6} are proposed to provide a high net bit rate scalability upon three bandwidth sizes multiple of an UWB bandwidth unit size set to 528 MHz. This size is selected to ensure baseband compatibility with the ECMA-WI-Media system characterized with the same RF logical subchannel size.

A pilot insertion ratio is adjusted to $ro = 1/7$ in a preamble based patterns. Scattered pilots should be added to carry out frequency synchronization. Tables 18.10 and 18.11 provide

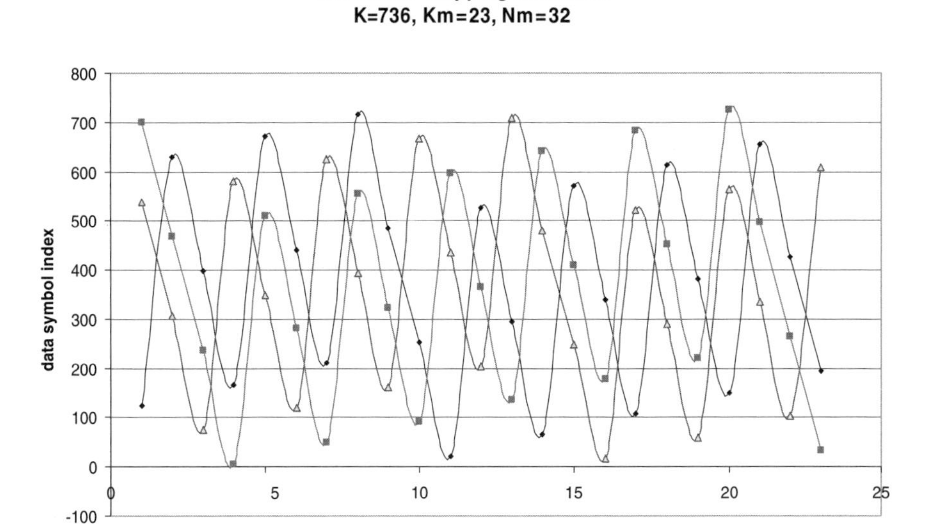

Figure 18.22 Subcarrier mapping allocation in subchannels.

Table 18.10 Net bit rate of the proposal system with $x = 8.17$

Bit rate (Mbps) $x = 8.17$, $r_o = 1/7$ (mode 1)									
	528 MHz			1056 MHz			1584 MHz		
Code rate	QPSK	16QAM	64QAM	QPSK	16QAM	64QAM	QPSK	16QAM	64QAM
1/3	255.1	510.3	765.4	510.3	1020.5	1530.8	765.4	1530.8	2296.1
1/2	386.6	773.1	1159.7	773.1	1546.2	2319.3	1159.7	2319.3	3479.0
5/8	483.2	966.4	1449.6	966.4	1932.8	2899.2	1449.6	2899.2	4348.7
2/3	515.4	1030.8	1546.2	1030.8	2061.6	3092.4	1546.2	3092.4	4638.7
3/4	579.8	1159.7	1739.5	1159.7	2319.3	3479.0	1739.5	3479.0	5218.5
5/6	644.3	1288.5	1932.8	1288.5	2577.0	3865.5	1932.8	3865.5	5798.3

Table 18.11 Net bit rate of the proposal system with $x = 5.44$

Bit rate (Mbps) $x = 5.44$ $r_o = 1/7$ (mode 2 and 3)									
	528 MHz			1056 MHz			1584 MHz		
Code rate	QPSK	16QAM	64QAM	QPSK	16QAM	64QAM	QPSK	16QAM	64QAM
1/3	241.9	483.8	725.7	483.8	967.6	1451.3	725.7	1451.3	2061.7
1/2	366.5	733.0	1099.5	733.0	1466.0	2199.0	1099.5	2199.0	3123.9
5/8	458.1	916.2	1374.4	916.2	1832.5	2748.7	1374.4	2748.7	3904.8
2/3	488.7	977.3	1466.0	977.3	1954.7	2932.0	1466.0	2932.0	4165.1
3/4	549.7	1099.5	1649.2	1099.5	2199.0	3298.5	1649.2	3298.5	4685.8
5/6	610.8	1221.7	1832.5	1221.7	2443.3	3665.0	1832.5	3665.0	5206.4

potential bit rates with an efficient bandwidth unit set to 506.18 MHz. In the case of doubled channel sizes, the bit rate is doubled. Several x values are considered, depending on the selected transmission mode. QPSK and 16-QAM modulated data subcarriers are considered. 64-QAM is mentioned for a new generation of millimeter-wave generation based on heterodyning techniques and photonic components.

Encoded bits associated to six OFDM symbols are block interleaved based on the turbo-based interleaving proposal described in [13, 49] and described in Section 18.5.2.2.

This novel interleaving processing which controls in an algebraic way the interleaving spreading between interleaved samples and preserves special data multiplexing provides very good results with regard to classical matrix interleaving [13, 50]. Gains are up to 10 dB in comparison with the first Wi-Media two-stage interleaving proposal, depending on the propagation scenario (see Section 18.5.3).

Interleaved encoded bits are mapped through QPSK or 16-QAM complex data symbols and forwarded to the OFDM multiplex after a dynamic interleaving processing carried out upon data symbols associated with successive OFDM symbols. The cyclic prefix is then appended at the beginning of each OFDM signal to form the OFDM symbol (Figure 18.23).

The dynamic subcarrier interleaving is envisioned as a simple mitigation technique against narrow band interferer for UWB-OFDM systems [49]. A subcarrier interleaving is performed at the OFDM multiplex upon data symbols in a dynamic way as an alternative to typical Multi Band (MB) processing of ECMA standard where Time Frequency Interleaving (TFI) patterns are defined to carry out dedicated frequency hopping between six adjacent OFDM symbols using three adjacent subchannels. The multi-band processing described by TFI patterns, is designed in a Simultaneously Operating Piconets (SOP) manner to ensure a minimum

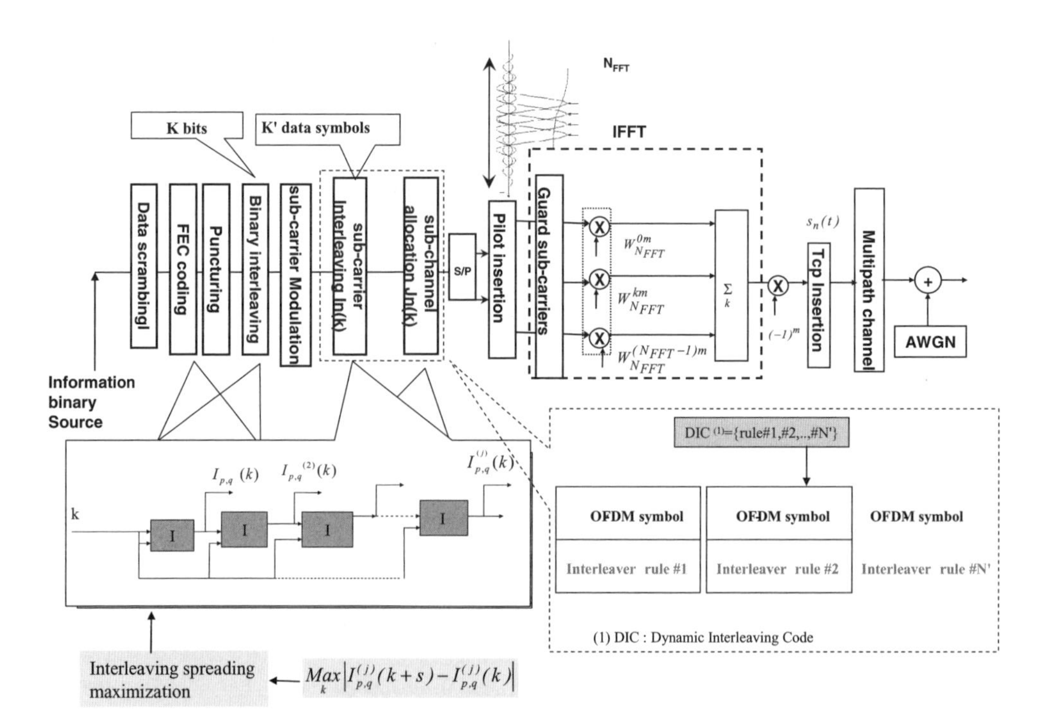

Figure 18.23 France Telecom R&D OFDM PHY layer proposal for 60 GHz WPANs.

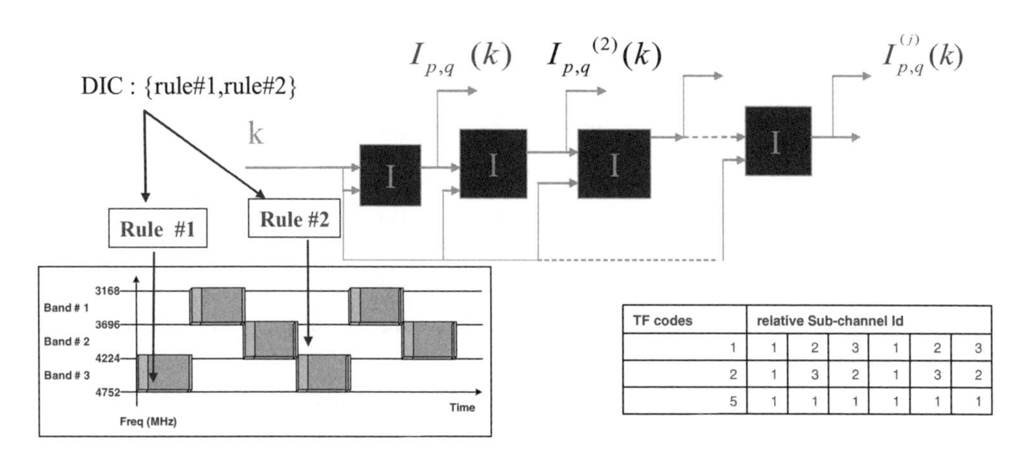

TF codes	relative Sub-channel Id					
1	1	2	3	1	2	3
2	1	3	2	1	3	2
5	1	1	1	1	1	1

Figure 18.24 Dynamic interleaving code implementation at the subcarrier level using turbo-based interleaver in Wi-Media like TFI mode.

collision between subchannels allocated to SOPs. Each SOP is assigned to a user. The dynamic subcarrier interleaving [49] achieves independent frequency interleaving patterns between adjacent OFDM symbols using a single subchannel. The selected interleaving patterns and the implementation order is described as *Dynamic Interleaving Code (DIC)* (cf. Figure 18.24). It introduces time-variant propagation realizations in the frequency domain as an alternative to MB. This concept improves the spectrum efficiency per user of TFI. It also enhances Fixed Frequency Interleaving (FFI) modes by introducing time-variant frequency diversity upon successive OFDM symbols. The processing assessed at 60 GHz, provides gains ranging from 2 to 4 dB in the face of static frequency interleaving (see Section 18.5.3).

At the receiver side, a zero-forcing equalization is performed, followed by a well-known ML decision. Metrics at the input of the decoder are quantized upon 12 bits and decoded with the Viterbi algorithm. An OFDM preamble, dedicated to channel estimation, is appended before each elementary frame composed of six OFDM symbols.

18.6.3 Link Level Performance

This section is dedicated to some BER results implementing the interleaving processing upon the UWB-OFDM system proposed by France Telecom R&D. The interleaving algorithm is the turbo-based structure described in Section 18.5.2.2.

Interleaving parameters $\{p, q, j\}$ are selected to increase the *interleaving spreading* $\Delta L(s)$ between adjacent samples and samples separated by $s - 1$ samples. S values are chosen depending on the interleaving implementation and interleaving spreading criteria. For the $s = \{s_0, s_1, s_q\}$ targeted values, the interleaving parameters $\{p, q, j\}$ are selected to generate the appropriate interleaving pattern that increase the interleaving spreading.

18.6.3.1 Binary Interleaving Benchmarking

Table 18.12 provides the selected interleaving parameters for the binary interleaving process associated with the transmission mode 1 and a transmission bandwidth size set to 528 MHz. The France Telecom R&D interleaving is compared with the Relative Prime (RP) interleaver issued

Table 18.12 Median value of the dispersion MPJ(s)

	p	j	$s=1$	$s=2$	$s=3$	$s=4$	$s=2944$	$S=5888$
Interleav_a	8	3	8305	1054	7251	2108	2944	5888
Interleav_b	16	3	8161	1342	6819	2684	2944	5888
Interleav_d	69	3	6899	3866	3033	7732	2944	5888
Interleav_e	128	1	2305	4610	6915	8444	2944	5888
Random			5161	5187	5122	5224	5191	5210
RP interleaving			5965	5734	231	6196	2944	5888
Wi-Media-like −2 stage			192	384	576	768	1	2

by Crozier [51, 52], a two-stage interleaver derived from the first Wi-Media specifications [5] and a random interleaver. An illustration of interleaving spreading is presented in Figure 18.25.

The interleaving spreading ΔL(s) is assessed as identical to the median of the dispersion given by $P_L(s,k) = |L(k+s) - L(k)|$ where $L(k)$ corresponds to the considered interleaving algorithm. The median value is denoted $MP_L(s)$. Dispersion and interleaving spreading are related by:

$$\begin{aligned} P_L(s,k) &= |L(k+s) - L(k)| \\ \Delta L(s) &= \underset{0\leq k\leq K-1}{\text{Min}}\{P_L(s,k)\} = \text{Median}\{P_L(s,k)\} \end{aligned} \tag{18.8}$$

The two-stage Wi-Media is composed of a two-step interleaving process with different interleaving sizes. The first stage performs an interleaving $I_{M1}(k)$ across six OFDM symbols using a matrix interleaver ($K_1 = N_{\text{CBPS}}x6$). N_{CBPS} is the number of coded bits per OFDM symbol. The second stage is also a matrix interleaving performed at the OFDM symbol level with an interleaving block size $K_2 = mN_{pm}$. The number of columns is approximately set to $\sqrt{N_{pm}}$ and the number of rows equal to $m\sqrt{N_{pm}}$. We select $N_r = 92$ and $N_c = 32$.

The Crozier interleaver generates an RP interleaver described by a modulo operation where p_c is prime factor with the interleaving block size which directly provides the interleaving spreading between adjacent samples and the interger s_c provides the offset of the first sample:

$$I_{RP}(k) = [s_c + k \cdot p_c]_K \tag{18.9}$$

Several realizations of the turbo-based interleaver are considered to sketch the sensitivity to interleaving spreading values.

Several interleaving algorithms are compared with each other:the novel algorithm, RP and random).

The interleaving spreading is the distance between interleaved samples separated by $s-1$ samples calculated from the sample position in the input sequence. An illustration is given in Figure 18.25.

The Bit Error Rate (BER) is expressed with respect to the useful energy per bit to the noise ratio (Ebu/No). The number of simulated data symbols is set to 25×10^6 to ensure a reliable BER estimation. QPSK BER performance is given in Figure 18.26. 16-QAM BER performance is illustrated in Figures 18.27 and 18.28 considering a code rate set to 5/8 and

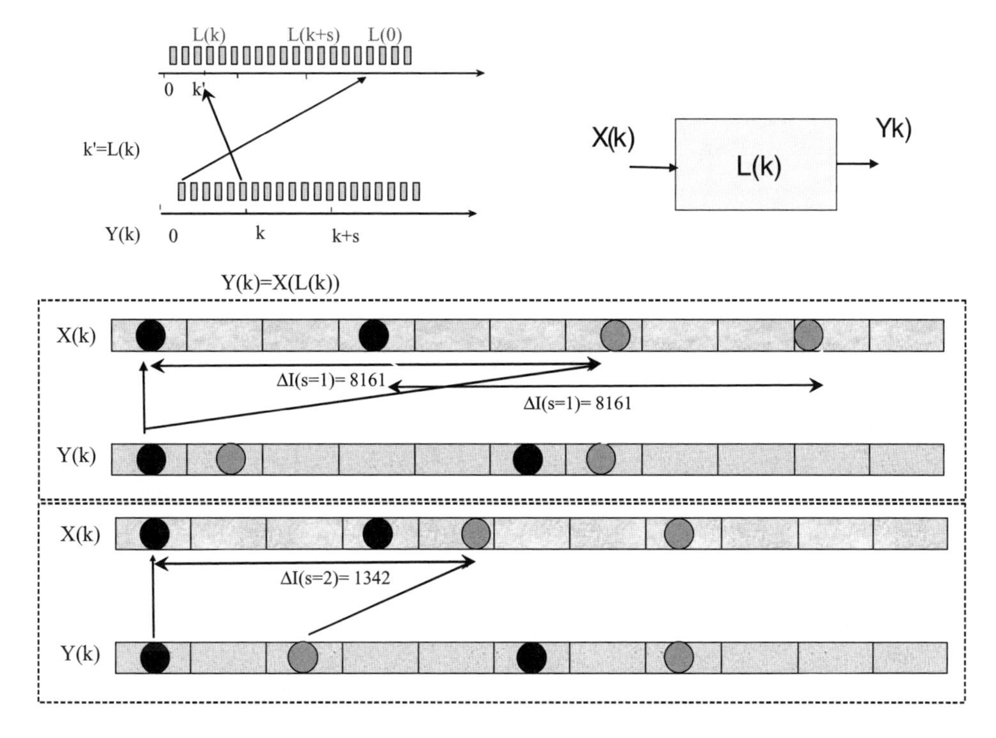

Figure 18.25 Interleaving spreading illustration.

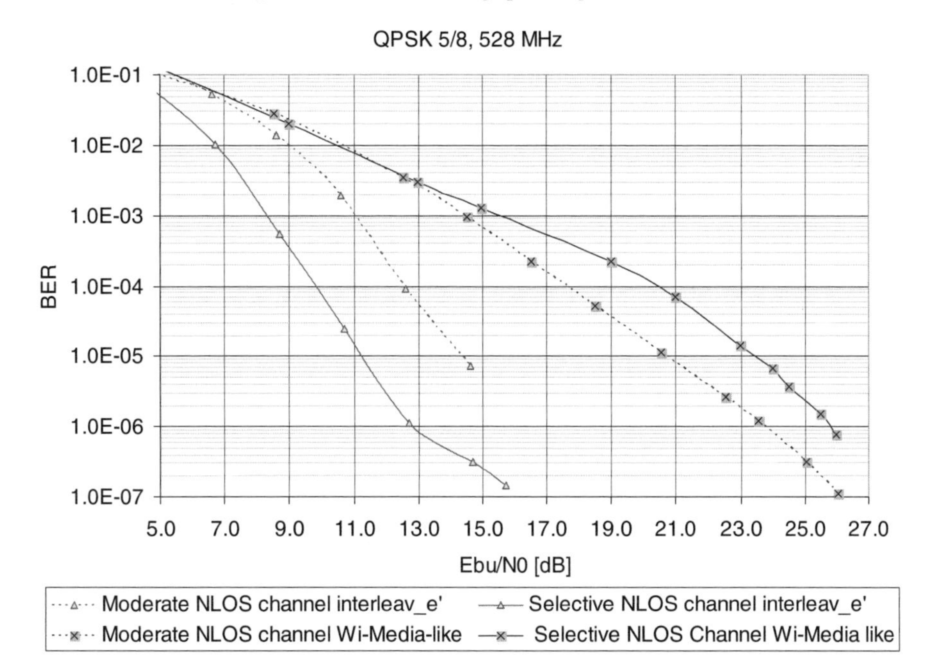

Figure 18.26 BER performance : QPSK 5/8.

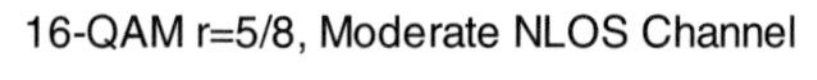

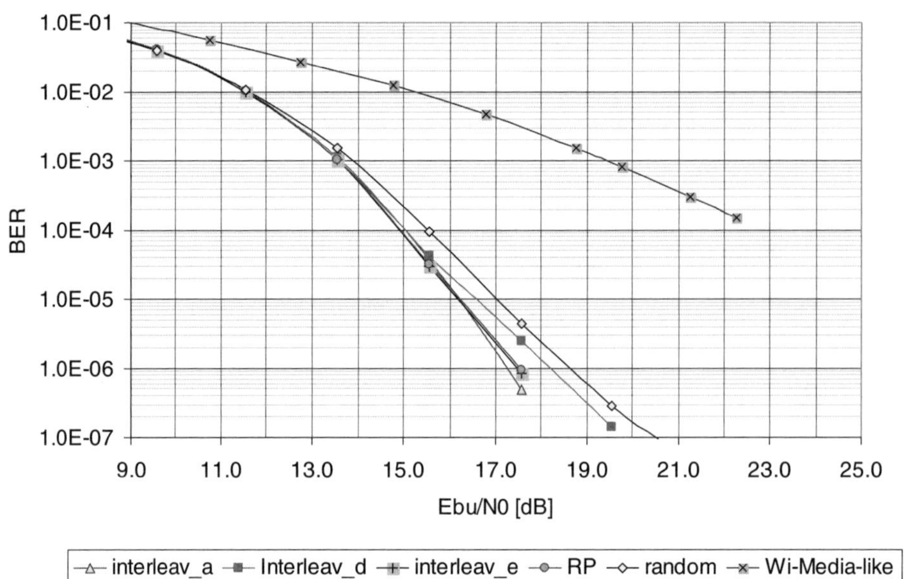

Figure 18.27 BER performance: moderate NLOS channel.

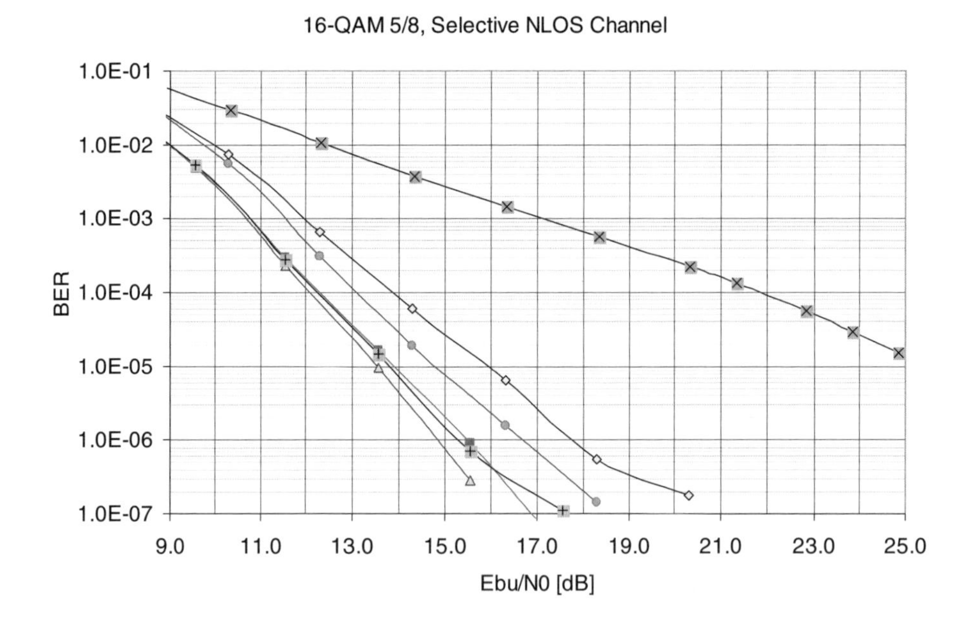

Figure 18.28 BER performance: selective NLOS channel.

16-QAM data symbols. The simulated propagation channel model is the CEPD model described in Section 18.4.2. Figure 18.29 shows plots of performance of sub-carrier mapping allocation. The selective NLOS channel is the CEPD-75 and the moderate NLOS channel is the CEPD_73 channel. The comparison is quantized with a BER target set to 10^{-5}.

QPSK Modulated Data Subcarriers: This configuration corresponds to a data rate near 480 Mbps and a transmission bandwidth size identical to Wi-Media (528 MHz, Logical RF subchannel size). Performance is compared with the highest Wi-Media data rate mode without spreading set-up [5]. In this version of Wi-Media, QPSK data subcarriers are implemented instead of dual carrier modulation.

The turbo-based interleaving process involves gain ranging from 6 to 10 dB compared with the two-stage Wi-Media interleaver. In the case of Wi-Media interleaver, BER degradations increase with the frequency selectivity of the channel. When the turbo-based interleaver is implemented, BER performance is enhanced when the frequency selectivity of the channel increases. This result shows the advantages of high interleaving spreading in the presence of high frequency selectivity of the channel. Bits are spread over almost three separate OFDM symbols ($\Delta\text{L}(s = 1) = 8161$, with $N_{\text{CBPS}} = 2944$). The system benefits from time and frequency interleaving process over three OFDM symbols, drawing benefits from time variations of the propagation channel. Furthermore, the Wi-Media interleaver induces neighboring bits when we consider bits issued from separate OFSM symbols ($s = 2944$, $\Delta L(s) = 1$).

16-QAM Modulated Data Subcarriers: The proposed algorithm is the most outstanding interleaver with gains up to 12 dB when considering the wi-Media-like interleaving and selective multipath channel. The comparison between the turbo-based algorithm and RP presents a gain of up to 1 dB over a severe multipath channel model. The RP interleaving spreading is large enough to spread bits over two OFDM symbols ($\text{DL}(s = 1) = 5965$). On the other side, $\text{DL}(s)$ is small for $s = 3$ with s smaller than the constraint length of the encoder. The random interleaving has degradation near 2 dB facing the turbo-based interleaving, even though the interleaving spreading is close to 5000 for all s values. We may attribute this result to the importance of high interleaving spreading between adjacent bits ($s = 1$).

The different turbo-based interleaving configurations present small variations. The turbo-based interleaver may significantly increase the radio coverage of short-range radio system with an advanced algorithm which does not introduce additional complexity.

18.6.3.2 Dynamic Subcarrier Mapping Allocation

This section is dedicated to the subcarrier allocation and interleaving process carried out at the subcarrier level. The interleaver is a frequency interleaver with an interleaver depth set to Npm, the number of data subcarrier per OFDM symbol. Three configurations are considered:

- Configuration 1 corresponds to the case where a single interleaving pattern rule is applied in a static manner.
- Configuration 2 corresponds to the case where two different rules are applied.
- Configuration 3 corresponds to the case where three different interleaving parterns are applied.

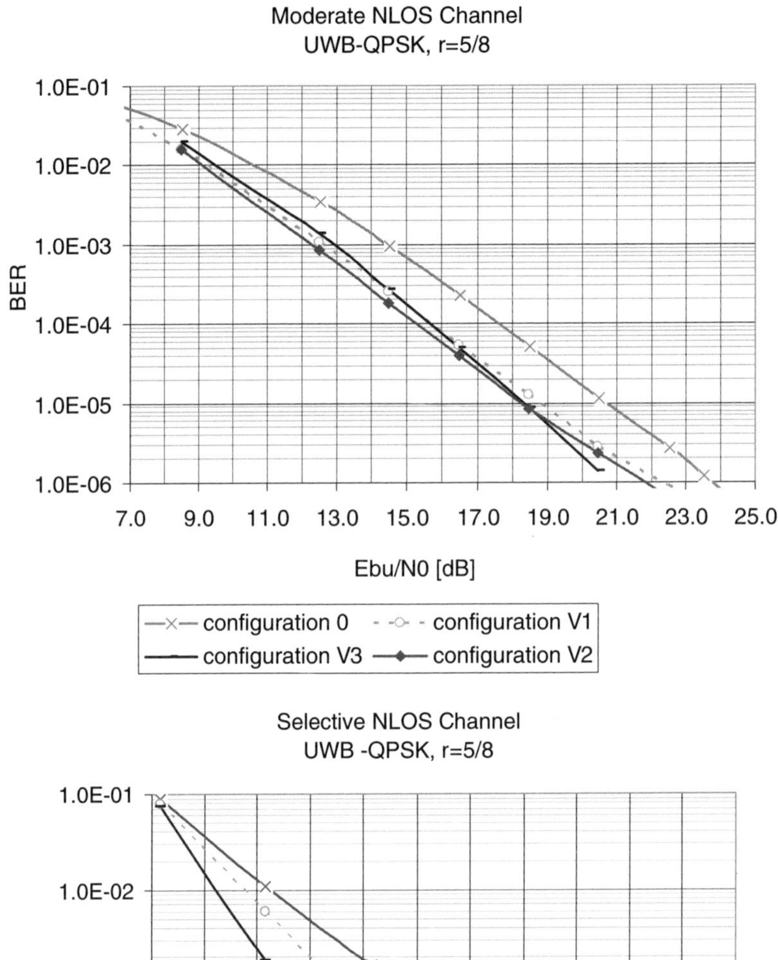

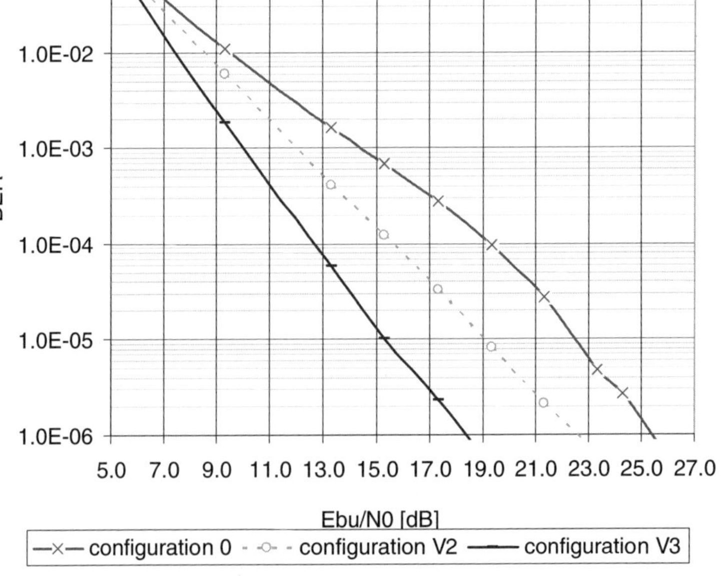

Figure 18.29 Sub-carrier mapping allocation performance.

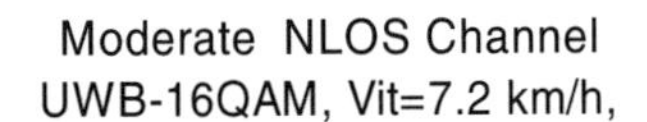

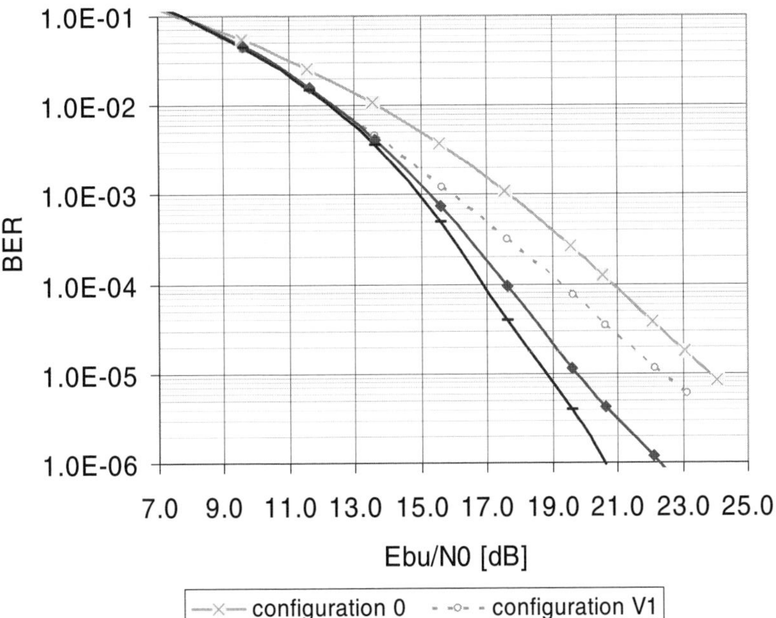

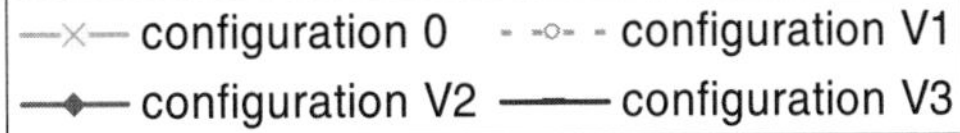

Selective NLOS Channel
UWB-16QAM, Vit=7.2 km/h,

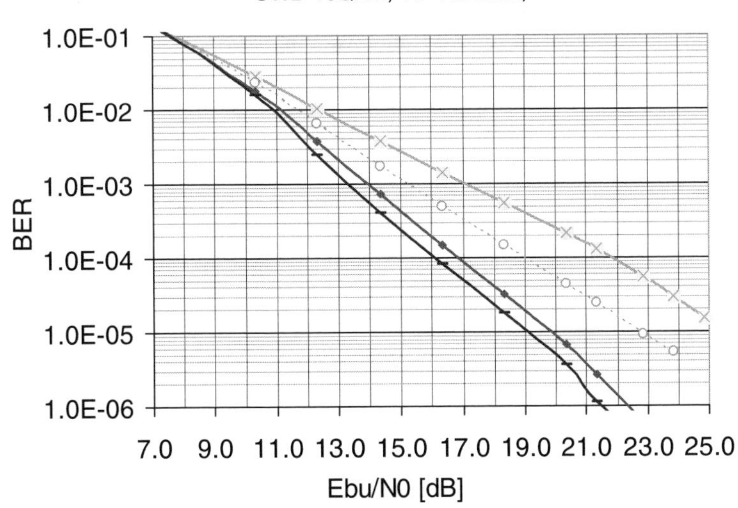

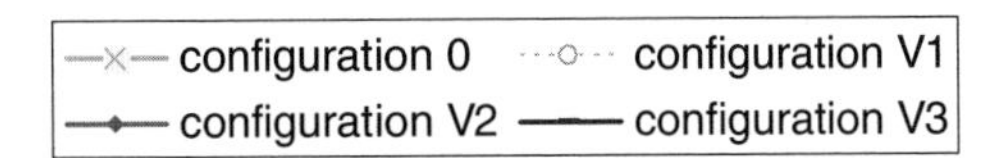

Figure 18.29 *(Continued)*

The interleaving spreading of the three selected interleaving patterns are given in the next table; q is arbitrarly set to two.

p	j	$\Delta I(s=1)$	$\Delta I(s=2)$
4	2	**231**	**274**
16	1	**193**	**350**
23	3	**367**	**2**

Link level performance is given for moderate NLOS and selective NLOS channels associated with the CEPD model described in the Section 18.4.2. QPSK and 16-QAM modulated subcarriers are considered.

Gain between static and dynamic interleaving data symbol mapping allocation is set to 4 dB. When we increase the number of permutation rules, gains are near 1 dB between two and three permutation rules. These gains allow limited radiated power and ensures improved coexistence between WPAN devices.

18.7 Conclusions

This chapter presents a complete 60 GHz PHY layer system intended for Multi-Gigabit Wireless applications dedicated to short range transmissions. After a presentation of regulation issues and state of art of millimeter wave systems, a propagation model is proposed on the basis of Ultra-Wideband measurements carried out in indoor environment. This model is based on a conversion rate of dedicated measurements and straightforward experimental observations on model realizations. It does not introduce any restricted stochastic assumptions on the model. The LOS/NLOS criterion significantly differentiates realizations of the channel, materialized by the time and frequency selectivity of the channel.

A scalable UWB-OFDM system is proposed, directedtowards a spectrum efficiency optimization considering appropriate OFDM dimensioning and advanced baseband processing. Furthermore, the baseband proposal may operate either in UWB spectrum mask or at 60 GHz. A novel and harmonized interleaving process is presented to carry out binary interleaving and subcarrier mapping allocation under OFDM-TDMA and OFDMA process. The interleaving algorithm is based on a turbo-structure and algebraic expression, maximizing interleaving spreading and performing dedicated data multiplexing in accordance with access mode. It has been implemented at the binary level, maximizing in a multilevel scale the interleaving spreading between adjacent bits and data symbols [13, 50]. Link level results are summarized in this chapter, highlighting gains up to 12 dB in the face of the two-stage Wi-Media-like interleaver.

The same algorithm is implemented at the subcarrier level in a dynamic manner to cope with narrow band interferers and strengthen the time and frequency diversity of the channel [49, 50]. The dynamic implementation may whiten colored noise in the presence of narrow band interferers and a selective propagation channel. The dynamic interleaving process introduces interference diversity; it is an alternative spreading coding technique, usually used to whiten noise. Furthermore, the additional time variations of the channel introduced by dynamic interleaving process is not carried forward on subcarrier orthogonality loss usually observed in the presence of a highly time-variant propagation channel. The main link level results are summarized using the proposed propagation channel. Dynamic implementation is also foreseen as an enhanced multi-band process where different interleaving patterns are implemented when using the same RF logical subchannels. Gains range from 2 to 4 dB.

Further studies will be focused on Advanced OFDMA process using properties of the algorithm. The second important point will be directed towards optimized combinations of binary interleaving process, subcarrier mapping allocation and subchannel distribution in multi-carrier transmissions.

Acknowledgments

This work has been initiated within the framework of the IST-FP6 MAGNET project, France Telecom R&D thanks the MAGNET consortium and the European Commission for their support in the achieved work carried out within the UWB-Multl-carrier cluster lead by Isabelle Siaud from France Telecom R&D. These studies are actually extended through internal projects and the IST-FP7 OMEGA project leading to novel and harmonized PHY layer concepts for short-range UWB-OFDM transmissions. This work has been the baseline for IEEE802.15.3c contributions prepared and promoted by Pascal Pagani from France Telecom R&D in IEEE802.15.3c standardization group. It will form the future evolution of the PHY layer proposal presented in this chapter.

References

1. FCC (2002) First Report and Order 02-48.
2. Fisher, R. and Barr, J. (2003) *TG3 mm Wave Interest Group Motion*, IEEE802.15-03/0317r1.
3. ECC Decision of 24 March 2006 on the harmonised conditions for devices using Ultra-Wideband (UWB) technology in bands below 10.6 GHz, (ECC/DEC/(06)04).
4. Draft revision of ECC Decision of 24 March 2006 on the harmonised conditions for devices usingUltra-Wideband (UWB) technology in bands below 10.6 GHz (ECC/DEC/(06)04), December 2006.
5. MBOA SIG (2004) *MultiBand OFDM Physical Layer Proposal for IEEE 802.15 Task Group 3a*, IEEE802.15-04.
6. Standard ECMA-368 (2005) *High Rate Ultra Wideband PHY and MAC Standard*, 1st edn.
7. Graas, E., Piz, M., Herzel, F. and Kraemer, R. (2005) *Draft PHY Proposal for 60 GHz WPAN*, IEEE802.15-05/0634r1.
8. Siaud, I. (2005) *Definition and Evaluation of Millimeter Short Range WPANs Physical layer Systems for Very High Data Rate Applications*, Wireless Word Research Forum, WWRF#13, Jeju Islands.
9. Siaud, I., Le Gouable, R. and Hélard, M. (2001) *On Multicarrier Transmission Techniques over Recorded Indoor Propagation Channel Models for Future Broadband RLANs at 60 GHz*, PIMRC'01; 30, San Diego.
10. Tensorcom, COMPA consortium (2007) *Merged proposal: New PHY Layer and Enhancement of MAC for mm Wave System Proposal*, IEEE802.15.3c WG, N°IEEE802.15-07/934r1.
11. Si-Beam Company (2007) *Proposal for HD AV and Data Support*, IEEE802.15.3c WG, N°IEEE802.15-07/942r2, Orlando.
12. Siaud, I. and Legouable, R. (2005) *Ultra WideBand MultiCarrier Spread Spectrum Techniques for Short Range Radio Communications at 60 GHz*, Wireless World Research Forum WWRF#14.
13. Siaud, I. and Ulmer-Moll, A.M. (2007) *Turbo-like Processing for Scalable Interleaving Pattern Generation: Application to 60 GHz UWB-OFDM Systems*, ICUWB'07, Singapore.
14. Siaud, I., Pagani, P. and Ulmer-Moll, A.M. (2007) *WPAN UWB Air Interface Benchmarking and Futures Strategies*, Wireless World research Forum WWRF#19, Chennai.
15. Grass, E., Siaud, I., Glisic, S. *et al.* (2008) *Asymmetric Dual-Band UWB/60 GHz Demonstrator*, PIMRC'08.
16. IPHOBAC Project IST-2006-35317 (2006) *Deliverable D2.1, Definition of End User Requirements.*
17. Moignard, M., Siaud, I. and Charbonnier, B. (2006) *Very High Data Rate Home Area Networks*, ISIS workshop, Boppard.

18. MAGNET project IST-507102 (2005) *User Centric Scenarios for PNs of a Valid Architecture*, deliverable No MAGNET/WP1.3/DTU /D1.3.1b/R/PU/001/1.0, December.
19. Sadri, A. *et al.* (2007) 802.15.3c Usage models, doc.IEEE802.15-06-0055-22-003c.
20. Munoz, L., Aguero, R., Choque, J. *et al.* (2004) Empowering next-generation wireless personal communication networks. IEEE Communications Magazine, **42** (5), 64–70.
21. MAGNET Beyond IST-507 102 (2006) *WP3 Internal Report, IR3.1.1, Radio-Engineering for Optimisation of PAN Devices.*
22. IST-MAGNET deliverable (2004) WP3, D3.2.1: Requirement Specification for PHY-Layer.
23. Sawada, H., Shoji, Y., Choi, C.S. *et al.* Antenna models proposal for each Usage Model Definition, doc.IEEE802.15-06.0427-04-003c.
24. CEPT Recommendation TR 22-03 (1990) Provisional Recommended Use of the Frequency Range 54.25–66 GHz by Terrestrial Fixed and Mobile Systems, Athens.
25. ERC Report 25, European Table of Frequency Allocations and Utilizations Covering the Frequency Band Range 9 KHz to 275 GHz, Lisboa 2002 revised Dublin 2003.
26. ETSI DTR/ERM-RM-049 V0.0.2 (2006) Electromagnetic compatibility and Radio spectrum Matters (ERM); System reference document; Technical characteristics of multiple gigabit wireless systems in the 60 GHz range, document ETSI BRAN No. bran44d033.
27. Smulders, P.F.M. and Correia, L.MI. Characterization of propagation in 60 GHz radio channels, *Electronics & Communication Engineering Journal*, April 1997.
28. Ikeda, H. and Shoji, Y. (2006) 60 GHz Japanese regulation 2, IEEE802.15.3c TG, IEEE802.15-05/525r3 document.
29. Wooyong, L., Jinkyeong, K., Kyeongpyo, K. and Yongsun, K. (2006) *Status of 60 GHz Unlicensed band Korean Regulation*, IEEE802.15.3c TG, IEEE802.15-05/0671r1 document.
30. Malhouroux, N., Siaud, I. and Guillet, V. (2004) PAN–PD/FD Channel Characterisation and SISO modelling at 17 and 60 GHz (Part I), IST-MAGNET D3.1.2a Deliverable.
31. Siaud, I., Malhouroux, N. and Guillet, V. (2005) *Flexible Ultra-WideBand versus Wideband PHY layer Design and Performance for WPANs at 60 GHz (IST MAGNET project)*, European Conference on Propagation and Systems, Brest-France.
32. Pagani, P., Malhouroux, N., Siaud, I. and Guillet, V. (2006) Characterization and modeling of the 60 GHz indoor channel in the office and residential environments, IEEE P802.15-06-0027-01-003c, January 2006-05-03.
33. Pagani, P., Malhouroux, N., Siaud, I. and Guillet, V. (2006) Analysis of the angular characteristics in the 60 GHz indoor propagation channel, IEEE P802.15-06-0028-01-003c, January 2006-05-03.
34. Pagani, P., Siaud, I. and Malhouroux, N. (2006) Adaptation of the France Telecom 60 GHz Channel Model to the TG3c framework", IEEE-15-06-0218-003c.
35. Hirose, T. (2003) Study of mm wave propagation modeling to realize WPANs, IEEE (0.802).15-03/0365.
36. Pollock, T., Skafidas, E., Liu, C. *et al.* (2006) Indoor 60 GHz Channel Model with AoA, IEEE (0.802).15-06-0103-01-003c.
37. Liu, C., Skafidas, E., Pollock, T. *et al.* (2006) NICTA Indoor 60 GHz Channel Measurements and Analysis update, IEEE802.15-06-0112-00-003c.
38. Li, H.-J. and Wang, Y.-Z. *et al.* (2006) Propagation Model and Channel Measurement for 60-GHz Indoor Wireless Communication, IEEE (0.802).15-06/145r1.
39. Smulders, P.F.M. and Correia, L.M. (1997) *Characterisation ofpropagation in 60 GHz radio channels. Electronics and Communication Engineering Journal.*
40. Siaud, I. (1997) Indoor Propagation Channel Simulation for the Performance Evaluation of HIPERLAN systems, ETSI EP BRAN document, N° wg3td46.
41. Siaud, I. (1996) *A Digital Signal Processing Approach for the Mobile Radio Propagation Channel Simulation with Time and Frequency Diversity applied to an indoor environment at 2.2 GHz*, Personal Indoor Mobile Radio Communications Conference, PIMRC'96, Taiwan.
42. Obara, K. (2004) *MM-Wave WPAN Meeting System Using OFDM*, IEEE802.15-04/141r2.
43. Toyoda, I., Suzuki, Y., Nishikawa, K. and Seki, T. (2006) *Single Carrier PHY for 60 GHz WPAN*, IEEE802.15-06/007r0.
44. Siaud, I. (2002) Combined Channel Coding and receiver antenna Pattern Influence on COFDM performance: Evaluation in a residential Indoor multipath environment at 60 GHz, International OFDM Workshop, InOW'2002.
45. IST Broadway project (2002) WP1, Deliverable D2 Functional System Parameters description, Motorola S.A, Farran, Intracom, TNO, UoA, TUD, IMST.

46. IST Broadway project (2002) WP 3, Deliverable D3 BROADWAY Baseband Requirements, Motorola S.A., TUD, Intracom.
47. IST MAGNET (2005) Update D3.2.2a Candidate Air Interfaces and Enhancements IST-507102, My personal Adaptive Global Net project, Deliverable D3.2.2b.
48. Kunhert, H., Lenk, F., Hilsenbeck, J. *et al. Low Phase Noise GaInP/HBT MMIC Oscillators up to 36 GHz*, http://www.fbh-berlin.de/ver01/th2a.
49. Siaud, I. and Ulmer-Moll, A.M. (2007) A Novel Adaptive subcarrier Interleaving application to millimeter-wave WPAN OFDM Systems (IST MAGNET project), IEEE portable (0.2007) conf., Orlando, USA.
50. Siaud, I. and Ulmer-Moll, A.M. (2007) Advanced interleaving algorithms for OFDM based millimeter wave WPAN transmissions, SCEE Seminar, France.
51. Crozier, S. and Guinand, P. (2001) *High-Performance Low-Memory Interleaver Banks for Turbo-Codes*, Proceedings of the 54th IEEE Vehicular Technology Conference (VTC 2001 Fall), Atlantic City, New Jersey, USA, pp. 2394–8.
52. Crozier, S., Lodge, J., Guinand, P. and Hunt, A. (1999) *Performance of Turbo Codes with Relative Prime and Golden Interleaving Strategies*, Proceedings of the 6th International Mobile Satellite Conference (IMSC '99), Ottawa, Ontario, Canada, pp. 268–75.

19

Modulation Techniques and System Architectures for Multi-Gb/s 60 GHz Radios

Alberto Valdes-Garcia and Troy Beukema
IBM. T.J. Watson Research Center, Yorktown Heights, NY

19.1 Introduction

Due to the relatively large amount of bandwidth available in the 60 GHz band and the propagation characteristics at these frequencies, two main usage scenarios are being considered for the introduction of broadband millimeter-wave technology to the market:

1. Directional, 'point-and-shoot' data-transfer for multi-media kiosk access and peer-to-peer communication among portable devices. For these applications a Gb/s, short distance (1–3 m) line-of-sight (LOS) link is required and the emphasis is on low cost, low complexity and low power consumption.
2. WPAN-like home or office network for high-definition video and/or data transmission. For this application, a nonline-of-sight (NLOS) link (considering radiation-absorbent objects between the receiver and transmitter) over a moderate distance (aproximately 10 m) is required. The emphasis is on robustness to multi-path signal dispersion and system throughput.

More details about specific cases of these scenarios can be found in [1]. Some different modulation options that can be considered for a Gb/s system at 60 GHz include low-complexity single-carrier systems such as on-of keying (OOK) and minimum-shift keying (MSK) [2] with clock-and-data recovery (CDR) based demodulators and higher complexity systems such as single-carrier (SC) quadrature-amplitude modulation (QAM) or multiple-carrier, orthogonal

Short-Range Wireless Communications Rolf Kraemer and Marcos D. Katz

frequency-division multiplexed (OFDM) QAM [3] with high precision A/D and complex digital-signal processing (DSP). In Sections 19.2 and 19.3, example system architectures for MSK and OFDM based systems are introduced with a discussion of their properties for each usage case. With basis on a survey of the current state-of-the-art for silicon integrated circuits, Section 19.4 presents an analysis of the HW requirements and potential power consumption for the high data-rate systems presented.

19.2 MSK-Based System for LOS Gb/s Communications

A directional channel with low time dispersion from multi-path and/or channel bandwidth limitation permits the use of a low-complexity signaling scheme to realize multiple Gb/s data rates in the 60 GHz band. MSK is an example low-complexity modulation which exhibits many desirable properties for use at these frequencies. A key advantage of this modulation is the constant envelope transmission which enables higher transmit power-amplifier (PA) efficiency compared with OOK or phase-shift-keyed based systems such as differential QPSK (DQPSK). Since no information is modulated on the carrier amplitude, a limiting receiver chain can be used, removing the need for automatic gain-control (AGC). Elimination of the need for receiver AGC enables the potential for fast, efficient packet synchronization.

The MSK signal can be demodulated using either a simple FM discriminator with a low-power clock-and-data recovery (CDR) data slicer or a higher complexity A/D + DSP approach. A block diagram of a MSK system using a FM discriminator/CDR based demodulator is shown in Figure 19.1. At the transmitter, a binary data stream modulates the phase of the 60 GHz carrier at a rate of $+\pi/2$ radians per bit period to transmit a 1 and $-\pi/2$ radians per bit period to transmit a zero. This signal is bandwidth limited, amplified and sent through a directional or otherwise low multi-path dispersion channel. At the receiver, the signal is mixed down to an IF, bandwidth limited and sent to a limiter/discriminator FM detector which produces a nonreturn-to-zero (NRZ) analog waveform. The NRZ waveform is decoded with a clock-and-data recovery slicer, avoiding the need for high-precision A/D while achieving data throughput in the Gb/s range.

The modulation also exhibits good bandwidth efficiency, as shown in Table 19.1, approaching 1 b/s/Hz for 99.5% power spectrum width when an appropriate bandpass filter is used to

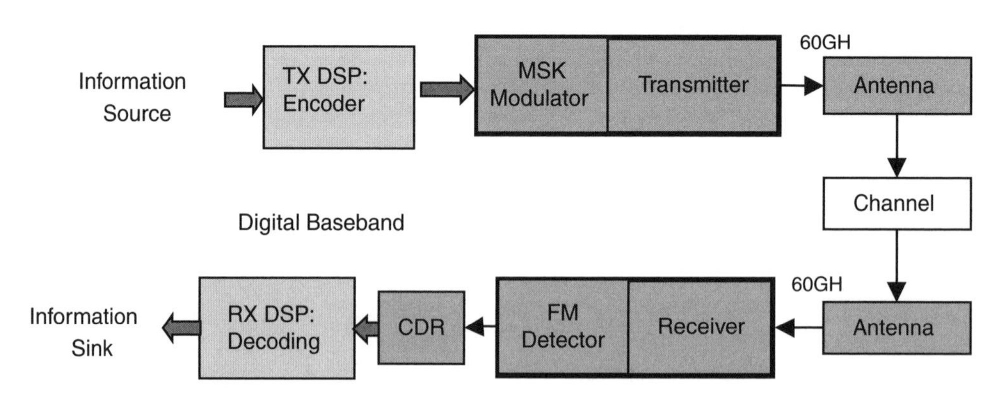

Figure 19.1 System architecture for an MSK-based system to operate in a LOS channel.

Table 19.1 Bandwidth requirements for MSK transmission at different data rates.

Data rate (Gb/s)	3 dB IF Filter BW	First side-lobe (dB)	99.5% BW (GHz)
1	Infinite (no filter)	−23	1.6
1	1 GHz	−30	1.1
2	Infinite (no filter)	−23	3.3
2	2 GHz	−30	2.2

attenuate the spectral side-lobes. At this spectral efficiency, MSK can achieve 2 Gb/s data rate while containing channel bandwidth use below 2.5 GHz.

Phase noise of the local oscillator in 60 GHz systems can be an important factor to consider in defining an appropriate modulation. A straightforward FM-discriminator demodulation of MSK generates residual-FM at the output from the carrier phase noise, but a combination of data run-length encoding and high-pass filtering the discriminated output can be used to mitigate the degradation from high $1/f$ phase noise near the carrier frequency. Other differential phase SC modulations such as DQPSK or DBPSK inherently filter out the lower-frequency carrier phase noise since the low-frequency noise is common to adjacent high-speed channel symbols. A MSK based modulation can also be demodulated using A/D + DSP in conjunction with a carrier tracking loop to increase the robustness of the system against carrier phase noise.

Possible applications for the MSK system include directional point-to-point links with high-gain antennas to realize wireless point-to-point LAN bridges, and short-range (3 m), directional links to realize power-efficient high-rate modulation in portable 'point-and-shoot' application devices. Without equalization, however, this modulation is not suitable for omnidirectional or low-gain 60 GHz multi-path channels which can exhibit RMS delay spreads up to 25 ns RMS or more.

19.3 OFDM-Based System for NLOS Gb/s Communications

A nondirectional or multi-path channel presents a significant challenge in realizing a robust multi-Gb/s data rate system at 60 GHz. Due to the high channel symbol rate required for data rates in the range 2 Gb/s and above, channel dispersion from multi-path must be addressed by either employing a multi-carrier modulation (OFDM), or using a SC modulation with direct-sequence spread-spectrum (DS-SS) RAKE receiver and/or a powerful multi-path equalizer in the receiver. Since OFDM is already in widespread use at 2.4/5 GHz and in UWB standards, it is considered here for use in 60 GHz broadband NLOS systems also. The primary benefit OFDM brings is reduction of the subchannel symbol rate to a point that the channel delay spread is an acceptable fraction of the symbol period to enable reliable data decoding. However, the OFDM modulation can introduce a significant degradation in PA efficiency, since peak/average power ratio for systems employing 64 subchannels and above can exceed 10 dB. Reduction of the peak/average power through coding or other approaches is a key challenge for deployment of OFDM at 60 GHz where it is difficult to realize high transmitter power.

A high-level block diagram of an OFDM-based 60 GHz system is shown in Figure 19.2. The system requires high rate D/A and A/D (in the range of 1–2 Gs/s) with a FFT-based digital modulator and IDFT based demodulator. Although high rate D/A could be realized with relatively low power and complexity, the A/D presents a challenge since it requires 5–8 bits

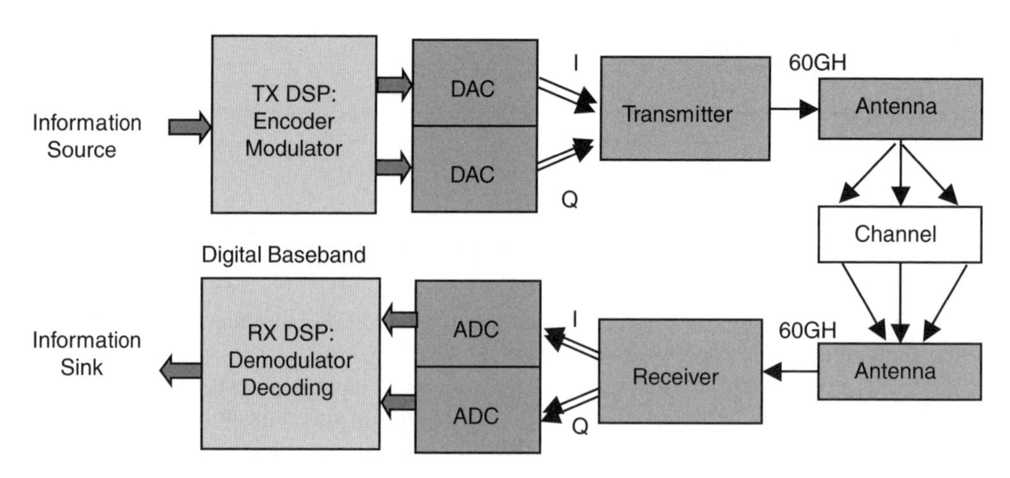

Figure 19.2 System architecture for an OFDM-based system to operate in a NLOS channel.

at a sampling rate in the range of 2 Gs/s to realize 2 Gb/s data rate systems. Although Nyquist sampling ($1 \times$ oversample) can be used to reduce A/D rate, elimination of oversampling reduces the ability of the OFDM system to filter away adjacent channel interference in the demodulator.

Packet synchronization is also an important issue in a broadband OFDM system employing time-division multiple-access protocol. Since the modulation is linear, the receiver must be given time to detect and apply automatic-gain-control (AGC) to the received signal without reducing the maximum throughput of the system significantly. Longer data slots can be used to mitigate this synchronization overhead problem, but may require that the OFDM demodulator implement a time drift tracking algorithm over the longer slot duration to minimize sample error degradation arising from asynchronous transmitter and receiver reference clocks.

Local oscillator phase noise is an important consideration in the design of 60 GHz OFDM systems. Although use of a larger number of subchannels can increase multi-path resistance, the LO phase noise eventually limits performance due to spectral smearing of the subchannels by the phase noise, and the inability to track phase noise at slow subchannel symbol rates. All OFDM demodulators must implement a coherent carrier phase recovery/tracking loop. Use of a lower number of subchannels enables the carrier tracking loop bandwidth to be higher, increasing the tolerance to LO phase noise. Phase noise also directly limits use of high-density QAM constellations in the subcarriers. As an example, a system phase noise level of approximately -90 dBc/Hz or lower at 1 MHz offset is needed to realize 16-QAM OFDM with 512 subchannels supporting a data rate in the range of 2 Gb/s.

A typical high-performance OFDM system employs frequency-domain symbol interleaving and forward-error correction (FEC) coding detected at the receiver with a soft-decision Viterbi algorithm (VA) decoder. Realization of the VA at the receiver is a large challenge for data rates in the 2 Gb/s range. Therefore, alternate coding methods which are robust to subchannel fading while providing lower computational overhead are of interest for development of 60 GHz OFDM based systems.

Applications for the OFDM system include channels with the potential for significant time dispersion from multi-path, including single-bounce delays which can introduce a deep notch in the frequency domain. The OFDM based modulation adds no benefit in a line-of-sight link, and in fact will not perform as well as SC based modulations for these applications due to its higher peak/average power (resulting in lower average power transmission for OFDM) and higher susceptibility to LO phase noise at 60 GHz.

19.4 Implementation Trade-Offs in Silicon Technologies

RF Front-End

Recent advances in silicon IC technology have enabled the demonstration of a plethora of devices, circuits and subsystems operating in the range of 60–100 GHz [4]. In pure CMOS processes, significant progress has recently been made with respect to the level of integration at millimeter-wave frequencies. The latest examples of integrated CMOS radio circuits operating at 60 GHz demonstrate receiver front-ends with a local oscillator [5–7]. In SiGe bipolar technology, a 60 GHz transmitter and receiver chipset with integrated modulation/demodulation capabilities has already been demonstrated [8]. MSK is one of the modulation formats supported by the chipset. In addition to the silicon chips, 60 GHz dipole antennas have been developed along with low-cost packaging techniques [9].

The packaged 60 GHz chip-set was used in a video system to demonstrate multi-Gb/s wireless communication [10]. A HDTV image from a camera was transmitted through the chipset using MSK modulation with coding and baseband functions implemented inside a field-programmable gate-array (FPGA). The uncompressed HDTV transmission across the 60 GHz channel was running at 2 Gb/s (1.485 Gb/s data plus additional overhead for coding and synchronization) over a range up to 3.5 m. The transmitted video was displayed on a HDTV monitor as shown in [10].

Thanks to its constant envelope, an MSK system can take advantage of an efficient nonlinear PA to reduce the transmitter power consumption. A 60 GHz switched mode PA has recently been demonstrated in SiGe [11], showing that higher-efficiency PAs can also be employed at these frequencies with modern silicon technologies. On the other hand, a OFDM systems require a linear PA which must accommodate a relatively high peak-to-average (PAR) ratio. This is an area where III–V based semiconductors have shown a consistent advantage over silicon. A recent PA design, however, shows that a SiGe integrated PA can achieve 20 dBm of saturated output power with a 1 dB compression point of around 12 dBm [12].

ADC and DAC

A crucial component of any 60 GHz radio using multiple carriers and/or frequency domain equalization will be a high-speed ADC. Table 19.2 presents a summary of the most recently reported ADCs with sample frequencies greater than 400 MHz [13–30]. Most of the listed designs were reported in the last 15 months. The designs are grouped by technology nodes to analyze the performance trends. The effective number of bits (ENOB) is calculated from the reported signal-to-noise-and-distortion ratio (SNDR) that corresponds to the highest reported input signal test frequency, which is listed as ERBW (effective resolution bandwidth). In other words, the ADC has at least ‘X’ ENOB over the signal bandwidth ERBW. To compare their

Table 19.2 Summary of current state-of-the-art in high speed ADCs.

Process (um)	ENOB	ERBW (GHz)	Power (W)	FOM (pJ/conversion)	Architecture	Reference
0.18	7.3	0.80	0.77	3.14	Folding-interpolation	[13]
0.18	3.4	0.65	0.07	4.97	Flash	[14]
0.18	2.9	0.25	0.01	2.14	Time-interleaved SAR	[15]
0.18	4.7	0.94	0.31	6.38	Time-interleaved Flash	[16]
0.18	5.1	0.80	0.35	6.56	Flash	[17]
0.18	4.9	0.40	0.11	4.28	Pipeline	[18]
0.13	5.0	0.30	0.01	0.27	Time-interleaved SAR	[19]
0.13	4.2	0.50	0.04	2.35	Flash w/active interpolation	[20]
0.13	5.2	0.60	0.16	3.65	Flash	[21]
0.13	8.5	0.50	0.25	0.68	Time-interleaved	[22]
0.09	5.3	0.50	0.06	1.37	Two-step sub-ranging	[23]
0.09	3.7	0.63	0.0025	0.16	Modified flash	[24]
0.09	3.6	1.00	0.23	9.18	Flash	[25]
0.09	4.9	0.33	0.01	0.52	Time-interleaved	[26]
0.09	5.2	0.30	0.12	5.48	Folding-interpolation	[27]
0.09	8.7	0.40	0.35	1.07	Time-interleaved	[28]
0.09	7.0	0.16	0.14	3.35	Pipeline	[29]
0.065	4.0	0.25	0.01	0.75	Time-interleaved SAR	[30]

conversion efficiency, the following figure of merit (FOM) is employed:

$$\text{FOM} = \frac{\text{Power}}{(2 \cdot \text{ERBW}) \cdot \left(2^{\text{ENOB}}\right)} \tag{19.1}$$

ADCs with sampling frequencies beyond 100 MHz often have a SNDR (which defines the ENOB) that decreases rapidly with input frequency due to the many design challenges (clock jitter, increased matching requirements, etc.). A better ADC keeps an ENOB closer to the designed number of bits over an ERBW closer to the Nyquist frequency. Some papers report a FOM calculated using the maximum sampling frequency instead of 2ERBW and the designed bits instead of ENOB. This leads to lower (better) FOM numbers which are, however, not meaningful to compare high-speed performance.

In order to have a better understanding of the trends in ADC efficiency for different technology nodes, Figures 19.3 and 19.4 present the FOM as a function of ENOB and ERBW for the designs listed in Table 19.3. The increase in ADC conversion efficiency from the 0.18 to 0.13 μm CMOS technology nodes is clear. However, a similar trend is not apparent for the shift from 0.13 to 90 nm. For 90 nm and beyond, improvements are more related to architecture innovations, where time-interleaved implementations show a trend in achieving higher efficiencies. At low ENOB (< 5 bit) and ERBW < 600 MHz there are several examples of ADCs achieving a FOM < 1 pJ/conversion. Nevertheless, as both speed and resolution increase, efficiency decreases exponentially. The ADC reported in [21] (2.5 mW for about 600 MHz of ERBW and 4ENOB) has the best conversion efficiency (0.16 pJ/conversion) among the published ADCs at these speeds and is a major advance in state-of-the-art.

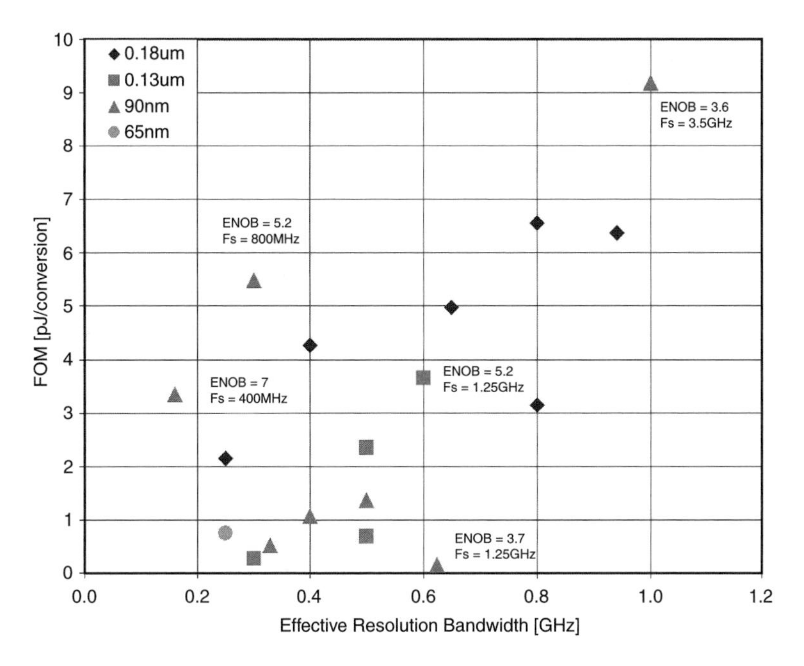

Figure 19.3 FOM vs effective resolution bandwidth for different technology nodes.

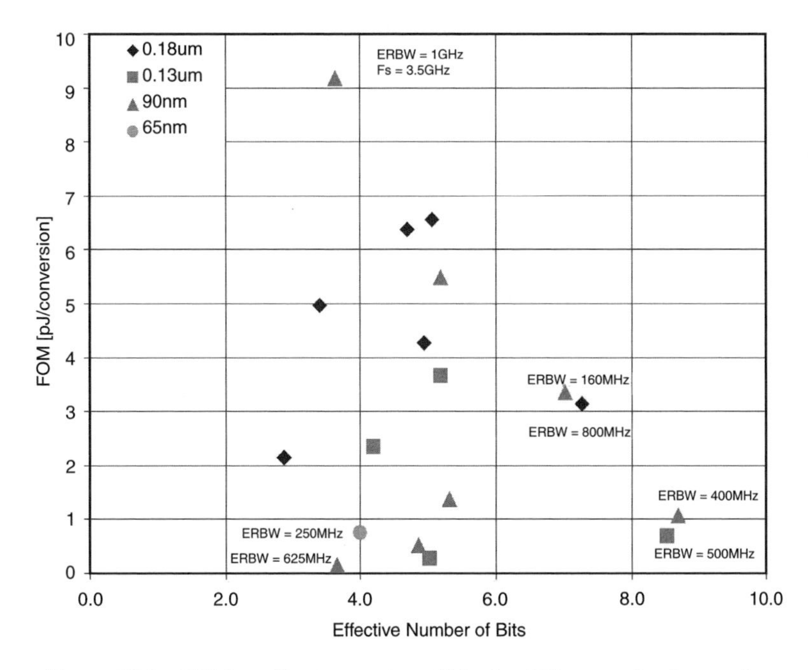

Figure 19.4 FOM vs effective number of bits for different technology nodes.

Table 19.3 Summary of current state-of-the-art in high-speed DACs.

Technology	Measured SFDR/output freq./Sample rate	Power (W)	Reference
0.25 um BiCMOS	63 dB/600 MHz/1.2 GHz	6	[31]
0.18 μm CMOS	67 dB SFDR/260 MHz/1.4 GHz	0.4	[32]
0.13 μm CMOS	54.5 dB SFDR/100 MHz/200 MHz	0.025	[33]
90 nm CMOS	61 dB SFDR/193 MHz/800 MHz	0.049	[34]

For a 2 Gb/s OFDM system assuming QPSK constellation, the ADC requirements are approximately 5ENOB and 1 GHz of ERBW. Such an ADC hasn't been reported yet. If a FOM of 3 pJ/conversion can be achieved for this performance in a robust fashion considering both process and temperature variations, that ADC would consume about 200 mW. If the architecture reported in [21] proves to be robust for implementation in more advanced CMOS technology nodes (i.e. 65 nm and beyond) and scalable for a higher speed and resolution, a FOM of 0.5 pJ/conversion could be achieved for the performance mentioned and each ADC would consume about 30 mW. For 3 Gb/s or higher using 16QAM OFDM over the same 1 GHz bandwidth the required resolution would be about 8 b (with basis on 802.11a designs). If an efficiency of 1 pJ/conv is ever achieved at this performance, each ADC would consume approximately 1 W.

In the area of DACs, fewer examples are found in the literature. Some of the relevant designs are listed in Table 19.4. No DAC was found in the literature with the required characteristics for a Gb/s OFDM design ($\sim$50 dB SFDR, sample rate $>$2 GHz). Fast, current-switched DACs employing conventional thermometer coding and/or segmented architectures will be needed to achieve the required dynamic range, linearity and sampling rates necessary for application in high-rate wireless systems.

Table 19.4 Summary of current state-of-the-art in digital base band ICs.

Technology	Characteristics	Power (W)	Reference
0.25 μm CMOS	Baseband and MAC processors including ADCs and DACs supporting 802.11a standard (OFDM) up to 54 Mb/s	0.778	[35]
0.18 μm CMOS	MIMO baseband and MAC processors including ADCs and DACs in 3 separate RX and TX chains supporting 802.11n draft standard (OFDM) up to 150 Mb/s	2.489	[36]
0.18 μm CMOS	128 point FFT/IFFT processor operating up to 1 G/s for UWB (OFDM) applications	0.175	[37]
0.25 μm CMOS	Reconfigurable FFT/IFFT processor. Highest throughput is 4096 points at 200 MHz clock	0.49	[38]
0.13 μm CMOS	Performs 2048 point FFT/IFFT operations at a 22.86 MHz clock supporting WiMAX	0.0173	[39]

Digital Baseband

A digital baseband chip, implementing the MAC, coding and decoding functions, as well as modulation and demodulation, is the last component to form a complete silicon solution (not necessarily single-chip) for a Gb/s 60 GHz wireless system. Driven by recent wireless standards, digital and mixed-signal wireless baseband ICs reported in the literature are limited to data-rates in the range of 20–500 Mb/s [35–39]. Table 19.4 summarizes the characteristics of recent notable examples of these ICs. The MIMO baseband processor reported in [36] is probably the closest to the requirements of a digital PHY for an OFDM based Gb/s radio. The power consumption of this type of PHY implementation (currently ~2.5 W for 150 Mb/s throughput) is expected to be reduced in deep submicrometer technologies, however the processed data rate must increase by a factor of 10.

19.5 Conclusions

The use of the 60 GHz band presents significant opportunities and challenges for different commercial applications providing wireless links at Gb/s data rates. MSK and OFDM are modulation formats which present convenient characteristics for systems operating at these frequencies in LOS and NLOS environments, respectively. Recent developments on silicon-based integrated millimeter-wave technology satisfy most of the requirements for the implementation of these systems from an RF perspective.

Due to its reduced complexity, an MSK-based Gb/s system for short-range LOS applications has moderate requirements for a digital-baseband section and it would be ready for commercial deployment in the near future,based on currently available hardware solutions. On the other hand, while the implementation of a 60 GHz OFDM based system still demands further improvements in the RF section (i.e. improved phase noise and increased PA linearity at these frequencies), the widest technological gap to be filled in this case is in the digital and mixed-signal (ADC, DAC) areas. Specifically, architecture and circuit-level improvements are required to increase both the ERBW and ENOB of GHz range ADCs without compromising conversion efficiency since it does not seem that these performance improvements will come naturally with technology scaling. Similarly, even though the throughput and power efficiency of the baseband processor will benefit from continued scaling in the deep submicrometer regime, it is not expected that 5 Gb/s OFDM processors will have low enough power for portable applications for technology nodes in the 45 nm range.

References

1. Sadri, A. *et al.* (2006) 802.15.3c Usage Model Document, IEEE 802.15 Working Group for WPAN, no. IEEE15-06-0055-06-003c.
2. Pasupathy, S. (1979) Minimum shift keying: A spectrally efficient modulation. *IEEE Communications Magazine*, **17** (4), 14–22.
3. van Nee, R. and Prasad, R. (2000) *OFDM for Wireless Multimedia Communications*, 1st edn, Artech House, p. 280.
4. Floyd, B., Pfeiffer, U., Reynolds, S. *et al.* (2007) *Silicon Millimeter Wave Radio Circuits at 60-100 GHz*, in Proceedings IEEE Topical Meeting on Silicon Monolithic Integrated Circuits in RF Systems, pp. 213–8 (Invited).
5. Wang, C.-H., Chang, H.-Y., Wu, P.-S. *et al.* (2007) *A 60 GHz los-power six-port transceiver for gigabit software-defined transceiver applications*, in *Proc. IEEE ISSCC*, pp. 192–3.

6. Emami, S., Doan, C.H. and Niknejad, A.M. (2007) *A highly integrated 60 GHz CMOS Front-End Receiver*, in *Proc. IEEE ISSCC*, pp. 190–1.
7. Mitomo, T., Fujimoto, R., Ono, N. *et al.* (2008) A 60-GHz CMOS receiver front-end with frequency synthesizer. *IEEE JSSC*, **43** (4), 1030–7.
8. Reynolds, S., Valdes-Garcia, A., Floyd, B.A. *et al.* (2007) Second generation transceiver chipset supporting multiple modulations at Gb/s data rates, in *Proc. IEEE BCTM*, pp. 192–7.
9. Pfeiffer, U. *et al.* (2006) A chip-scale packaging technology for 60-GHz wireless chipsets. *IEEE TMTT*, **54** (8), 3387–97.
10. Katayama, Y. *et al.* (2008) *Multiple-Gbps Wireless Systems over 60-GHz SiGe Radio Link with BW-Efficient Noncoherent Detection*, in *Proc. IEEE ICME*, pp. 513–16.
11. Valdes-Garcia, A., Reynolds, S. and Pfeiffer, U.R. (2006) A 60 GHz class-E power amplifier in SiGe, in *Proc. IEEE ASSCC*, pp. 199–202.
12. Pfeiffer, U.R. (2006) A 20 dBm fully-integrated 60 GHz SiGe power amplifier with automatic level control, in *Proc. ESSCIRC*, pp. 356–9.
13. Taft, R.C., Menkus, C.A., Tursi, M.R. *et al.* (2004) A 1.8-V 1.6-G sample/s 8-b self-calibrating folding ADC with 7.26 ENOB at nyquist frequency. *IEEE JSSC*, **39** (12), 2107–15.
14. Koo, J.-H., Kim, Y.-J., Park, B.-H. *et al.* (2006) A 4-bit 1.356 Gsps ADC for DS-CDMA UWB system, in *Proc. IEEE A-SSCC*, pp. 359–62.
15. Ginsburg, B.P. and Chandrakasan, A.P. (2007) Dual time-interleaved successive approximation register ADCs for an ultre-wideband receiver. *IEEE JSSC*, **42** (2), 247–57.
16. Jiang, X. and Chang, M.-C.F. (2005) A 1-GHz signal bandwidth 6-bit CMOS ADC with power efficient averaging. *IEEE JSSC*, **40** (2), 532–5.
17. Hung, C.-K., Shiu, J.-F., Chen, I.-C. and Chen, H.-S. (2006) A 6-bit 1.6 GS/s flash ADC in 0.18 um CMOS with reversed-reference dummy, in *Proc. IEEE A-SSCC*, pp. 335–8.
18. Shen, D.-L. and Lee, T.-C. (2007) A 6-bit 800-MS/s pipelined A/D converter with open-loop amplifiers. *IEEE JSSC*, **42** (2), 258–68.
19. Chen, S.-W.M. and Brodersen, R.W. (2006) A 6-bit 600-MS/s 5.3-mW asynchronus ADC in 0.13 um CMOS. *IEEE JSSC*, **41** (12), 2650–7, 2669–80.
20. Viitala, O., Lindfors, S. and Halonen, K. (2006) A 5-bit 1-GS/s flash-ADC in 0.13 um CMOS using active interpolation, in *Proc. IEEE ESSCIRC*, pp. 412–25.
21. Sandner, C., Clara, M., Santner, A. *et al.* (2005) A 6-bit 1.2-GS/s low-power flash-ADC in 0.13 um digital CMOS. *IEEE JSSC*, **40** (7), 1499–1505.
22. Gupta, S.K., Inerfield, M.A. and Wang, J. (2006) A 1-GS/s 11-bit ADC with 55-dB SNDR, 250-mW power relaized by a high bandwidth scalable time-interleaved architecture. *IEEE JSSC*, **41** (12), 2650–7.
23. Figueiredo, P.M. *et al.* (2006) A 90 nm CMOS 1.2 V 6b 1 GS/s two-step subranging ADC, in *Proc. IEEE ISSCC*, pp. 2320–1.
24. der Plas, G.V., Decoutere, S. and Donnay, S. (2006) A 0.16 pJ/conversion-step 2.5 mW 1.25 GS/s 4b ADC in a 90 nm digital CMOS process, in *Proc. IEEE ISSCC*, pp. 2310–1.
25. Park, S., Palaskas, Y., Ravi, A. *et al.* (2006) A 3.5 GS/s 5-b Flash ADC in 90 nm CMOS, in *Proc. IEEE CICC*, pp. 489–92.
26. Draxelmayr, D. (2004) A 6b 600 MHz 10 mW ADC array in digital 90 nm CMOS, in *Proc. IEEE ISSCC*, pp. 264–55.
27. Makigawa, K., Ono, K., Ohkawa, T. *et al.* (2006) A 7 bit 800 Msps 120 mW folding and interpolation ADC using a mixed-averaging scheme, in *Proc. IEEE Symposium on VLSI Circuits*, pp. 15–17.
28. Hsu, C.-C., Huang, F.-C., Shih, C.-Y. *et al.* (2007) An 11b 800 MS/s time-interleaved ADC with digital background calibration, in *Proc. IEEE ISSCC*, pp. 464–5.
29. Peach, C.T., Ravi, A., Bishop, R. *et al.* (2005) A 9-b 400 M sample/s pipelined analog-to-digital converter in 90 nm CMOS, in *Proc. IEEE ESSCIRC*, pp. 535–8.
30. Ginsburg, B.P. and Chandrakasan, A. (2006) A 500 MS/s 5b ADC in 65 nm CMOS, in *Proc. IEEE Symposium on VLSI Circuits*, pp. 140–1.
31. Jewett, B., Liu, J. and Poulton, K. (2005) A 1.2 GS/s 15b DAC for precision signal generation, in *Proc. IEEE ISSCC*, pp. 110–1.
32. Schafferer, B. and Adams, R. (2004) A 3 V CMOS 400 mW 14b 1.4 Gs/s DAC for multi-carrier applications, in *Proc. IEEE ISSCC*, pp. 360–1.

33. Clara, M., Klatzer, W., Seger, B. *et al.* (2007) A 1.5 V 200 MS/s 13b 25 mW DAC with randomized nested background calibration in 0.13 um CMOS, in *Proc. IEEE ISSCC*, pp. 250–1.
34. Cao, J., Lin, H., Xiang, Y. *et al.* (2006) A 10-bit 1G sample/s DAC in 90 nm CMOS for embedded applications, in *Proc. IEEE CICC*, pp. 165–6.
35. Thomson, J. *et al.* (2002) An integrated 802.11a baseband and MAC processor, in *Proc. IEEE ISSCC*, pp. 250–1.
36. Petrus, P. *et al.* (2007) An integrated draft 802.11n compliant MIMO baseband and MAC processor, in *Proc. IEEE ISSCC*, pp. 266–7.
37. Lin, Y.-W., Liu, H.-Y. and Lee, C.-Y. (2005) A 1-GS/s FFT/IFFT processor for UWB applications. *IEEE JSSC*, **40** (8), 1726–35.
38. Zhong, G., Xu, F. and Willson, A.N. (2006) A power-scalable reconfigurable FFT/IFFT IC based on a multi-processor ring. *IEEE JSSC*, **41** (2), 483–95.
39. Chen, Y., Lin, Y.-W. and Lee, C.-Y. (2006) A Block Scaling FFT/IFFT Processor for WiMAX Applications, in *Proc. IEEE A-SSCC*, pp. 203–6.

20

System Concepts and Circuits for a 60 GHz OFDM Transceiver

Eckhard Grass, Chang-Soon Choi, Klaus Tittelbach-Helmrich, Maxim Piz, Marcus Ehrig, Klaus Schmalz, Yaoming Sun, Srdjan Glisic, Frank Herzel and Christoph Scheytt

IHP GmbH, Frankfurt (Oder)

20.1 Introduction

Due to the rising demand for wideband applications and services, a large increase of the data rates for future mobile radio systems is expected. Whereas for wired transmission via optic fibre there are no technological bottlenecks, even for data rates in the region of many GBit/s, these rates are a big challenge for wireless systems. They can only be tackled by an increase of spectral efficiency or by the development of unused frequency resources within the micrometer-and millimeter- wave bands.

For applications that require high data rates in conjunction with relatively small ranges and only low mobility, the use of frequency bands in the region of millimeter-waves is very promising. It is expected that in near future, the band around 60 GHz, with several GHz bandwidth, will be available worldwide for unlicensed radio transmission. The current situation is shown in Figure 20.1. Frequency regulation in Europe is currently in progress. The band from 63 to 64 GHz is reserved for intelligent transportation systems (ITS). It is hoped that a similar amount of bandwidth as in other regulatory regions will be made available in Europe eventually.

The main advantage of this frequency band is the large available bandwidth. In addition, small antenna dimensions – also in the case of multi-antenna systems – result from the short wavelength. A disadvantage is the free space path loss, which is approximately 20 dB higher, compared with conventional mobile radio systems. The high penetration loss prevents the waves from propagating through walls, but it also reduces the interference between adjacent cells and other radio systems. In comparison with UWB techniques the latter is an important advantage in terms of frequency-reuse in a densely populated environment.

Short-Range Wireless Communications Rolf Kraemer and Marcos D. Katz

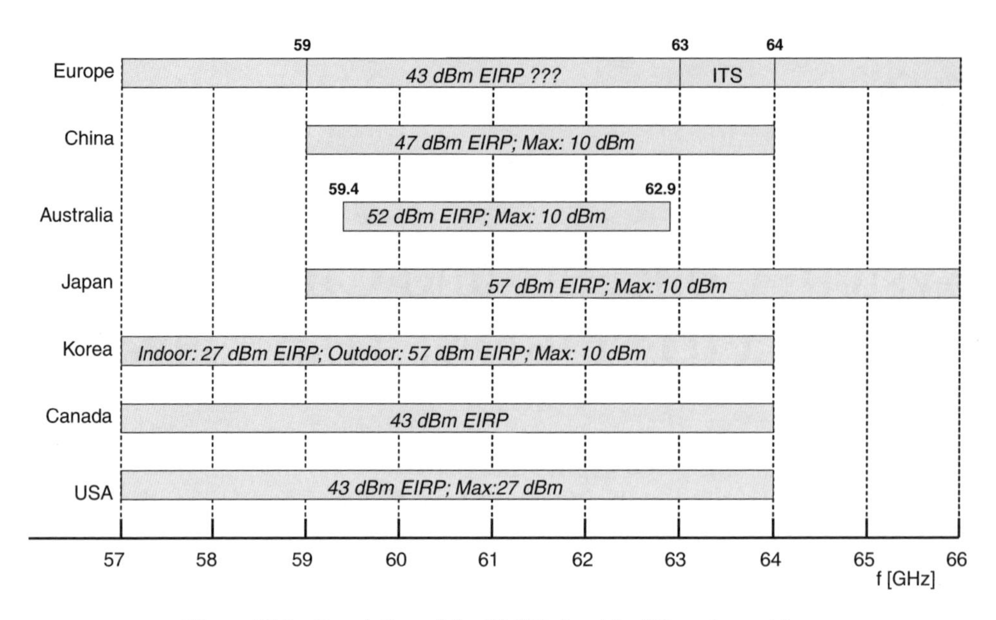

Figure 20.1 Regulation of the 60 GHz band in different countries.

Currently, the 60 GHz standardization within IEEE802.15.3c is in progress. Furthermore, within the new study group VHT (very high throughput) a 60 GHz PHY is also considered in IEEE working group 802.11. A 60 GHz PHY in the standard IEEE802.11VHT has the potential to become *the* next generation wireless LAN standard, allowing for data rates of several gigabits per second.

Additionally, a 60 GHz standard is being developed within the ECMA standardization group. Currently, at ECMA similar PHY parameters are considered as in IEEE802.15.3c. However, the number of different PHY modes in the IEEE standard is significantly higher. This may be counterproductive for potential system designers.

The interest in the 60 GHz band is also reflected in a number of international projects. One of them is the German WIGWAM project (http://www.wigwam-project.de/) which has recently been followed by the project EASY-A (http://www.easy-a.de/) with 15 main partners. Other projects dealing with aspects of 60 GHz communication are Broadway (http://www.ist-broadway.org/), Magnet (http://www.ist-magnet.org/) and OMEGA (http://www.ict-omega.eu/). The number and size of these projects is an indication of the economic potential of this new technology.

20.2 System Design Aspects

20.2.1 System Architecture

20.2.1.1 Channel Plan

As introduced in Chapter 1, there are several activities to define a 60-GHz standard for gigabit wireless personal area network (WPAN) applications. Although their system approaches and main target applications are different from each other, there is a common agreement that

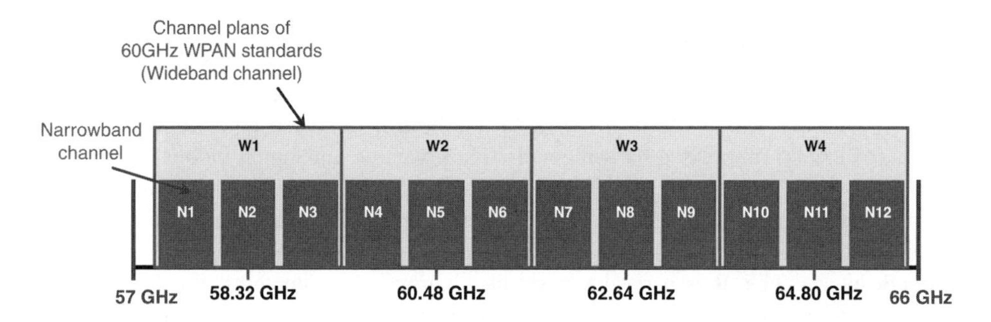

Figure 20.2 Channel plans for wideband channel and narrowband channel.

60 GHz WPAN devices have the same channel plan, regardless of what 60 GHz standard a device supports [1, 2]. From 57 to 66 GHz, over 9 GHz bandwidth, four independent channels were defined, centred on the frequencies of 58.32, 60.48, 62.64 and 64.80 GHz. Each channel has the maximum channel bandwidth of 2.16 GHz. Within the boundary of the unlicensed 60 GHz frequency-band allocated by a governmental agency of a specific region, a 60 GHz device can select a channel as long as it adheres to the defined centre frequencies and maximum channel bandwidth. All centre frequencies and maximum channel bandwidths were decided with the assumption that a phase-locked loop (PLL) utilizes a 19.2 MHz crystal oscillator. Since it is widely used for cellular phones, it is expected to be the basis for cost-effective 60 GHz transceivers.

This channel plan is illustrated in Figure 20.2 as labeled from W1 to W4. It allows allocating several narrow-band channels in each frequency channel, which can be used not only as data channels but also as return channels or control channels. This narrow-band channel plan is believed to be very useful in wireless local area networks (WLAN) applications where one access point needs to support a much higher number of device users than for wireless PANs. For future applications of gigabit WLAN, narrow-band channels which have a maximum bandwidth of 720 MHz are proposed and incorporated with 'wide-band channels' defined for 60 GHz WPAN as shown in Figure 20.2 labeled as N1 to N12. The narrowband channel plan provides not only high flexibility to accommodate many users, but also full compatibility to the channel plan of 60 GHz WPAN standards. All centre frequencies of narrow-band channels can be generated from a PLL based on a 19.2 MHz crystal oscillator, which was originally designed for WPAN applications. In addition, this proposed scheme promises a reasonable number of available channels, regardless of regions.

20.2.1.2 60 GHz Channel Characteristics

There are many activities to characterize 60 GHz indoor channels and make a channel model based on measured data. Presently, 60 GHz channel models which were built up on the basis of measurement data are available for indoor residential and office environments in the IEEE 802.15.3c standard group [3].

Unlike IEEE standards dealing with below-6 GHz operation, it includes several channel models according to the transmitter antenna beamwidth in the same environment. 60 GHz wireless systems have the potential to utilize higher antenna gain than 'low-frequency' systems without excessive antenna size. This is advantageous not only to compensate high atmospheric loss of 60 GHz bands, but also to reduce the influences of multipath-interference with the help of narrow beamwidth. For instance, it was observed that the rms delay spread decreased from 18 to 4.7 ns by changing from an omnidirectional antenna to a high-directivity antenna with 3 dB beamwidth of 10° [4]. In addition, it has been reported that multipath interference can be suppressed by using a circular polarized antenna in indoor environments [4, 5]. The experimental result shows that using a circular polarized antenna provides shorter rms delay spread compared with the case of using linear polarized antenna. This makes the wireless system free from complex digital equalization of intersymbol interference caused by multipath interference, thus it results in simpler hardware architectures. For these reasons, these types of antenna are preferred for very short-range file-download applications (such as Kiosk Usage Model 5 in IEEE 802.15.3c [6]) where mobile devices such as digital cameras and portable multimedia players require simpler hardware architectures and lower power consumption.

However, high-directivity antennas will result in smaller coverage which is a serious limiting factor for mobile communication systems such as multi-user WLANs and WPANs. Moreover, it requires additional modifications of the Medium Access Control (MAC) and physical layer architectures in order to support point-to-multipoint communication and operation of piconets. To provide better flexibility and link adaptability facilitating point-to-multipoint communications, moderate-directivity antennas or omnidirectional antennas are simple solutions in the design of 60 GHz wireless systems, especially for gigabit WLAN applications.

For the specification of the main PHY parameters line-of-sight (LOS) channel environments, an obstructed LOS (OLOS) channels were characterized. OLOS measurements were carried out by a person partially or totally blocking the LOS path [7]. Figure 20.3(A) and (B) shows the examples of channel characteristics for LOS and OLOS measured in small conference room environments. We can clearly observe that even a LOS channel suffers from strong multipath propagation and becomes highly dispersive and frequency selective in indoor environments. In obstructed channel environments, this becomes more serious by power reduction of the LOS direct path signals with more than 20 dB. Figure 20.4 summarizes the rms delay spreads

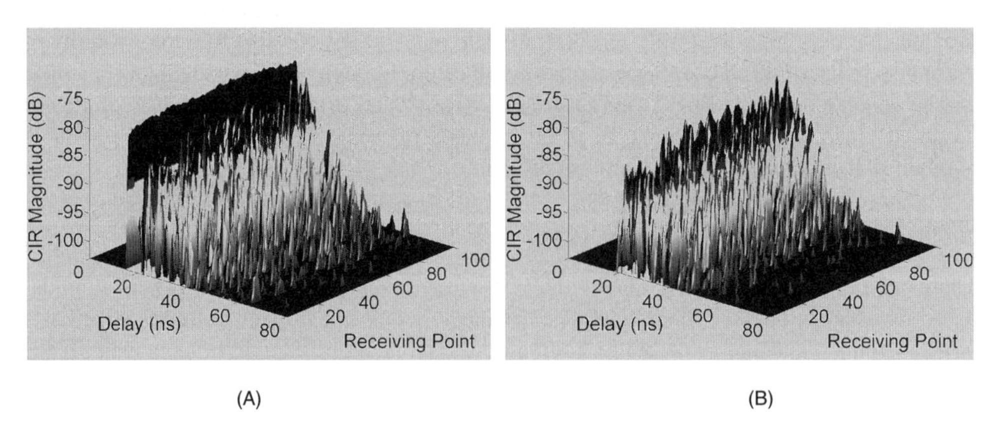

Figure 20.3 60-GHz channel measurement time-domain results for: (A) LOS; (B) OLOS (obstructed LOS) conditions.

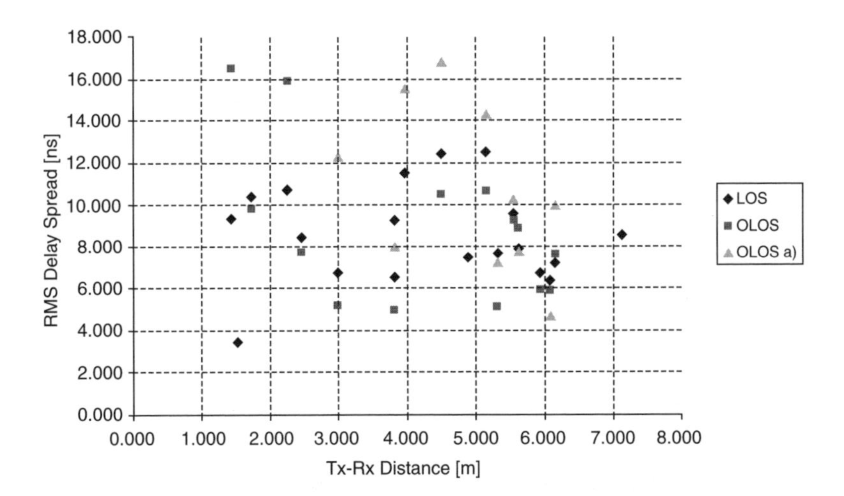

Figure 20.4 Measured rms delay spreads as a function of transmission distance for LOS and OLOS (obstructed LOS) conditions.

measured at different transmission distances for LOS and OLOS channels. They range from 6 to 10 ns. It can be concluded that, in this type of 60 GHz indoor channel environments, wireless systems should be equipped with a countermeasure for strong multi-path interferences, for example digital equalization techniques and OFDM transmissions.

There is another important 60-GHz channel characteristic that one should consider for indoor WPAN systems. It is most likely to happen in desktop environments which are one of the more widely used channel environments for WPAN applications. The reflected signals from the desk can cause serious signal fading since they have almost the same signal amplitude, but can be out of phase from directly transmitted signals [8]. This phenomenon can be expressed with the two-path model for LOS components in channel characteristics. Assuming that L is the transmission distance and h_{T} and h_{R} are the heights of transmitter and receiver, respectively, the path-loss given by the two-path model can be expressed as

$$\frac{P_r}{P_t} = \left(\frac{\lambda}{4\pi L}\right)^2 \left|1 + \Gamma \exp\left[j\frac{2\pi}{\lambda}\frac{2h_1h_2}{L}\right]\right|^2 \tag{20.1}$$

where Γ is the reflection coefficient of the desk and λ is the wavelength.

Figure 20.5 shows the two-path responses as a function of the receiver height for 60-GHz desktop channel environments. The points are measured data and the solid line is result based on Equation (20.1). It is clearly observed that serious signal fading occurs around 143 mm receiver height and it agrees well with the calculation results. To avoid these signal fading phenomena in indoor environments, this two-path response should be taken into account and a solution, for example a receiver space-diversity system, needs to be considered for system design [9]. In IEEE 802.15.3c channel models, the two-path response was also considered and set to be the statistical value for the reduction of simulation scenarios.

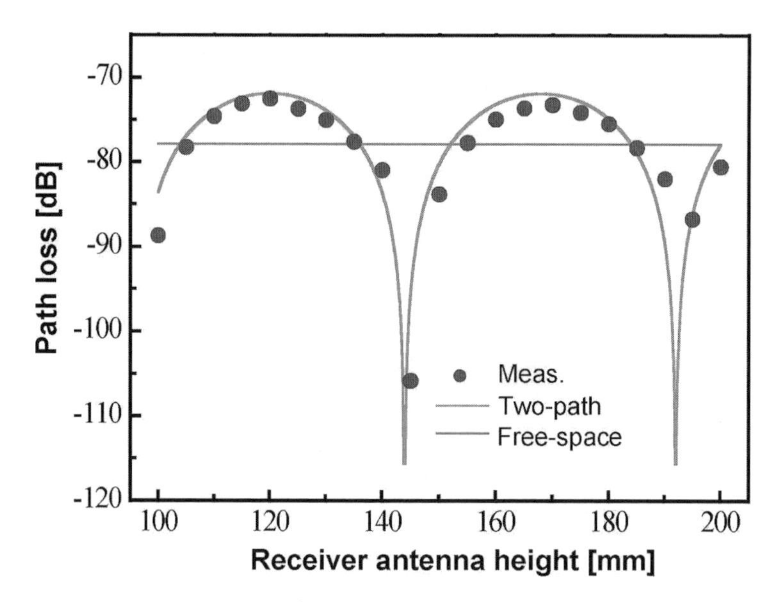

Figure 20.5 Measured (circles) and calculated (solid line) two-path responses as a function of receiver antenna height.

20.2.1.3 Baseband Modulation: OFDM versus Single Carrier

For 60 GHz WPAN system design, several baseband modulations are being considered. Most of them can be categorized into three major schemes: OFDM, single-carrier and single carrier with frequency domain equalization. In Table 20.1, the main properties are compared.

OFDM provides the strongest robustness to multipath interference and the best spectral in conjunction with ensuring the smallest out-of-band radiation. This is the foremost reason why most of modern wireless communication systems employ OFDM baseband modulations, for example, Digital Video Broadcasting, Indoor WLAN (IEEE 802.11a/g) and Mobile WiMax (IEEE 802.16e) applications. It is also expected to be used for the fourth generation (4G) of cellular communication systems. However, the hardware requirement for the RF analogue front-end is most stringent for OFDM transmission system. For example, high peak-to-average power ratio (PAPR) characteristics of OFDM modulated signal seriously limit output power level of RF high power amplifiers (HPA) in order to ensure their linearity. This output backoff

Table 20.1 Comparison of Single carrier modulation versus OFDM for 60 GHz

	OFDM	Single Carrier	Single Carrier-FDE
Spectral efficiency	++	−	+
PAPR	− −	+	+
Multipath robustness	++	− −	+
Out of band radiation	++	−	−
Complexity of baseband processor	−	+	−
Complexity of AFE	−	+ +	+

is typically required to be of the order of 6–12 dB, which in turn degrades the power efficiency of the HPA. In 60 GHz wireless systems, this drawback becomes more serious because 60 GHz RF front-end technologies as well as that of HPAs have not been matured yet in comparison with below-6 GHz operations. In addition, the target transmission rate for a 60 GHz wireless system should be higher than 1 Gbps. To realize such high-rate baseband processor and analog-to-digital converter (ADC) which can support OFDM transmission is a still challenging task from the viewpoint of hardware implementation.

On the other hand, single carrier modulation makes it possible to reduce the problem of hardware requirement. However, it occupies a larger transmission bandwidth than OFDM, which limits the number of usable frequency channels in the 60 GHz-band. Moreover, it can be very sensitive to multi-path fading phenomena unless a digital equalization technique is used. These features make it only suitable for short-range and LOS communication systems in which high-directivity antennas are used for the suppression of multi-path interference.

Frequency domain equalization techniques have been proposed for single carrier modulation to combat multi-path fading. They are known under the term, single carrier with frequency domain equalizer (SC-FDE). On the receiver side, digital equalization of transmitted signals can be done in the frequency domain with FFT and IFFT modules. This makes it possible to use broad beamwidth antennas for ensuring wide coverage communication. Since this SC-FDE system basically transmits single-carrier signals, PAPR and susceptibility to RF phase-noise are almost equivalent to those of 'pure' single-carrier systems. This results in less stringent requirements for RF analogue front-ends. However, SC-FDE is not so advantageous in terms of hardware requirements for a digital baseband processor and ADC because it needs almost the same components as an OFDM baseband processor.

In other words, each modulation has its own structural advantages. Therefore, it is beneficial to choose the baseband modulation depending on application scenarios. As discussed previously, single carrier modulation is preferred for file-download applications where wireless systems guarantee short-range and LOS links with the requirement of low-power consumption. For fixed and nomadic applications such as WLAN and HD video transmission systems free from power constraints, OFDM is believed to be the best choice to provide high performance in indoor environments. To come up with both demands, IEEE 802.15.3c will support both baseband modulations and they will be independently optimized according to specific application.

20.2.1.4 60 GHz Analog Front-End Architectures

In modern communication systems, three RF architectures are mainly used, which are schematically shown in Figure 20.6.

The first is called the direct conversion scheme which converts RF signals into the baseband in a single step for the receiver and *vice versa* for the transmitter. This is apparently simpler and requires fewer RF components than any other approach, thus it leads to lower power consumption. In addition, it facilitates monolithic one-chip integration due to the absence of external filters for image-rejection or channel selection. However, DC offset and LO leakage/pulling have been always a limiting factor for its extensive use. These problems become much more serious for 60-GHz operation due to immature RF technologies compared with microwave bands. Nevertheless, several trials have been conducted for short-range

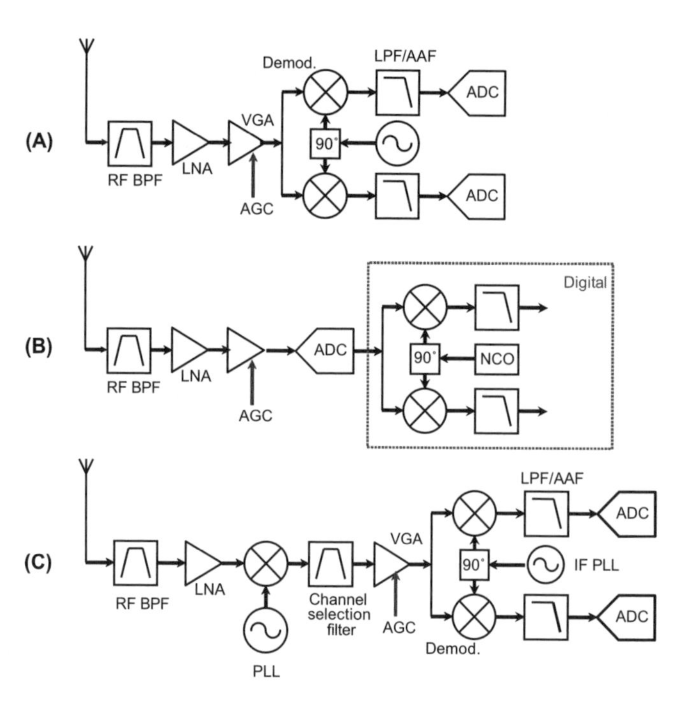

Figure 20.6 Schematic illustration for (A) direct conversion receiver (B) low-IF receiver and (C) superheterodyne receiver.

file-download applications that require a low-cost transceiver with relatively lower power consumption.

A variant of this direct conversion scheme is a low-IF scheme which is also called a digital-IF scheme. RF signals are not directly down-converted to baseband, but to the low-IF-band. Then, these low-IF signals are sampled with an ADC and I/Q demodulation is done in the digital domain. This was originally proposed to overcome the problems of a direct conversion transceiver while keeping the simple RF analogue front-end structure. However, it inherently requires high sampling rates of ADCs and this is hardly realizable in gigabit 60 GHz wireless system where the sampling rate of ADCs is already a bottleneck for hardware implementation. Therefore, low-IF schemes for 60 GHz systems are currently not in the mainstream of 60 GHz system development.

The final architecture is a superheterodyne scheme which has been most widely used in wireless communication systems. IF, by definition, is the intermediate frequency to which RF signals are first down-converted with a high frequency local oscillator (LO). After channel selection filtering, these IF signals are down-converted into the baseband with an IF LO. This provides not only higher channel selectivity, but also better immunity to DC offset, LO leakage and I/Q mismatch which are critical problems in a direct conversion scheme. For its practical use in the 60 GHz band, the IF should be carefully selected considering not only the tradeoff between image-rejection capability and channel selectivity, but also the availability of RF components. In the literature, different IFs have been suggested: 5, 9 and 12 GHz. The advantage of a 5 GHz is the compatibility to the 802.11a WLAN standard, facilitating the design of a 60 GHz system, especially when combined with a 5 GHz backup system. This is still attractive where dual-band operation is being considered, such as gigabit WLAN systems which are due to be defined in IEEE 802.11 VHT. A high IF makes it easy to remove image-signals by the inherent frequency response of a low noise amplifier (LNA) and power amplifier (PA) without inserting a lossy image rejection filter.

Alternatively, the IF can be generated from LO by frequency division, which is called the sliding IF scheme. This removes an additional crystal oscillator and phase-locked loop for IF-conversion and results in less power consumption and less complexity while maintaining the advantage of the superheterodyne scheme. In this approach, RF channels between 57 and 66 GHz are selected by tuning the LO which also results in IF tuning. This makes it difficult to design a channel selection filter operating at IF since the IF varies according to the channel number.

20.2.1.5 Multiple Antenna Technologies

One of the key advantages of 60 GHz wireless systems with respect to 'low-frequency' systems is to have smaller antenna size. This is due to the short wavelength. It enables multiple antenna configurations without increasing the size of the transceiver too much. With the rapid evolution of Si-based integration processes, these multiple antenna configurations become easily to realize and to implement practically. In particular, they are expected to provide an effective solution to the problems that 60 GHz wireless systems suffer from, for example, LOS constraint and high path loss.

Switched diversity is the simplest way to alleviate the LOS constraint and provides diversity gain in obstructed and blocked LOS conditions. In addition, two-path signal fading in a desktop environment can be simply solved by employing this diversity scheme. For these reasons, IEEE 802.15.3c and ECMA standards have already adopted a transmitter switched diversity scheme and an optimized frame structure, as shown in Figure 20.7 [1].

There are several reports that multiple-input-and-multiple-output (MIMO) techniques in 60 GHz systems are not so effective as those in below the 6 GHz-band due to the LOS constraint and strongly correlated multiple transmitters and receivers. However, these MIMO

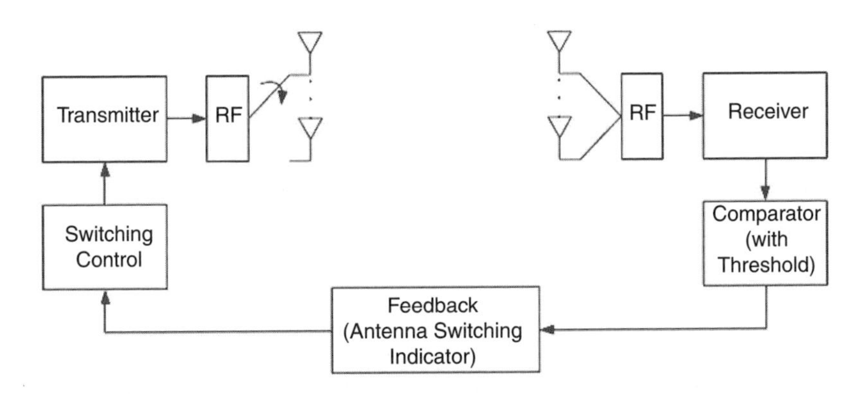

Figure 20.7 The structure of transmit switched diversity system defined in IEEE 802.15.3c draft.

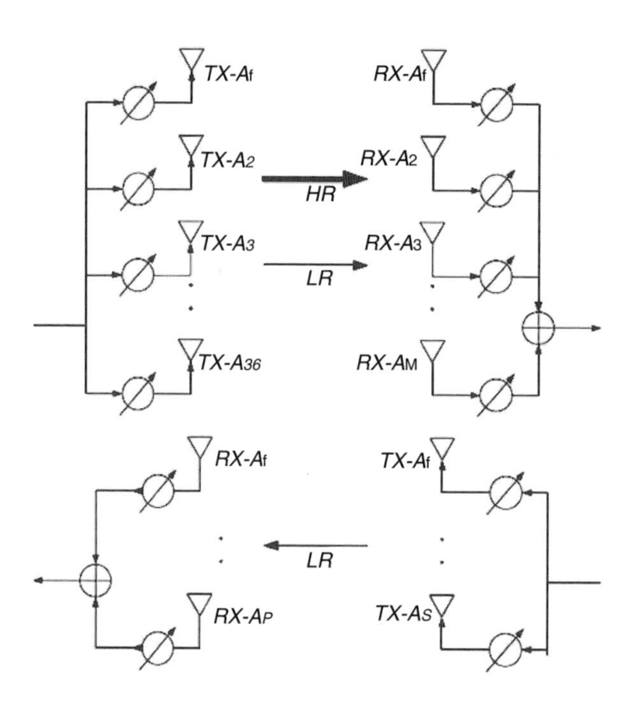

Figure 20.8 General wireless HD antenna array configuration.

techniques are still beneficial in strong multipath indoor environments and their processing gain can be improved by hardware optimization of 60 GHz wireless systems.

Beamforming is also an attractive multiple antenna technology for 60 GHz wireless systems. With phased-array transceivers, the beam can be steered to find the best path from a transmitter to a receiver. This process can be done when the link condition changes, for example, the LOS link blocked by a person in an indoor environment. Moreover, it offers improved antenna gain and suppressed multi-path fading with narrower antenna beamwidth. The WirelessHD consortium has already declared its intention to use this beamforming technology for uncompressed HD video transmission in 60 GHz-link. IEEE 802.15.3c also supports it and provides frame structure and a training process. Figure 20.8 shows the phased antenna array configuration proposed by WirelessHD.

20.2.2 60 GHz RF Analog Front-Ends: State-of-the-Art and Future Development

20.2.2.1 The Choice of Device Technology

IHP has developed 60 GHz wireless systems mainly for multi-gigabit WLAN applications as well as HD video distribution applications in indoor environments. As explained in Section 20.2.1, these applications have to provide wider coverage and less susceptibility to multi-path fading without limiting power consumption too much. From this point of view, OFDM has been preferred to single carrier modulation and developed with the target of multi-gigabit transmission in the 60 GHz-band. However, it requires a 60 GHz RF front-end to possess higher performance, including higher linearity of the power amplifier and lower phase noise of PLL. Furthermore, the target applications are required to radiate high transmission power in order to have wider coverage compared with WPAN applications.

For these 60 GHz applications, III–V semiconductor technologies such as GaAs and InP were mostly used in the past. However, the 60 GHz circuits based on these III–V technologies are expensive and hard to integrate with other circuitries. On the other hand, RF CMOS technologies have not matured yet to support the required performance for OFDM transmission, even though they have made great progress recently. They may be useful in short-range file-download applications which require low power consumption rather than long-range WLAN applications. From our point of view, SiGe BiCMOS technology is believed to be the most promising solution in terms of cost and performance, therefore it has been used for our 60 GHz RF analogue front-ends.

20.2.2.2 60 GHz RF Front-Ends: Previous Results

Figure 20.9(A) and (B) schematically show the 60 GHz analog receiver and transmitter front-ends developed so far. These chipsets were fabricated in IHP SiGe:C 0.25 μm BiCMOS technologies with f_T of 200 GHz.

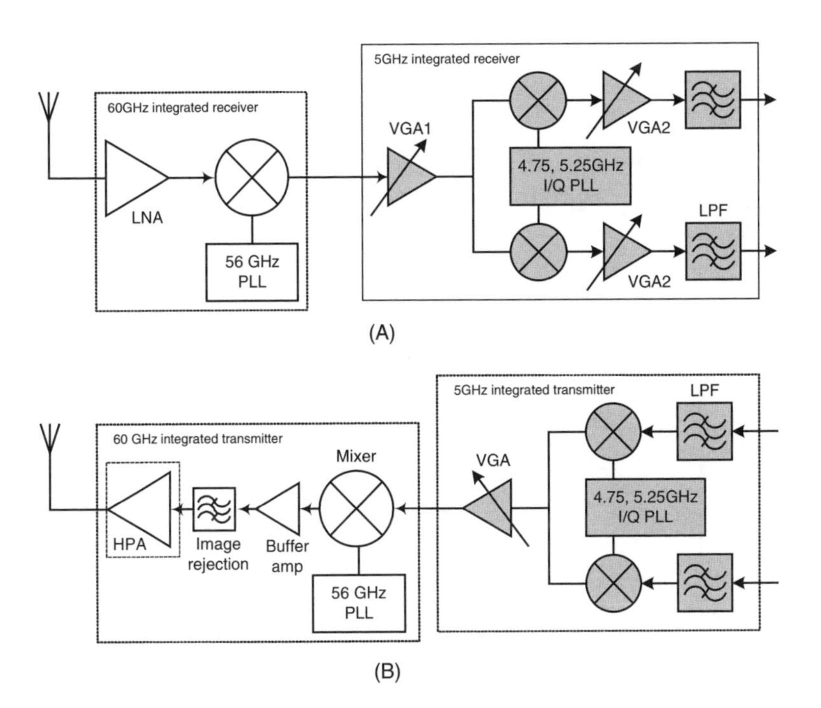

Figure 20.9 (A) 60 GHz receiver analogue front-end; (B) 60 GHz transmitter analogue front-end developed with IHP 0.25 μm SiGe:C BiCMOS technology.

The 60 GHz receiver front-end consists of three building blocks, an LNA, a down-conversion mixer and a 56 GHz PLL. The LNA is based on a three-stage common-emitter differential topology. In order to obtain a symmetrical design, a single-ended LNA was laid out first, then it was copied upside down. The emitters of differential pairs are connected directly to the chip ground. By doing so, the LNA is single-ended measurable. It has a measured gain of 18 dB and a simulated noise figure of 6.8 dB at 60 GHz. The measured and simulated reflection is below −17 dB. The detailed design has been presented in [10].

A Gilbert cell is used as the frequency down-conversion mixer core. The output buffer is realized by two emitter-followers. Two series capacitors and a shunt inductor are used to match the 5 GHz IF-port to 100 Ω differential output impedance. The middle of the inductor represents a virtual ground. This output matching structure is a second-order high-pass topology, optimized to have a band-pass characteristic. It has a 1-dB bandwidth of 2 GHz, which is sufficient for our Gbps communication system, regardless of modulation scheme. The measured and simulated conversion gains for the LNA and mixer combination are shown in Figure 20.10. For these measurements, the LO frequency was fixed at 56 GHz.

An RF PLL has been regarded as one of the critical components that significantly affect system performance of OFDM signal transmission. In the first prototype of our RF front-end, a 56 GHz PLL is implemented using a fourth-order topology and fully integrated in both receiver and transmitter. It consists of a voltage-controlled oscillator (VCO) with a differential output, a frequency divider, a charge pump and a loop filter. The 512 divider is realized with nine consecutive 1 : 2 divider stages. A programmable RF PLL would bring on an excessive divider ratio, which induces the phase noise caused by the reference signal to be unacceptable. Therefore, in this prototype demonstrator, we fixed the dividing factor in the PLL and channel

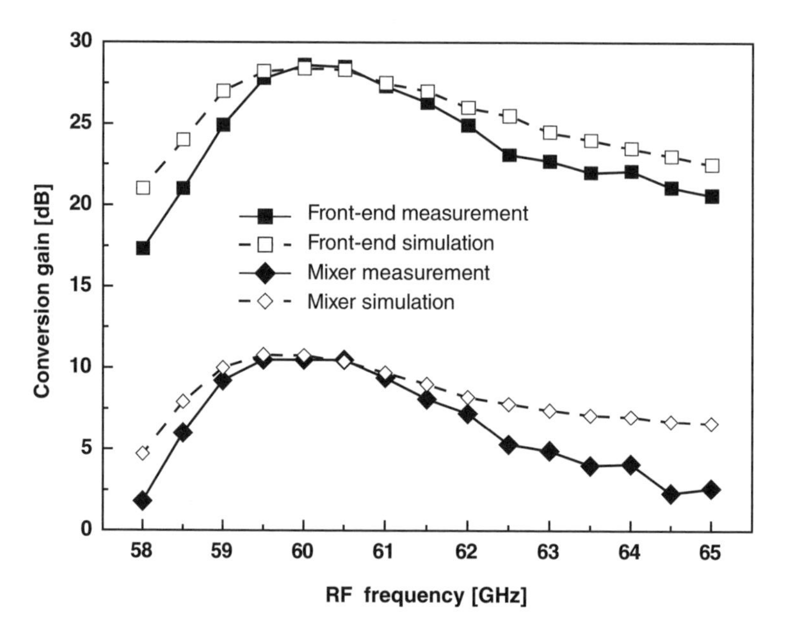

Figure 20.10 Measured and simulated conversion gain for the frequency down-conversion mixer and the combination of LNA/mixer with a fixed 56 GHz LO input.

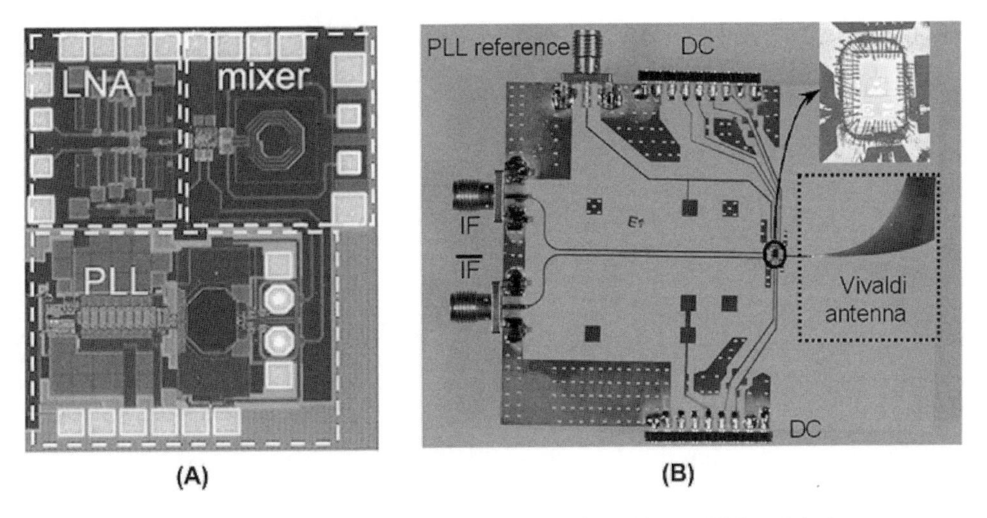

Figure 20.11 Photographs of: (A) 60 GHz receiver chipset; (B) board design.

selection is assumed to be achieved in the IF stage. The detailed design of a 56 GHz PLL we employed in the RF front-end can be found in [11]. The measurement shows that the PLL has a bandwidth of 4.5 MHz and a spur level below −55 dBc. It consumes 80 mA from a 3 V supply and 60 mA from a 2.6 V supply. The VCO phase-noise at 1 MHz offset from the carrier is between −90 and −95 dBc/Hz. Figure 20.11A shows the photograph of the developed 60 GHz receiver front-end where an LNA, a down-conversion mixer and a 56 GHz PLL are integrated into one chip. Figure 20.11B shows the board design for the 60 GHz receiver front-end.

The developed 60 GHz RF transmitter front-end consists of a frequency up-conversion mixer, a preamplifier, an image rejection filter and a power amplifier, as shown in Figure 20.9. This frequency up-conversion mixer is based on a Gilbert-cell mixer and is optimized for high linearity and output power. The simulation result shows the maximum output 1 dB compression point (1 dBCP) of −15 dBm. In order to make this output signal acceptable in terms of input range of a following power amplifier, a preamplifier with 18 dB gain is employed. It is also used in order to compensate the high insertion loss given by the image-rejection filter [12].

The power amplifier features a multi-stage differential cascode topology. One cascode stage provides high gain of around 11 dB with IHP SiGe BiCMOS technology, and therefore three cascode stages were employed to get more than 25 dB amplification in the power amplifier. It was designed fully differentially for better common mode noise rejection and robustness. Figure 20.12(A) shows the measured and simulated S-parameters for a power amplifier. The single-ended output 1 dBCP of 10.5 dBm at 61 GHz and saturation power of 14.9 dBm were measured. A photograph of our 60 GHz transmitter chipset with the power amplifier integrated in the RF circuits is shown in Figure 20.12(B).

After down-conversion from the 60 GHz-band into the 5 GHz-band with a 60 GHz LNA and mixer, these IF signals are demodulated with a 5 GHz-band I/Q demodulator. As illustrated in Figure 20.9, it consists of an IF variable gain amplifier (VGA), a wideband quadrature PLL, frequency down-conversion mixers, two baseband VGA, and low-pass filters with tunable cut-off frequency. A 5 GHz-band I/Q modulator has similar components except that no baseband VGA is employed.

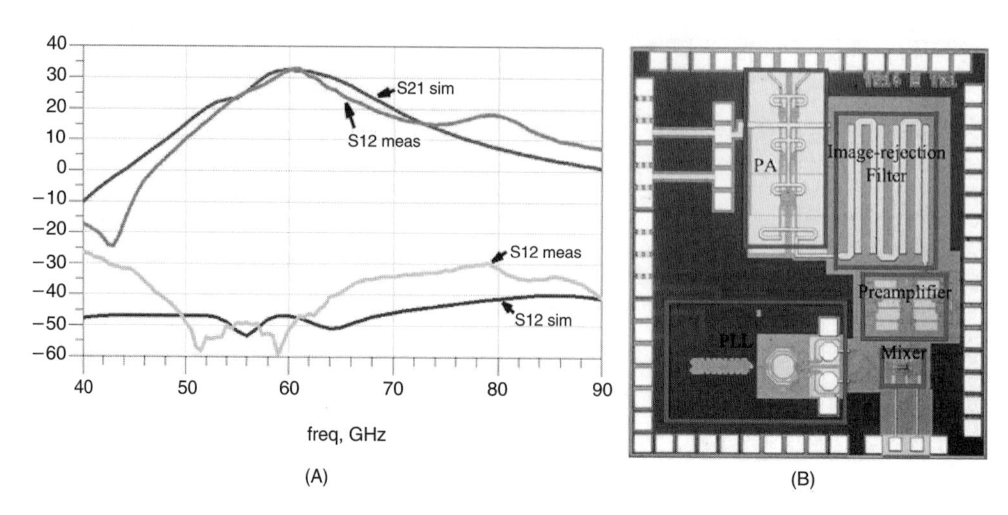

Figure 20.12 (A) Measured and simulated S-parameters for power amplifier; (B) photographs of the developed 60 GHz transmitter chipset where a power amplifier is integrated.

In this prototype demonstrator, two 500 MHz channels from 60.5 to 61.5 GHz can be processed. Therefore, the PLL in the IF stage was designed to provide 4.75 and 5.25 GHz I/Q signals. To generate them, a 10 GHz VCO in conjunction with a bipolar divide-by-two circuit is used. With the help of a large internal signal swing, improved phase-noise performance of a MOS-VCO was achieved with low power consumption of SiGe frequency dividers. This I/Q PLL consumes 57 mA at 2.5 V supply voltage.

A modulator employs an IF VGA cell which is realized with a transadmittance stage followed by a transimpedance stage with internal AC coupling. In the case of a demodulator, two cascaded VGA cells are used as an IF VGA stage. The detailed description on design methodology of I/Q modulator/demodulator can be found in [13]. We observed that the spurious level for the modulator and demodulator was as low as −73 dBc although the large PLL tuning range of approximately 1 GHz was designed and used. The phase noise at 1 MHz offset is −112 dBc/Hz for modulator and demodulator. Figure 20.13 shows the single-tone signal output power of the modulator as a function of I/Q baseband signal amplitude. The gain provided by the VGA is controlled by an external bias signal. The output 1 dBCP is −7 dBm at the maximum gain of the modulator. OFDM signal modulation was also tested, and the output spectrum and signal constellation of 64QAM OFDM signals are show in Figure 20.14.

20.2.2.3 60 GHz RF Front-Ends: Future Development

For the next generation 60 GHz RF front-ends, a sliding IF architecture is being considered because it provides simpler architecture and less power consumption than a conventional superheterodyne transceiver [14]. With a 48 GHz-band PLL incorporating a static 1 : 4 frequency divider, a 12 GHz-band IF is generated and used for IF-conversion [15]. This eliminates the use of an additional PLL for IF generation, resulting in simpler architecture. Moreover, the PLL operation frequency can be reduced to 80% of RF, which is another factor we can expect to reduce power consumption and complexity. This high IF of more than 10 GHz is beneficial

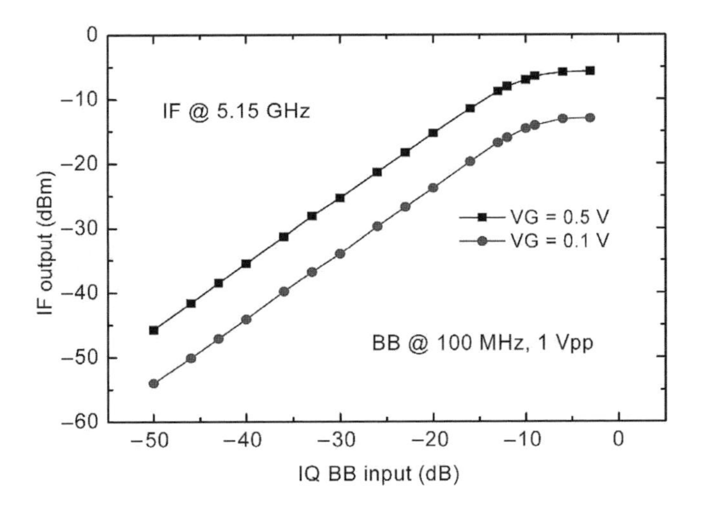

Figure 20.13 Measured IF output power of the designed variable gain amplifier as a function of input baseband signal amplitudes.

to mitigate image-rejection problems of superheterodyne transmissions. Figure 20.15 shows the schematic diagram for a sliding-IF receiver for future development.

This sliding-IF transceiver requires a new design of PLL and it is important to support the channelization plan that most of 60 GHz WPAN standards, including IEEE 802.15.3c, ECMA and Wireless HD agreed on. The channelization plan requires the generation of four different RF frequencies: 58.32, 60.48, 62.64 and 64.8 GHz, which support four independent channels over the 9 GHz bandwidth. The channel spacing is 2160 MHz which is an integer multiple of 9.6 MHz. This suggests the usage of a low-cost 19.2 MHz crystal oscillator with a 1 : 2 frequency divider as described in the Section 20.2.1.1. Such a low comparison frequency at the phase-detector results in a large noise amplification of the reference noise within the PLL

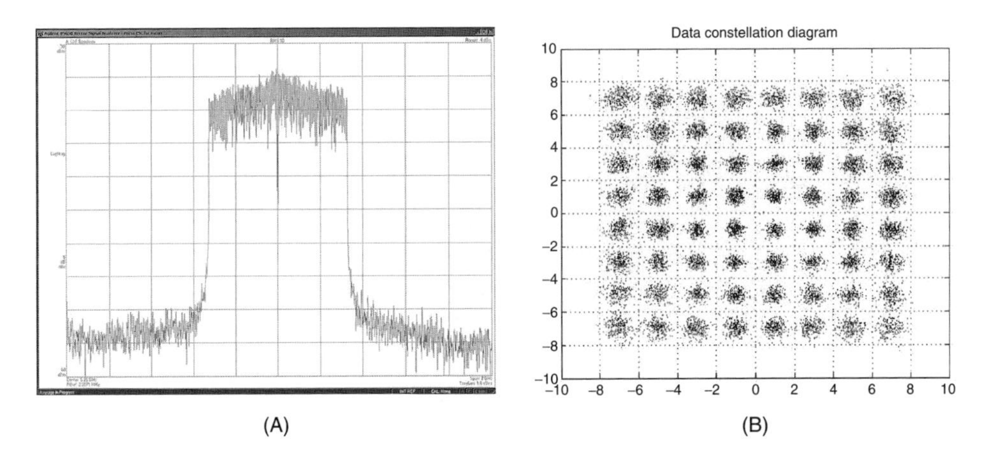

Figure 20.14 (A) RF spectrum; (B) signal constellation of OFDM signal at 5 GHz-band measured at the output of a 5 GHz I/Q modulator.

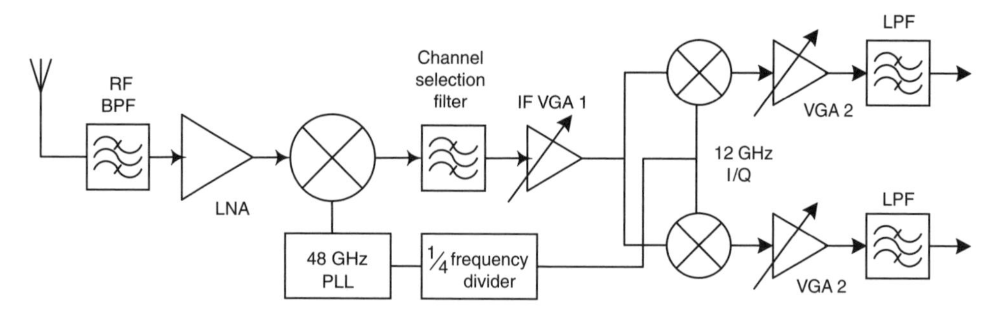

Figure 20.15 Sliding-IF architecture for next-generation 60 GHz RF analogue front-end.

bandwidth. For a 48 GHz PLL with an input frequency of 9.6 MHz, the phase-noise multiplication within the PLL bandwidth is about 20 log(48 GHz/9.6 MHz) = 74 dB. Therefore, further division of crystal oscillator reference is not desirable since the noise multiplication factor would rise by another 6 dB. The sampling frequency which is supposed to be used for OFDM transmission is 2592 MHz. Since this is also an integer multiple of 19.2 MHz, the same crystal can be used for generation of the sampling frequency for DAC and ADC [16].

A new 48 GHz PLL based on a 19.2 MHz crystal oscillator has been designed and used to down-convert 60 GHz RF signals to a 12 GHz-band. A sliding-IF of about 12 GHz which is generated from the 48 GHz PLL using a 1 : 4 frequency divider is used to down-convert this IF signal into baseband with an I/Q demodulator. For this PLL, the channel spacing at the output is reduced to 0.8×2160 MHz = 1728 MHz. Since this is an integer multiple of 9.6 MHz, the cost-effective 19.2 MHz crystal oscillator followed by a 1 : 2 frequency divider can be used to drive the PLL, as shown in Figure 20.16. The phase-frequency detector (PFD) compares the 9.6 MHz signal from the divided VCO frequency with the divided crystal frequency. The PFD output is connected to a charge pump (CP) and a low-pass filter (LPF) which is connected to the VCO tuning input. Figure 20.16 schematically shows the designed 48 GHz-band PLL architecture as suggested in Figure 20.16.

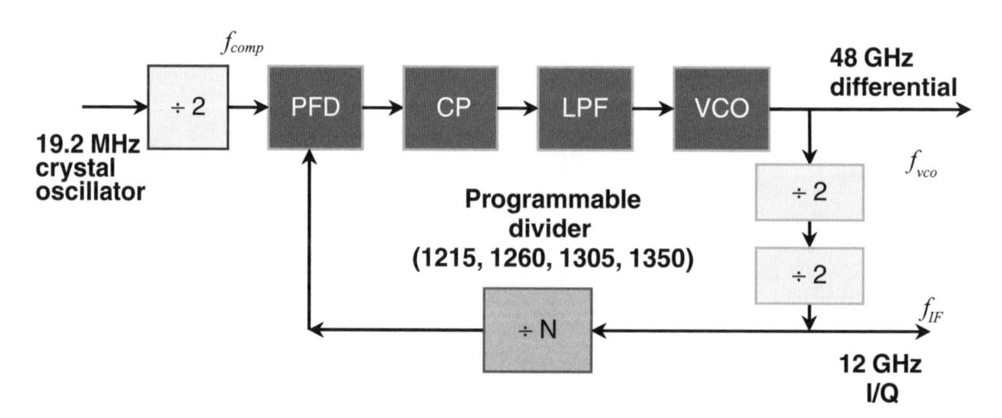

Figure 20.16 48 GHz-band PLL architecture providing the compatibility to IEEE 802.15.3c channelization plan.

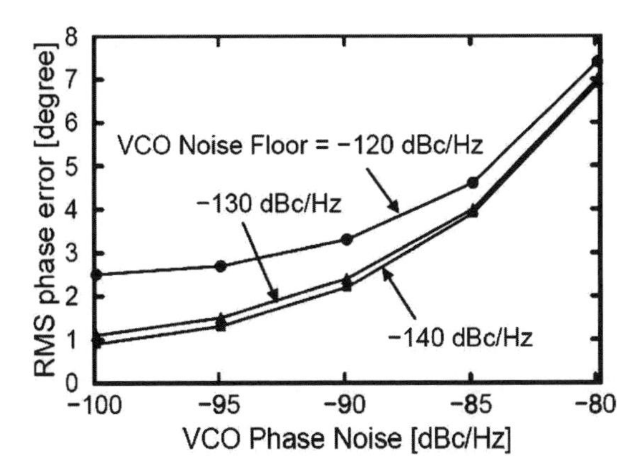

Figure 20.17 RMS phase error as a function of VCO phase-noise at 1 MHz offset for different VCO noise floors.

The phase noise of this PLL mainly depends on three parameters: the noise floor of the 9.6 MHz reference, the VCO phase-noise at 1 MHz offset (characterizing the −20 dB/decade range of the phase-noise spectrum) and the noise floor of the VCO. We have varied the two VCO noise parameters while setting the noise floor of the reference to −153 dBc/Hz. This value is based on the crystal reference noise floor of available 19.2 MHz crystals, taking into account the division of the crystal frequency by two. Then, the rms phase-error was calculated, adjusting the bandwidth by optimizing the charge pump current for a low-phase error. Figure 20.17 shows the calculated rms phase-error as a function of VCO phase-noise at 1 MHz offset for different VCO noise floors. For our parameter range, rms phase-error mainly depends on the VCO phase-noise in the −20 dB/decade range, provided that the VCO noise floor is below −130 dBc/Hz. From the result, VCO phase-noise below −90 dBc/Hz at 1 MHz offset is highly desirable to obtain an rms phase error around 3° which is acceptable for 16QAM OFDM transmission. This remains one of the most difficult tasks if advanced CMOS technology is used. The low supply voltage of CMOS makes it difficult to achieve acceptable phase-noise levels since it prevents a large internal voltage swing in the VCO.

As described in Section 20.2.1.4, it is difficult to optimize the channel selection filter in the sliding-IF configuration. Nevertheless, it is indispensable for indoor WLAN applications because we cannot avoid adjacent channel interferences. This also facilitates coexistence with legacy WPAN devices which use the above-mentioned channelization plan. One option for a channel selection filter for sliding- IF is a tunable bandpass filter. However, integration of a small-size and high-quality factor bandpass filter operating at 12 GHz is quite challenging. Therefore, for simplicity, a fixed channel selection filter with wider bandwidth than one channel is first employed although it cannot perfectly eliminate adjacent channel signals. This problem can be alleviated by rejecting the residual adjacent channel with a baseband low-pass filter. For the new PLL, IF varies from 11.664 to 12.96 GHz. According to the proposals in IEEE 802.15.3c, the 3 dB bandwidth of signals is 1.728 GHz. Therefore, the required 3 dB bandwidth for a fixed channel selection filter becomes 3.024 GHz (1.728 + 1.296 GHz) as shown in Figure 20.18. It is observed that maximum interference in the pass-band of the channel selection filter does not exceed 1/3 of the bandwidth even in the worst case where the first channel is used with the adjacent channel in the second channel.

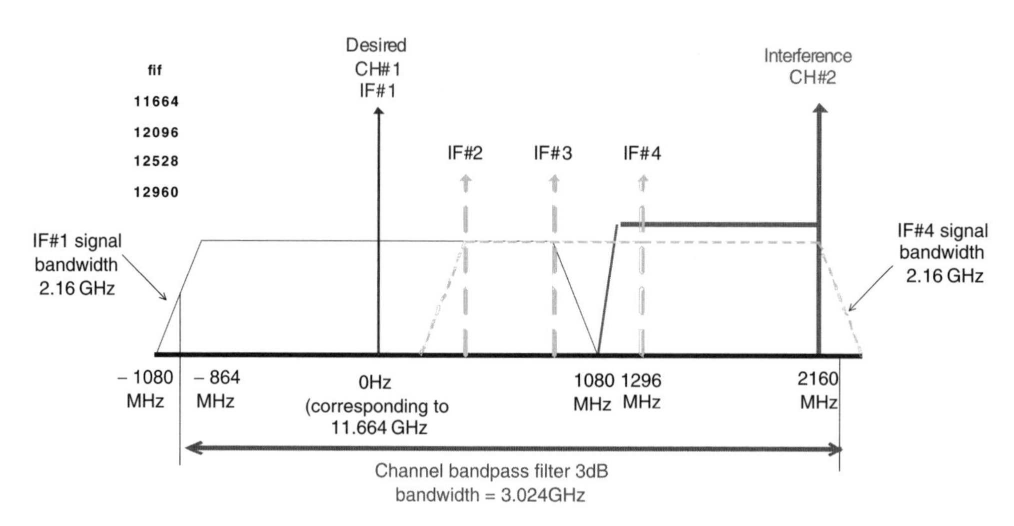

Figure 20.18 Illustration for required 3 dB bandwidth of IF channel selection filter.

20.3 OFDM Baseband Processor Architecture and Simulation Results

This chapter describes the PHY layer of the IHP demonstrator developed within the WIGWAM project (http://www.wigwam-project.de/). It also deals with synchronization and link performance published in [17, 18].

20.3.1 Concepts of the Physical Layer

The defined OFDM system for the IHP demonstrator can be viewed as an up-scaled version of the 802.11a standard. In this WLAN standard, OFDM is used with convolutional codes and a bit interleaver with a block length equal to the amount of coded bits in an OFDM symbol. Modulation ranges from BPSK to 64-QAM, with and without puncturing for proper rate matching to link quality. Details can be found in [19].

The defined PHY layer for the demonstrator differs in several respects from the old standard. The basic OFDM PHY parameters had to be adapted to the higher bandwidth and different channel characteristics. They are given on the left-hand side of Table 20.2. The selected

Table 20.2 Parameters of OFDM-PHY for narrowband and wideband system

FFT bandwidth	400 MHz	Pilot subcarriers	16	FFT bandwidth	800 MHz	Pilot subcarriers	30
FFT size	256	Zero sub-carriers	5	FFT size	512	Zero sub-carriers	5
Subcarrier spacing	1 5625 MHz	Symbol duration	800 ns	Subcarrier spacing	1 5625 MHz	Symbol duration	800 ns
Data sub-carriers	192	Cyclic prefix	160 ns	Data sub-carriers	384	Cyclic prefix	160 ns

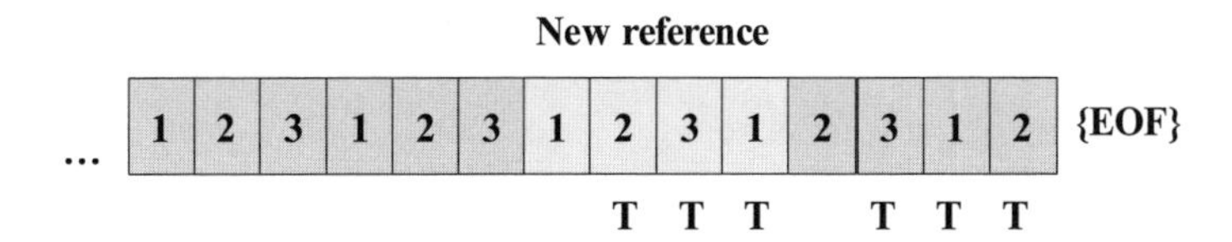

Figure 20.19 Multiplexing of coding streams and dual mode operation.

parameters, mainly the choice of the FFT size and guard length, can be seen as a trade-off between phase noise sensitivity, calling for higher subcarrier spacing, and tolerated channel delay spread, calling for longer symbol durations and therefore, lower subcarrier spacing. With the specified guard time, the usual indoor channel conditions with RMS delay spreads below 25 ns are well covered. It had been observed that the performance loss due to phase noise of the IHP VCO is around 1 dB for 16-QAM modulation with a code rate of 1/2.

The PHY layer had to be defined to support data rates up to 1 GBit/s and on the same time be realizable on slow FPGA devices. The standard convolutional codes were retained, but depending on data rate, the data stream is split and encoded into several streams to allow parallel processing on slow processors. The different coded streams are transmitted in different OFDM symbols (Figure 20.19). In addition, a new scheme to facilitate channel tracking at low hardware complexity has been applied. Dual mode operation has been introduced, where after each block of N regular OFDM data symbols transmitted with the nominal frame modulation, four data symbols are inserted, which are restricted to BPSK or QPSK. These symbols are used to re-estimate the channel in a decision-feedback manner. The restriction to BPSK or QPSK ensures fixed power in the constellation, leading to the same estimation accuracy as for the preamble, and the hardware for the initial preamble-based channel estimation can be reused. One further advantage is that the spacing (the N regular symbols) between such 'reference' data blocks can be adapted to the coherence time of the channel.

20.3.2 Coded Modulation Performance

The performance of the specified coded OFDM scheme in some typical indoor environment has been investigated in (Figure 20.20) under the premise of perfect synchronization

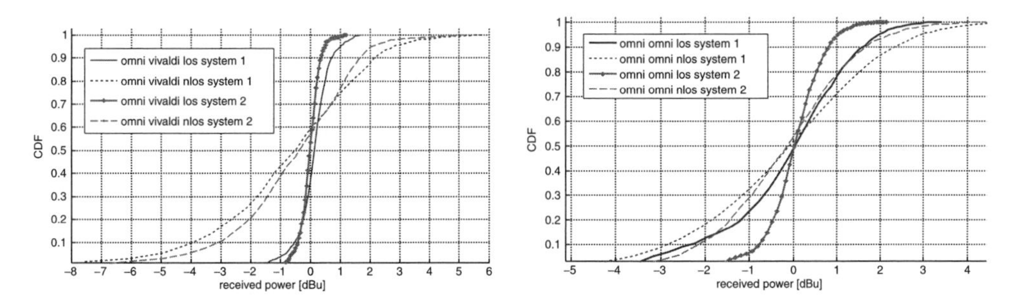

Figure 20.20 CDF of the channel models.

without considering dual mode transmission. Code performance depends on the channel model and chosen interleaving pattern. The channel model is based on measurements made by the Heinrich Hertz Institute, Berlin for a room of size $8.70 \times 5.70 \times 3\,\text{m}$, using our specific antennas. A large set of measurements was created and has been taken as the basis for simulation. The measurements were done for line-of-sight and non-line-of-sight conditions for two antenna constellations: in the first case, both the transmitter and receiver are equipped with an omnidirectional antenna, and in the second case the receiver uses a directional antenna in form of a Vivaldi antenna with a gain of $G = 12\,\text{dBi}$ and a 3 db-beamwidth of 30°, the type of antenna used for the demonstrator. The Vivaldi antenna was positioned to point towards the transmitter. The measurements were performed at different room positions for the receiver with antenna separations ranging from 2.5 to 6 m. The data of each subset of channel responses belonging to the same location was normalized for a mean expected channel gain of $G = 0\,\text{dB}$, so that the fluctuation of received power due to fast fading is preserved.

The cumulative distribution functions (CDF) for the received signal level were derived for our OFDM system comprising a bandwidth of 333 MHz and for an alternative system, where the transmission bandwidth B is roughly doubled, $B = 655$ MHz, see Table 20.2. The gain distributions shown in Figure 20.20 indicate the fluctuation of received power seen by each of the systems. For the omni-omni-link, considerable frequency diversity is seen on average. With increased distance, the dominance of the LOS-path ceases leading to lower coherence bandwidth. This effect is much lower for the Vivaldi-antenna if it always points towards the transmitter. The effect of the low beamwidth of the Vivaldi antenna is flattening of the channel, because reflections are likely to arrive from the side and are highly attenuated.

The aim of the simulations was to investigate the coded modulation performance of the narrowband and wideband OFDM system and in this way to find out if the smaller bandwidth delivers sufficient frequency diversity. Only a line-of-sight link was considered. Perfect channel estimation and synchronization was established and no pilot carriers were deployed. The energy contained in the guard interval was not counted. The frame length was arbitrarily set to 8192 bits. Simulation results are shown in Figure 20.21.

To establish a reasonable code performance, interleaver and puncturer have to be optimized for the channel. We used a similar permutation rule as for 802.11a, depending on just one parameter, the subcarrier step size Δ for adjacent code bits. This step size was set to $N1 = 12$ subcarriers for the narrowband system and $N2 = 24$ subcarriers for the wideband system.

Both systems show similar performance in case of the receiver Vivaldi antenna, but for the LOS-link based on omnidirectional antennas, the wideband OFDM outperforms the narrowband system as should be expected due to increased diversity. The higher performance is clearly seen for unpunctured transmission modes. In case of more frequency selectivity for the omni-omni-link, punctured transmission modes show only good performance for a rate of 2/3 and bad performance for a rate of $^3/_4$. Moreover, QAM-64 with a rate of $r = {}^1/_2$ outperforms QAM-16/R = 3/4 and QAM-16/R = 1/2 outperforms BPSK-3/4 and QPSK-3/4.

20.3.3 Synchronization Algorithm

We now consider the task of synchronization and channel estimation. In a conventional OFDM system, a preamble is used to estimate carrier frequency offset, find a good frame timing position and estimate channel coefficients, which constitute the channel transfer function at the used subcarrier locations in the frequency domain.

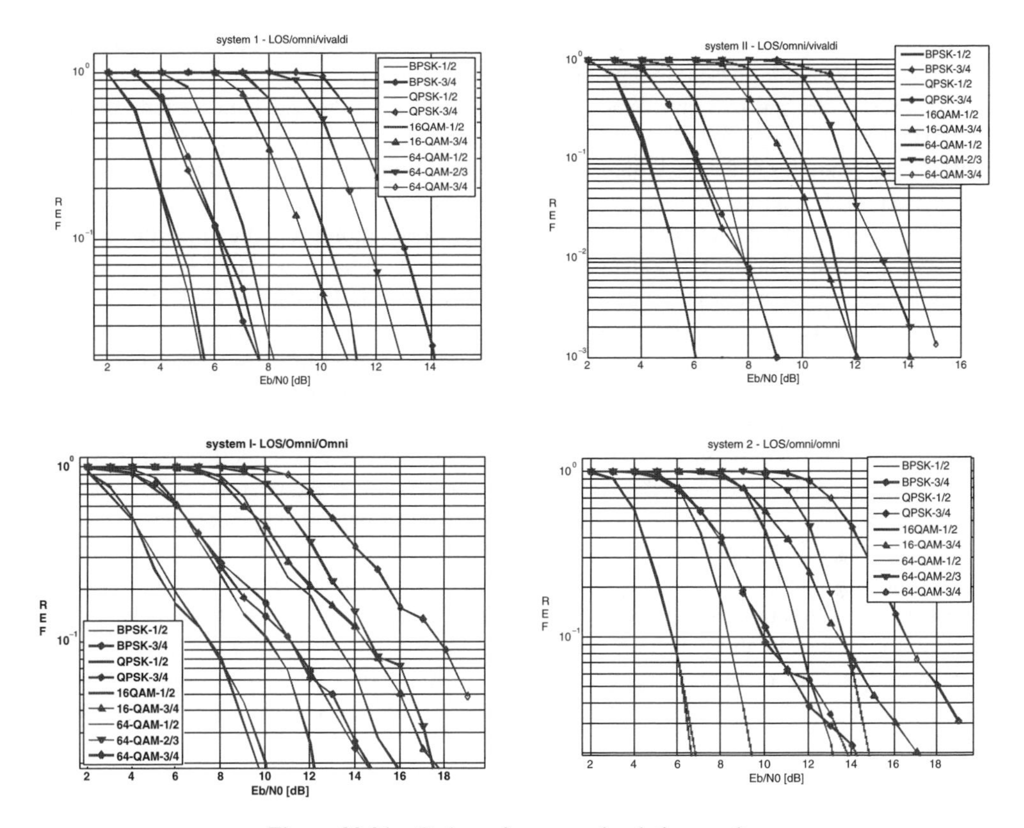

Figure 20.21 Code performance simulation results.

For the demonstrator, a new synchronization scheme has been developed and published in [18]. The preamble structure is shown in Figure 20.22. The preamble of 11 symbols is much longer compared with 802.11a. It enables reliable synchronization even at high phase noise and very low SNR. The justification for such a long preamble is that synchronization and channel estimation can be improved without sacrificing efficiency if long frames (beyond 100 OFDM symbols) can be reliably transmitted. The preamble structure differs from the old standard in that the first part is composed of a noninverted and an inverted A-sequence. The sign flip is used for coarse frame detection and makes it possible to avoid hardware costly cross

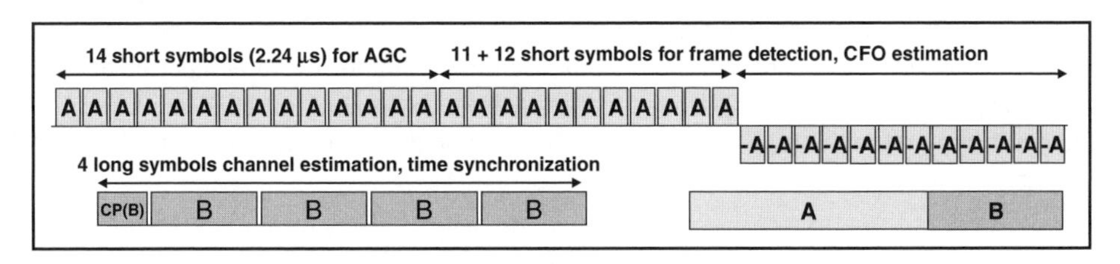

Figure 20.22 Defined preamble structure.

correlation. More precise frame timing estimation is accomplished using the second preamble part. This is done with phase unwrapping after FFT operation.

In contrast to the first part the system is further described for the wideband system specified on the right-hand side of Table 20.2. Sequence A has the length of the cyclic prefix used for data symbols (128 samples at the basic sampling rate of the OFDM system) and B the length of the FFT period. Both A and B sequences are created by the inverse Fourier transform of specified QPSK patterns. This type of sequence has good autocorrelation properties regardless of the choice of symbols. Through an exhaustive search algorithm, the two sets of QPSK-symbols were selected so as to obtain a low peak-to-average-power ratio (PAPR), namely $\mathrm{PAPR_A} = 4$ dB for the short and $\mathrm{PAPR_B} = 5$ dB for the long sequence.

The synchronizer may tolerate carrier frequency deviations up to two subcarrier spacings. Since phase noise causes a fluctuation in the instantaneous carrier frequency, the maximum combined absolute frequency deviation $\Delta f_{\max}$ of transmitter and receiver VCO must be sufficiently low to leave a margin for the phase noise. Phase noise was modelled with a Wiener process. The degradation of the subcarrier SNR due to phase noise has been analysed in various papers. In addition, the CFO estimation performance *itself* is degraded by phase noise, leading to a higher residual frequency error. Details are given in [18].

Two autocorrelators are used in parallel, a short one for frame detection and a long one for frequency correction. Both operate on the first preamble. In order to achieve a high level of robustness in the detection of the frame, a special normalized autocorrelation is used, which is easy to implement. The normalized max-norm based finite autocorrelation function on complex waveform $z(n)$ for a distance D and an average window L_{w} can be defined as

$$ACF_z(D, L_{\mathrm{w}}, n) = \frac{C_z(D, L_{\mathrm{w}}, n)}{\max(E_z(L_{\mathrm{w}}, n - D), E_z(L_{\mathrm{w}}, n))}$$

$$E_z(n, L_{\mathrm{w}}) = \sum_{k=0}^{L_w - 1} |z(n-k)|^2$$

$$C_z(D, L_{\mathrm{w}}, n) = \sum_{k=0}^{L_w - 1} \bar{z}(n - k - D) z(n - k)$$

where $\max(a, b)$ is the mutual maximum function for two numbers a, b and $\bar{z}(n)$ denotes the complex conjugate of $z(n)$. It is easy to show that $|ACF_z(D, L_{\mathrm{w}}, n)| \leq 1$ always holds and equality only applies, if the first sequence has the same shape as the second: $z(n - k - D) = \alpha \cdot z(n - k)$ for $0 \leq k < L_{\mathrm{w}}$.

The received signal $\tilde{z}_{RX}(n)$ of the first preamble just after channel convolution retains its periodicity of $N_{\mathrm{s}} = 128$ samples for a certain period of at least 11 short A$'$-sequences and another 11 inverted A$'$-sequences if the AGC has settled within the presumed period and if the channel response $h(n)$ is shorter than the cyclic prefix (the duration of an A-sequence). This means only the first inverted A-sequence will be corrupted. Short ACF $\Psi_{\mathrm{s}}(n)$ and long ACF $\Psi_{\mathrm{L}}(n)$ are defined as

$$\Psi_{\mathrm{S}}(n) = ACF_{z_{RX}}(D = 4N_{\mathrm{s}}, L_{\mathrm{w}} = 3N_{\mathrm{s}}, n)$$
$$\Psi_{\mathrm{L}}(n) = ACF_{z_{RX}}(D = N_{\mathrm{s}}, L_{\mathrm{w}} = 8N_{\mathrm{s}}, n)$$

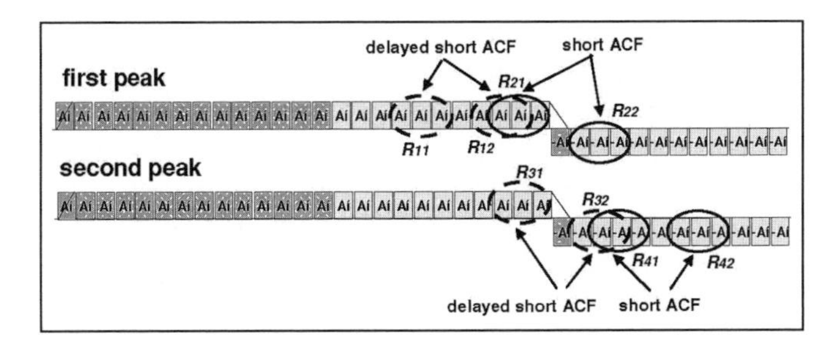

Figure 20.23 Operation of short autocorrelator.

The short ACF is used to detect the frame exploiting the sign-flip of the A′-sequences. For explanation of the detection mechanism, white noise and phase noise are neglected for the moment so that only the carrier frequency offset rotation remains after channel convolution. Then, for n_{12}, n_{22}, n_{32}, n_{42} defined as the last samples within the Regions R_{12}, R_{22}, R_{32}, R_{42} (see Figure 20.23), the following approximations apply:

$$\begin{aligned}
&z_{RX}(n) \approx z_{RX}(n-4N_s)\cdot \exp(j\Delta\Omega_0 \cdot 4N_s) && \text{for} \quad n \in R_{12} \\
&z_{RX}(n) \approx -z_{RX}(n-4N_s)\cdot \exp(j\Delta\Omega_0 \cdot 4N_s) && \text{for} \quad n \in R_{22} \\
&z_{RX}(n) \approx -z_{RX}(n-4N_s)\cdot \exp(j\Delta\Omega_0 \cdot 4N_s) && \text{for} \quad n \in R_{32} \\
&z_{RX}(n) \approx z_{RX}(n-4N_s)\cdot \exp(j\Delta\Omega_0 \cdot 4N_s) && \text{for} \quad n \in R_{42} \\
&\Rightarrow \Psi_S(n_{22}) \approx -\Psi_S(n_{12}) = -\Psi_S(n_{22}-5N_s)
\end{aligned} \tag{20.2}$$

$$\begin{aligned}
&\Rightarrow \Psi_S(n_{12}) \approx \exp(j\Delta\Omega_0 \cdot 4N_s) \\
&\Rightarrow \Psi_S(n_{22}) \approx -\exp(j\Delta\Omega_0 \cdot 4N_s) \\
&\Rightarrow \Psi_S(n_{32}) \approx -\exp(j\Delta\Omega_0 \cdot 4N_s) \\
&\Rightarrow \Psi_S(n_{42}) \approx \exp(j\Delta\Omega_0 \cdot 4N_s) \\
&\Rightarrow \Psi_S(n_{42}) \approx -\Psi_S(n_{32}) = -\Psi_S(n_{42}-5N_s)
\end{aligned} \tag{20.3}$$

Due to the noise contributions, the ACF signal will not have exactly the opposite phase with respect to the delayed signal as stated above and will also be decreased in amplitude. The synchronizer works in that way so as to identify all time instants, where the short ACF signal satisfies *to some extent* the antiphase condition $\Psi_S(n) \approx -\Psi_S(n-5N_s)$. The short ACF signal is related to the delayed signal as follows:

$$\tilde{\Psi}_s(n) := -\Psi_s(n)\cdot \exp(-j\cdot \angle\{\Psi_s(n-5N_s)\})$$

Here $\angle z$ denotes the angle of the complex number z in radians. The samples, which satisfy the following three conditions, belong to a set M:

$$|\Psi_s(n-5N_s)| > \alpha_1 \tag{20.4}$$

$$\mathrm{Re}\{\tilde{\Psi}_s(n)\} > \alpha_2 \tag{20.5}$$

$$\Rightarrow n \in M$$

$$\alpha_3 \cdot \mathrm{Re}\{\tilde{\Psi}_s(n)\} > \left|\mathrm{Im}\{\tilde{\Psi}_s(n)\}\right| \tag{20.6}$$

α_1, α_2 and α_3 are appropriate positive threshold parameters. Condition (20.4) ensures a high amplitude for the delayed ACF and conditions (20.5) and (20.6) ensure that the rotated ACF $\tilde{\Psi}_s(n)$ has a peak in the direction of the axis $Z = 1$ in the complex plane for $\alpha_3 \ll 1$. The next step consists in the separation of the acquired indices $n \in M$ into different cluster sets M_i. If these indices are grouped in ascending order $m_k \in M, m_1 < m_2 < m_3 < \ldots$, then the assignment is done so that adjacent indices, which belong to the same cluster M_i will differ in no more than d_1 positions: $(m_k \in M_i) \wedge (m_{k+1} \in M_i) \Leftrightarrow m_{k+1} - m_k \leq d_1$. Under normal circumstances, two index clusters will arise covering a region around n_{22} and n_{42}, where the antiphase condition is met. The width of the clusters depends on the chosen threshold parameters and the length of the channel response. The parameter d_1 is set to a few samples. We define a peak point p_i of cluster M_i as the truncated mean of the lowest and highest index within M_i:

$$p_i = \left\lfloor (\min_{n \in M_i}\{n\} + \max_{n \in M_i}\{n\})/2 \right\rfloor$$

Two peaks must be found within the frame and the distance between them must be within some tolerance range. A frame is detected, if the following condition holds: $d_2 - \Delta \leq p_{i+1} - p_i \leq d_2 + \Delta$, where p_i is taken as a first time reference. The long autocorrelation, which is used for carrier frequency estimation, is then evaluated in a save region at $p_i - d_3$ before the sign flip and at $p_i + d_4$ after the sign flip where the A′-sequence doesn't change for the full length of the ACF. The estimated CFO in radians per sample is taken from the averaged long ACF at the defined positions:

$$\Psi_{L_i} = [\Psi_L(p_i - d_3) + \Psi_L(p_i + d_4)]/2 \quad \Delta\hat{\Omega}_{0,i} = \frac{\angle\psi_{L_i}}{N_s}$$

$$\text{The frame is finally accepted, if } \left|\psi_{L_i}\right| > \alpha_4. \tag{20.7}$$

After the frame has been detected satisfying conditions (20.4)–(20.7), and the carrier frequency offset has been corrected using a numerical controlled oscillator (NCO), the four long symbols of the second preamble are evaluated in the frequency domain, applying the FFT at positions $p_i + d_5 + k \cdot N_{\text{FFT}}$, $k = 0, 1, 2, 3$ within the second preamble.

As in the 802.11a standard, a doubled cyclic prefix is used for the second preamble to tolerate a wider range for the FFT without raising intersymbol interference. The four symbols $S_{k,i}^{\text{Pr}}$ obtained for the ith subcarrier are averaged to achieve an estimation gain in the order of 6 dB and rotated by the complex conjugate of the ith reference QPSK symbol to attain an estimate of the channel coefficient for the ith subcarrier.

$$\hat{H}_i = \left(S_{0,i}^{\text{Pr}} + S_{1,i}^{\text{Pr}} + S_{2,i}^{\text{Pr}} + S_{3,i}^{\text{Pr}}\right)/4 \cdot \bar{S}_k^B/\sqrt{2}$$

A phase unwrapping method is used to estimate the beginning of the FFT window for data OFDM symbols. This method works well as long as there is a dominant line-of-sight path in the impulse response and still gives acceptable performance for non-line-of-sight conditions if the guard interval allows synchronization errors within a few samples. All active subcarriers are used to estimate the mean phase increase from one subcarrier to the next according to

$$\Delta\varphi_{\text{mean}} = \left(\sum_{i=-209\ldots-4, i=3\ldots208} [\angle(H_{i+1}\bar{H}_i)] \right) \Big/ 412$$

The actual FFT position with respect to the second preamble is then estimated as

$$\Delta n = N_{\text{FFT}} \cdot \Delta\varphi_{\text{mean}}/(2\pi) = 512 \cdot \Delta\varphi_{\text{mean}}/(2\pi)$$

For a better understanding, if the channel response were consisted of just one component at time $t = \text{mT}$, that is $h(n) = A \cdot \delta(n - m)$, then

$$H_l = A \exp(-j2\pi ml/N_{\text{FFT}}),$$

$$\Delta\phi_{\text{mean}} = \left(\sum_{i=-209\ldots-4, i=3\ldots208} [-2\pi m/N_{\text{FFT}}] \right) \Big/ 412 = -2\pi m/N_{\text{FFT}}, \quad \Rightarrow \Delta n = -m,$$

that is the exact FFT position would be obtained in this case. Finally, the start position for the data frame is chosen such as to have a constant predelay d_6 to the strongest path:

$$p_{\text{start}} = p_i + d_5 - \Delta n - d_6 + \Delta_{\text{preamble}}$$

Here Δ_{preamble} denotes the delay from the beginning of the first B-sequence to the first data symbol after the cyclic prefix.

20.3.4 Synchronizer Performance

We investigated the synchronizer performance for different scenarios summarized in Table 20.3. The specified thresholds and delay parameters were chosen as follows: $\alpha_1 = 0.42$, $\alpha_2 = 0.42$, $\alpha_3 = 1$, $\alpha_4 = 0.5$, $d_1 = 10$, $d_2 = 640$, $\Delta = 32$, $d_6 = 30$. The detection performance is shown in Figure 20.24A. At a high phase noise level of link 3 and link 6, a failure rate of 1.5% remains. For link 2 and 5, the failure rate approaches zero. The SNR is related to a noise bandwidth of 800 MHz. Figure 20.24B demonstrates the severe effect of the phase noise on CFO estimation which is independent of the type of link. For links 2 and 5, the CFO is of the order of 1% of the subcarrier spacing. Figure 20.24C shows the quality of the mean channel estimation performance per subcarrier, which was compared with an ideal system. For phase noise free transmission we observe an estimation gain of 7 dB arising from averaging over four symbols and an additional processing gain of $G = 0.9$ dB, because the used bandwidth is only 81%. On the other hand, the SNR is bounded by the phase noise for higher signal-to-noiseratios. For SNR = 20 dB, a performance degradation of 1 dB for $L = -95$ dBc/Hz at 1 MHz offset and 4 dB for $L = -90$ dBc/Hz was observed. Because of the small room size, the channel responses are of the order of 100 ns and were always covered within the specified guard time of 160 ns, even if the timing decision of the synchronizer has a spread of a few samples. The detection failure rate approaches zero for an SNR above 6 dB and for phase noise values below $L = -95$ dBc/Hz at 1 MHz offset. We also showed that the

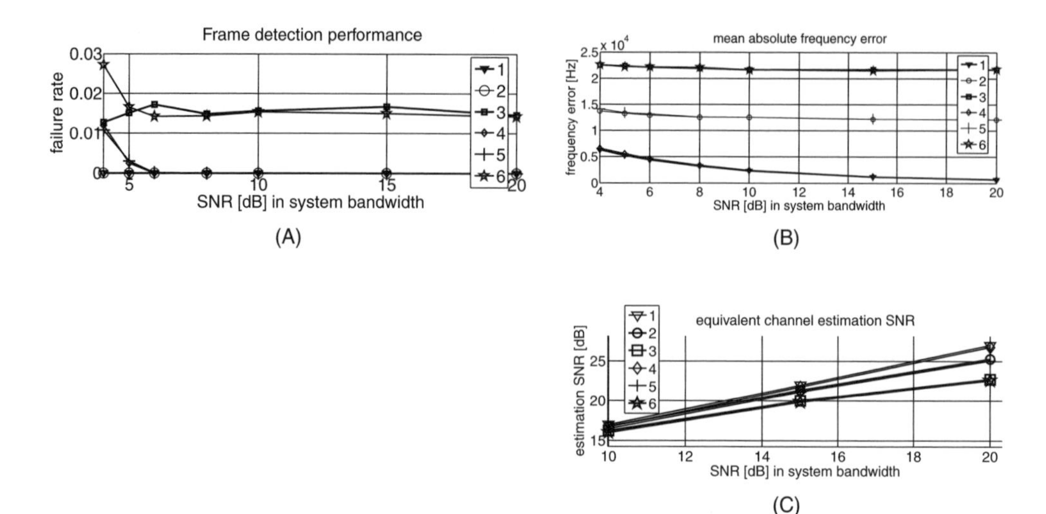

Figure 20.24 Frame detection performance (A), CFO estimation performance (B) and channel estimation SNR (C).

Table 20.3 Link scenarios

Link	Scenario
1	LOS-link, no phase noise
2	LOS-link, $L = -95$ dBc/Hz at 1 MHz
3	LOS-link, $L = -90$ dBc/Hz at 1 MHz
4	NLOS-link, no phase noise
5	NLOS-link, $L = -95$ dBc/Hz at 1 MHz
6	NLOS-link, $L = -90$ dBc/Hz at 1 MHz

phase noise of receiver and transmitter VCO level must be at least below −95 dBc/Hz to allow higher modulation modes. No false alarm occurred during simulations.

20.4 Medium Access Control (MAC) Implications

A Wireless LAN system providing data rates of several Gigabit per second poses new challenges to the Medium Access Control (MAC) protocol. Many basic MAC properties are independent of the special nature of the underlying physical layer (e.g., acknowledgement policies), but some MAC parameters (such as the transmit ↔ receive switching time) have a strong correlation to properties of the 60 GHz PHY. In this section we will discuss MAC issues for ultra- high data rate systems with special emphasis on WLAN systems in the 60 GHz band.

20.4.1 MAC Design Requirements

The most important design problem for the MAC protocol of an ultra-high rate WLAN system, independent of the special nature of the physical layer, is that the PHY and MAC protocol

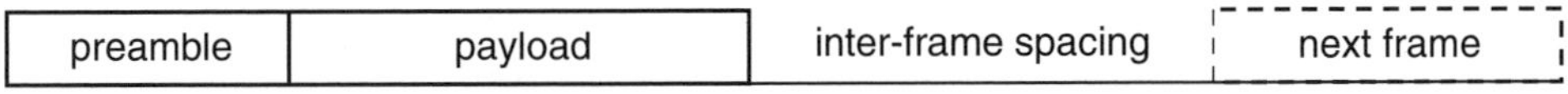

preamble	payload	inter-frame spacing	next frame

Figure 20.25 Basic structure of a typical WLAN frame.

overheads usually scale much slower than the data rate. This applies in particular to frame preambles, PHY and MAC headers, backoff durations and frame acknowledgements. For a well-designed, efficient system, the overhead is generally much smaller than the user payload. An ideal value would be around 1–5%, but up to 20–30% overhead would be tolerable. Typical IEEE 802.11 WLAN systems may achieve a MAC goodput (in Distributed Coordination Function, DCF) of about 80% of the channel capacity [20], that is an overhead of 20%.

A typical WLAN frame has the basic structure shown in Figure 20.25. Normally, the duration of the preamble and the interframe spacing is independent of the data rate and the payload size. The preamble in this sketch would also comprise something like a 'signal field' (as in 802.11a and 11g frames), which also has a fixed duration.

When considering the transmission of a single frame, the overhead consists of the preamble and the interframe spacing. Its relative amount, that is the goodput performance, will improve when the payload becomes larger (or the payload's data rate becomes smaller). A simulation of the goodput of single frames without (immediate) acknowledgement as a function of the payload size is presented in Figure 20.26. The overhead is 8.68 μs (7.68 μs preamble [incl. 1 μs signal field] + 1 μs interframe spacing), according to the standardization proposal for a 60 GHz PHY layer within the IEEE 802.15.3 standard [21].

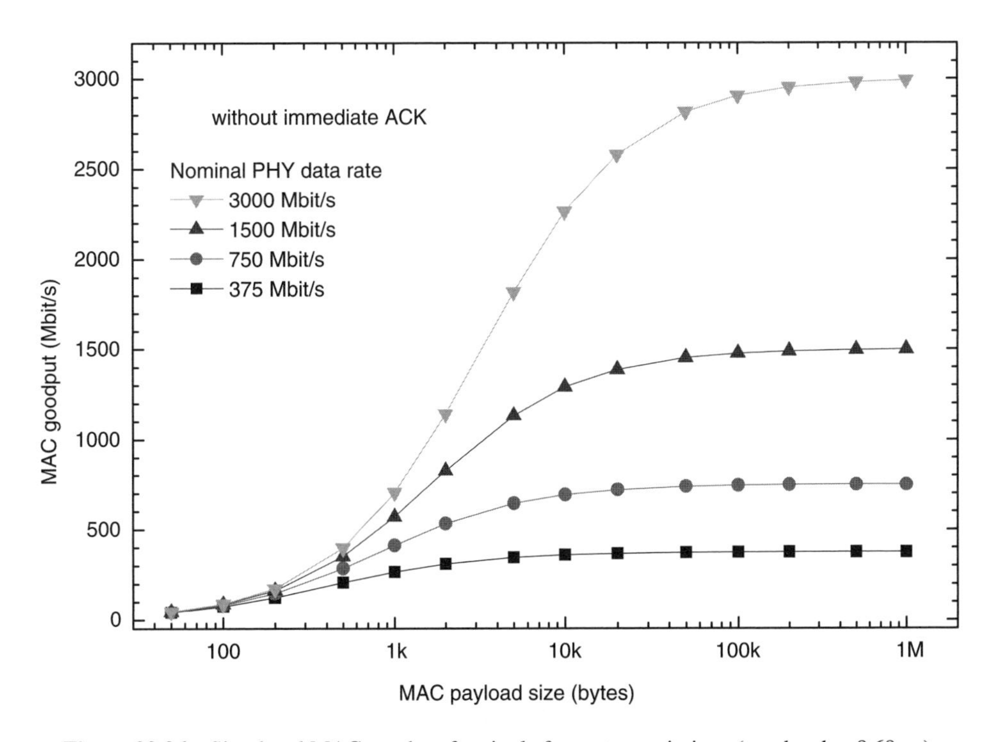

Figure 20.26 Simulated MAC goodput for single frame transmissions (overhead = 8.68 μs).

The figure shows that the MAC goodput is rather low and nearly independent of the nominal PHY data rate when the frame size is less than about 1 kbyte, which is the size of a typical IP packet. In other words, the increased nominal PHY rate has only little effect. The 80% level for the highest rate of 3 Gbit/s is reached at a MAC payload size of about 10 kbyte. When used within a traditional TCP/IP environment, one would need packet aggregation methods to achieve such payload sizes.

The overhead is further increased when the receiving station immediately acknowledges each frame. First, the acknowledgement frame itself is extremely inefficient with respect to goodput (complete frame overhead for 1 bit of information). Second, the interframe spacing is usually increased since both stations must switch between receive and transmit direction, and there must be enough time to check the received frame for transmission errors and to generate the acknowledgement. In the 60 GHz proposal mentioned previously, the respective overhead is 31.36 μs (2×7.68 μs preamble [including signal field] $+ 2 \times 8$ μs interframe spacing). A corresponding goodput simulation is shown in Figure 20.27.

The goodput for frames of 1 kbyte is below 250 Mbit/s for all data rates. The 80% level for a nominal PHY data rate of 3 Gbit/s is reached at about 50 kbyte payload size.

The backoff procedure, which is commonly used to control channel access in CSMA/CA-based protocols, induces additional overhead and further deteriorates the throughput (depending on the backoff window and the traffic statistics). Consequently, high data rate MAC protocols should avoid backoff processes (e.g., by assigning transmit opportunities to stations) and immediate acknowledgements (replacing them with delayed or group acknowledgements) as far as possible.

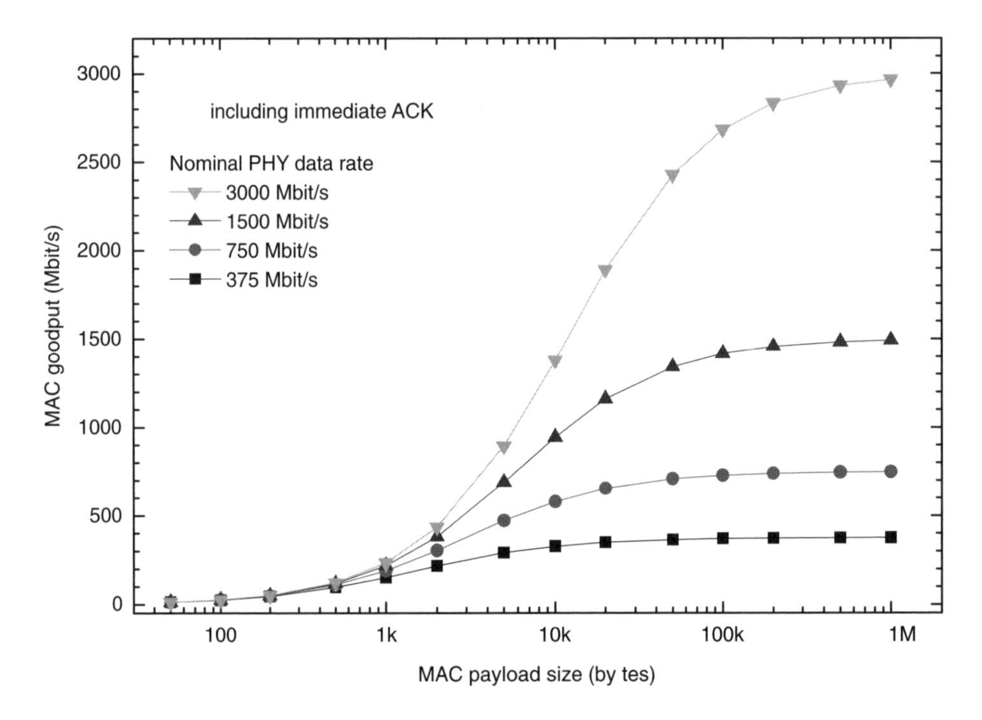

Figure 20.27 Simulated MAC goodput for frame plus acknowledgment (overhead = 32.68 μs).

20.4.2 MAC Implementation Aspects

Generally, the MAC protocol may be implemented in hardware or software or a combination of both. Pure hardware solutions suffer from a lack of flexibility with respect to upgrades and bug fixes. Moreover, modern MAC protocols are too complex to be favourably implemented in hardware.

Pure software implementations, on the other hand, require very powerful processors to meet the tight timing requirements (generation of an acknowledgement within a few microseconds) and to provide enough data throughput. A processor, which needs three clock cycles for copying a 32 bit data word (read word, write word and decrement counter), has to run at 300 MHz only in order to copy frame data at 3 Gbit/s. Such processors are available for PCs. It is, however, hard (or even impossible) to integrate the full Gbit/s MAC protocol into the PC's operating system because of the timing requirements. A separate general-purpose processor would be too power-hungry for a WLAN module. Thus, the combination of a medium-performance processor, executing the non-timing-critical MAC functions in software, with attached dedicated hardware for timing-critical operations (CRC checks, acknowledgement, timers, etc.) should be the MAC implementation platform of choice.

The MAC protocol representing the external interface of the WLAN module must be integrated into a computer system or (if it is acting as access point) connected to some high-speed network. This interface also must allow for several Gbit/s of throughput. This is not easily achievable on conventional PC interfaces. The limit for CardBus or standard PCI, for example, is 1 Gbit/s in burst mode. Thus, enhanced standards such as 64 bit PCI, high-speed PCI (66 MHz) or PCI Express are required. For access points, fibre optics would be a good solution.

Another task is due to the poor performance of a Gbit/s protocol for small packets. The size of IP packets, which are the 'backbone' of all Internet (TCP/IP) data transfer, is usually limited to about 1500 byte. From Figures 20.25 and 20.26 it is clearly visible, that these would be transferred with an extremely inefficient performance. So, we need intelligent packet aggregation at some suitable place in the network stack. Such a procedure has already been specified in the IEEE 802.11n High-Throughput standard.

20.5 Architecture and Performance of System Demonstrators

20.5.1 Architecture of 60 GHz Demonstrator

The claim of a demonstrator is to show the high-performance interconnection of all components involved in the 60 GHz system. The demonstrator includes all necessary components such as the Data Source (PC), the Medium Access Control (MAC), the Baseband Processor (BB) and the Analog Frontend (AFE). People want to see a *visible and pursuable* highspeed data transmission, for example a (large) file transfer from one PC to another or a high quality video transmission, for example HDTV. The performance of the demonstrator should also verify/confirm the robustness of the used algorithms and proposed hardware components.

For easy working with the demonstrator conventional PCs were used. The Operating System (OS) provides a complete TCP/IP stack for communication applications. File transfer and streaming capabilities were built into the higher network layers. Therefore, two PC will connect through a *wireless ethernet cable*. The data were transferred over a full duplex link.

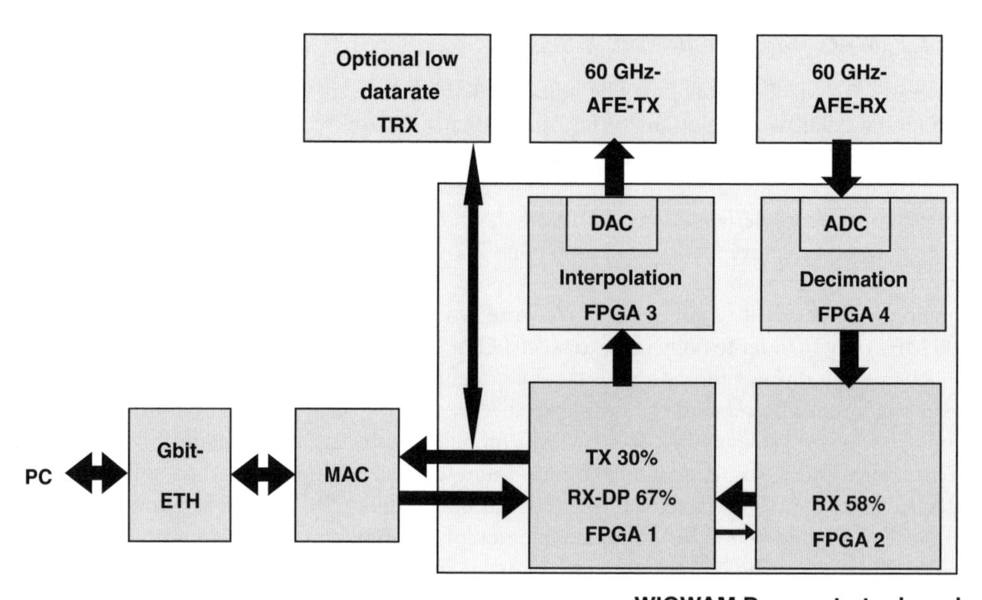

Figure 20.28 Block diagram of IHP 60 GHz demonstrator.

In this case two identical demonstrators were needed. Figure 20.28 shows a block diagram of the IHP 60 GHz demonstrator.

The main components (MAC, BB, AFE) were implemented separately on different hardware platforms. The PC sends data packets through the GBit-Ethernet cable to the other PC and doesn't know about the wireless connection.

The MAC hardware consists of a processor for packet management and a FPGA Hardware Accelerator for time critical calculations (CRC). It collects and encapsulates the Gbit-Ethernet packets with a header and CRC, and gives the packets to the BB. The MAC board and the BB boards are connected with a LVDS cable.

The BB hardware was completely implemented on four FPGAs. Two of these were assembled on a bigger main board for the implementation of the OFDM transceiver. The transmitter uses 30% of the resources of one FPGA and the receiver 125%. That is why the receiver was partitioned into these two FPGA. The other two FPGA were needed for the connection of the DAC and the ADC with the main BB board. The BB processes the Forward Error Correction (FEC) and interleaving of the data. The digital modulation scheme (OFDM) prepared the data for the AFE component. The two I/Q-signals (single ended) of the DAC and ADC are connected to the AFE with four short $50\,\Omega$ coaxial cables. The AFE component modulates the BB signal onto the carrier and amplifies it for transmission. The antennas are directly mounted on the AFE board.

At the receiver side, the AFE amplifies the antenna signal and transfers it to a BB signal. The BB processor, which includes the frame synchronization, the channel estimation, the channel equalization and the FEC-decoder, demodulates the received signal and sends the rebuilt data stream to the MAC. The MAC processes the data to the Gbit-Ethernet interface of the data sink PC.

The whole baseband transmitter implemented on FPGA consumes about 50 W. A thermal cooling solution had to be implemented to use the hardware under stable conditions. The two different voltages (3.3 and 5 V) were provided by a high-performance power supply.

In the case of the HDTV transmission, only a unidirectional link is necessary. An optional back channel can be established with a lower datarate transceiver, for example UWB up to 480 Mbps or WLAN up to 108 Mbps. This reduces the complexity of the whole transceiver. More aspects of this scenario are presented in Section 20.5.2.

20.5.1.1 Performance Measurements

In Section 20.5.1 it was mentioned that a high-speed data rate transmission is an ideal way to show the functionality of the demonstrator. It is possible to measure the data throughput of files with different lengths or to appraise the quality of video transmissions with different streaming data rates. This means that the data throughput can be measured at the PC side, but without direct results from the lower MAC and PHY layers. If the connectionless UDP is used and MAC retransmission failed, then lost packets (including the discarded erroneous packets from PHY) can be counted. TCP hides the error correction of the MAC, due to request retransmissions of the lost packets. In the worst case, the connection will close.

The quality parameter of the channel was measured with the Bit Error Rate (BER) or Packet Error Rate (PER). If bit errors happen then the FEC was not successful. To measure the BER a well-known bit sequence was transmitted and compared at the receiver side. The PER is calculate by simply counting the wrong and lost packets.

The measurement of the performance should be done in more incremental steps, so the successful coexistence of the additional layer can be checked and validated. Every step adds a layer and extends the used components. For a test of the interference of additional layers between the Ethernet links, we have to measure first only the data throughput between the PC. Normally 500 Mbps throughput should be possible in every direction. In the next step the MAC layer will be added on both sides. This inserts additional delays and latencies of the data packets in the link. During the measurement no error and data rate break should occur. If this test finishes successfully, the baseband component will be added. New delays and latencies are added and no errors should occur. The last step is the adding of the AFE. This introduces the noisy fading channel. Now packet errors and lost packets occur and influence the performance. Due to various effects the data rate may slow down or connection may be lost.

The BER and PER can be measured with a counter unit in the baseband component. The visualization of the results helps to establish the link. An additional output of the constellation diagram optimizes the setup of the demonstrator. The maximum data rate of our demonstrator can only be reached with an internal data generator at BB level.

The insert of additional components (MAC, BB, AFE) into a direct Ethernet connection may decrease the performance and influences the behaviour of the system. The MAC and the baseband introduce additional delays and latencies. Further simulations and measurements should be done to investigate the impact of the higher protocol layers, for example TCP (Slow Start at timeout). This can be critical if timeouts arise in the higher protocols and retransmissions were requested by both layers (TCP and MAC). So the system timing has to be recalibrated against the new timing behaviour of the inserted components or a connectionless protocol with an additional protocol and application layer have to be used.

20.5.2 Dual Band Demonstrator

On the basis of the demonstrator described in Sections 20.5.1 and 20.5.1.1, IHP developed, together with France Telecom, a combined UWB/60 GHz prototype. This prototype consists of a commercially available WiMedia compliant UWB development system, extended with IHPs 60 GHz radio front-end. The UWB system uses the frequency band from 3.1 to 5-GHz. The three centre frequencies of the UWB-OFDM transmission are: $f_1 = 3.432$ GHz, $f_2 = 3.96$ GHz and $f_3 = 4.488$ GHz. These centre frequencies are relatively close to the 5 GHz IF, of our 60 GHz up- and down-converters. Hence, it appeared conceivable to simply up-convert the UWB signal from the 3–5-GHz band to the 60 GHz band using our existing up-converter. Empirical experiments showed that it is possible to directly connect the 60 GHz up- and down-converters to the antenna output or input of the UWB development system. To show that the same OFDM signal can be transmitted both in the 3–5-GHz band as well as in the 60 GHz band, we decided to use the 60 GHz band for the downlink and 3–5-GHz (UWB) for the uplink. For this purpose, the frequency hopping was switched off. The architecture of our system is shown in Figure 20.29. For each of the two transceivers, a transmit/receive switch is needed. The switches are controlled with a TX/RX signal derived from the UWB development board.

After up-conversion, the transmit signal is band-filtered and directly fed to a Vivaldi antenna. The up-converter contains an integrated power amplifier which delivers about 3 dBm average output power.

The receiver reciprocates the architecture of the transmitter. Here, the signal received by a Vivaldi antenna is fed directly to the receiver chip. It contains an integrated LNA and the complete down-converter to the 5 GHz/UWB spectrum. The resulting IF signal is used as an input to the UWB development board. The architecture of this dual band demonstrator is shown in Figure 20.29.

This demonstrator allows a transparent asymmetric TCP/IP link. The maximum data rate of 480 Mbit/s was achieved within a range of up to 1 m. At the lowest data rate of 53 Mbit/s, a range of 9 m in direct line of sight conditions was possible. More details on the architecture and performance of this dual band demonstrator can be found in [22].

This demonstrator can be the basis for UWB and 60 GHz dual band systems. A strategy whereby the 60 GHz downlink is used whenever possible and the UWB system comes in when 60 GHz connectivity is not available has significant advantages. It delivers high data

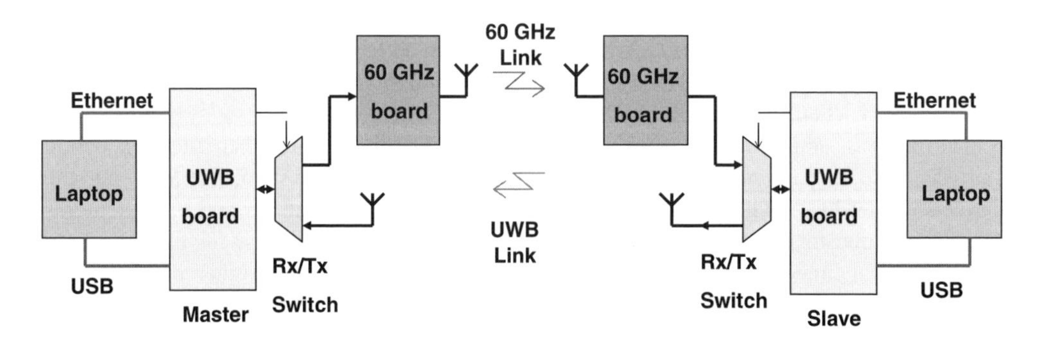

Figure 20.29 Architecture of dual band UWB/60 GHz demonstrator.

rate when possible and has a high coverage and robustness under difficult channel conditions. Valuable UWB spectrum is preserved for QoS-critical applications. A combination of 60 GHz transceivers with <6 GHz WLAN technologies such as IEEE802.11a,b,g,n is possible and has a similar effect.

20.6 Conclusions and Future Work

Currently, there is a multitude of international activities to develop communication systems for the 60 GHz band. This chapter has covered several aspects of this new technology.

Based on the characteristic of the 60 GHz wireless indoor channel, system parameters have been discussed. Details of physical layer algorithms and baseband processor parameters have been derived and investigated. The implications for the medium access control processor (MAC) have been described.

Finally, two different versions of 60 GHz OFDM demonstrators for data rates up to 1 Gbit/s have been developed and implemented. The main performance figures of our demonstrators have been presented. The analogue front-end is based on circuits implemented in a 0.25 μm SiGe BiCMOS technology. For the specification of the PHY parameters, a MATLAB model was used. After successful simulation of the PHY, the implementation was undertaken. Currently the baseband processor is implemented on an FPGA platform. This allows both, real-time operation and a hardware-in-the-loop approach for verification of the analogue front-end. We will continue our work to further optimize the performance of the demonstrator. The final version is expected to deliver a data rate in excess of 2 GBit/s.

The 60 GHz technology enables new services and user applications, which cannot be made available with present solutions. In combination with existing wireless technologies and standards many further user expectations and requirements in terms of data rate and coverage can be met.

Acknowledgements

Parts of this work were supported by the German Federal Ministry of Edurcation and Resaerch (BMBF) within the framework of the WIGWAM project (www.wigwam-project.de) under the grant reference number 01BU371 (WIGWAM-IHP).

Furthermore, we are grateful for the support of Wilhelm Keusgen and Michael Peter from the FhG Heinrich Hertz Institute for their support in the context of channel measurements and modelling.

References

1. IEEE 802.15.3c standard draft, *Amendment 2: Millimeter-Wave Based Alternative Physical Layer Extension.*
2. ECMA TC48 6th standard draft, *PHY and MAC Layers for 60 GHz Wireless Network.*
3. IEEE 802.15.3c contribution document, number: 15-07-0584-00-003c, *TG3c Channel Modeling sub-Committee Final Report.*
4. Manabe, T., Miura, Y. and Ihara, T. (1996) Effects of antenna directivity and polarization on indoor multipath propagation characteristics at 60 GHz. *IEEE J. Select. Comm.*, 441–8.
5. IEEE 802.15.3c contribution document, number: 15-06-0375-00-003c, *Propagation Model Using Circular Polarized Antenna.*

6. IEEE 802.15.3c contribution document, number: 15-06-0055-22-003c, *802.15.3c Usage Model Document (UMD), Draft.*
7. IEEE 802.15.3c contribution document, number: 15-06-0476-00-003c, *60 GHz Channel Measurements for Video Supply in Trains, Busses and Aircraft Scenario.*
8. IEEE 802.15.3c contribution document, number: 15-06-0109-00-003c, *Two-Path Channel Model for Wireless Desktop Applications.*
9. Choi, C., Shoji, Y. and Ogawa, H. (2007) Analysis of receiver space diversity gain for millimeter-wave selfheterodyne transmission techniques under two-path channel environment. IEEE Radio and Wireless Symposium, Long Beach, CA, pp. 75–78.
10. Sun, Y., Glisic, S. and Herzel, F. (2006) *A Fully Differential 60 GHz Receiver Front-End with Integrated PLL in SiGe:C BiCMOS*, in Proceedings of the 1st European Microwave Integrated Circuits Conference (EuMIC2006), Manchester, UK, pp. 198–201.
11. Winkler, W., Borngraeber, J., Heinemann B. and Herzel, F. (2005) A fully integrated BiCMOS PLL for 60 GHz wireless applications. IEEE International Solid-State Circuits Conference. ISSCC Dig. Tech Papers, **1**, 406–407.
12. Glisic, S., Sun, Y., Herzel, F., Winkler, W., Piz, M., Grass, E. and Scheytt, C. (2008) A fully integrated 60 GHz transmitter front-end with a PLL, an image rejection filter and a PA in SiGe. 34th European Solid-State Circuits Conference (ESSCIRC 2008), Dig. Tech. Paper.
13. Schmalz, K., Grass, E., Herzel, F. and Piz, M. (2007) An Integrated 5 GHz wideband quadrature modem for OFDM Gbps transmission in SiGe:C BiCMOS. *International Journal of Microwave Science and Technology*, 1–8.
14. Choi, C.-S., Grass, E., Herzel, F. *et al.* (2008) 60 GHz OFDM hardware demonstrators in SiGe BiCMOS: state-of-the-art and future development. Proc. of the IEEE 19th International Symposium on Personal, Indoor and Mobile Radio Communications (PIMRC), Cannes.
15. Herzel, F., Glisic, S. and Winkler, W. (2007) Integrated frequency synthesizer in SiGe BiCMOS technology for 60 and 24 GHz wireless applications. *Electronics Letters*, **43**, 154–6.
16. Herzel, F., Choi, C.-S. and Grass, E. (2008) Frequency synthesis for 60-GHz OFDM transceiver. Proc of the 1st European Wireless Technology Conference (EuWiT2008), Amsterdam, pp. 77–80.
17. Piz, M., Grass, E. and Peter M. (2006) A simple OFDM physical layer for short-range high-data-rate transmission at 60 GHz. 11th International OFDM-Workshop (InOWo'2006), Hamburg, Germany.
18. Piz, M. and Grass, E. (2007) A synchronization scheme for OFDM-based 60 GHz-WPANs. Proc. of the IEEE 18th International Symposium on Personal, Indoor and Mobile Radio Communications (PIMRC 2007), Athens, Greek.
19. IEEE Standard 802.11a-1999. Wireless LAN Medium Access Control (MAC) and Physical Layer (PHY) specifications: High-speed Physical Layer in the 5 GHz Band. LAN/MAN Standards Committee of the IEEE Computer Society, USA.
20. Bianchi, G. (2000) Performance analysis of the IEEE 802.11 distributed coordination function. *IEEE Journal on Selected Areas in Communications*, **18**, 535–547.
21. IEEE 802.15.3c proposal document, number: 15-07-0688-00-003c, *France Telecom – IHP Joint Physical Layer Proposal for IEEE 802.15 Task Group 3c*. IEEE 802.15 Working Group for Wireless Personal Area Networks (WPANs).
22. Grass, E., Siaud, I., Glisic, S. *et al.* (2008) Asymmetric Dual-Band UWB/60 GHz Demonstrator. Proc. of the IEEE 19th International Symposium on Personal, Indoor and Mobile Radio Communications (PIMRC 2008), Cannes.

21

Enabling Technologies for 60 GHz Communications: Front-end Friendly Air Interface Design, Full CMOS Integration and System-in-a-package

André Bourdoux, Piet Wambacq, Geert Carchon, Steven Brebels and Walter De Raedt
IMEC

21.1 System-level Strategies for the Physical Layer: Air Interface Design and Signal Processing Solutions

21.1.1 Introduction

Multi-gigabit/s communications are possible in the 60 GHz band thanks to the wide available bandwidth and the possibility of frequency reuse over small distances. This motivates the current standardization within the IEEE802.15.3c Task Group (http://www.ieee802.org/15/pub/TG3c.html). This band is also attractive because of the less aggressive multipath environment and the small wavelength, allowing low-profile antenna arrays. However, nonidealities of the 60 GHz Radio Frequency (RF) front-end (FE) have a much larger impact than at lower frequencies, because of the higher carrier frequency. A suitable air interface for low-cost, low-power 60 GHz transceivers should thus use a modulation technique that has a high level of immunity to FE nonidealities, especially phase noise (PN), analog-to-digital converter (ADC) quantization and clipping and nonlinearity of the power amplifier (PA). To alleviate the high back-off requirements of OFDM modulation and, at the same

Short-Range Wireless Communications Rolf Kraemer and Marcos D. Katz

time, allow low-complexity equalization, cyclic-prefixed block transmission is advantageous. Cyclic-prefixed transmission can be combined with modulations such as QPSK, O-QPSK, 8-PSK and even with the large class of (nonlinear) continuous phase modulation (CPM). The reason for choosing PSK or CPM modulation is that these modulations feature a low peak-to-average power ratio, thereby easing the power amplifier, digital-to-analog converter (DAC) and ADC requirement.

21.1.2 Link Budget

We will highlight the difficulty of achieving very high bit rates by considering a typical link budget at 60 GHz (Table 21.1). In order to meet a coded bit rate of 2 Gb/s without antenna gain, the range must be limited to about 1 m for a bit error rate (BER) of 10^{-5}. The same BER is achieved at about 9 m if 10 dB antenna gain is available at both the TX and RX sides. This motivates the use of beamforming (antenna gain), hence MIMO techniques are not advisable at 60 GHz. Antenna gain is fortunately easy to achieve at 60 GHz since the wavelength is only 5 mm in free space, and somewhat smaller when the relative dielectric constant is taken into account. Finally, since the position of the transmit and receive antenna is not predictable, it is preferable – though not mandatory – to resort to adaptive beamforming so that the antenna gain can be achieved independently of the relative positions and orientations

Table 21.1 Exemplary link budget.

Frequency			
Carrier frequency	60 000	60 000	MHz
Bandwidth	1750	1750	MHz
Tx			
Tx power	10.00	10.00	dBm
Tx antenna gain	0.00	10.00	dB
Channel			
Distance	0.91	8.95	m
LOS loss	67.17	87.04	dB
Oxygen attenuation	0.01	0.14	dB
Rx antenna gain	0.00	10.00	dB
Thermal noise	−81.57	−81.57	dBm
Rx			
Noise figure	6.00	6.00	dB
Other RX losses	−5.00	−5.00	dB
Coding rate	0.75	0.75	
Coding gain	1.25	1.25	dB
Processing gain	0.00	0.00	dB
SNR and Margins			
SNR QPSK (BER = 1e − 5)	12.64	12.64	dB
Bit rates			
Overhead	0.80	0.80	%
Rw spectral efficiency (QPSK)	2.00	2.00	bits/Hz
Net bit rate (QPSK)	2100	2100	bits/s

of the antennas. This, of course, puts additional challenges on the system design. We must also mention that beamforming is usually implemented by means of phase shifting and combining the signals from different antenna elements in an antenna array. If only beamforming gain is targeted, a resolution of 3 bits (i.e. 45°) is sufficient for the individual phase shifters. For spatial interference rejection, however, a higher phase resolution might be needed (typically 6 bits) in order to provide sufficient null depth.

21.1.3 Candidate Air Interfaces

The channel characteristics at 60 GHz have not been as extensively studied as at lower frequencies. However, several measurement campaigns have shown that the 60 GHz channel can exhibit multipath propagation in some cases. It is therefore necessary to design an air interface that is suitable for both line-of-sight (LOS) and non line-of-sight (NLOS) conditions. Single-carrier modulations are well suited for LOS conditions, but require a time-domain equalizer that can be costly in terms of implementation. On the other hand, orthogonal frequency division multiplexing (OFDM) is an elegant and efficient method for NLOS conditions but has an unnecessary implementation overhead in LOS conditions. A good compromise is therefore to resort to *cyclic-prefixed single-carrier transmission* [1], whereby the receiver has the choice of conventional single-tap time-domain equalization in LOS conditions or single-tap frequency-domain equalization in NLOS conditions. This results in the lowest complexity and power consumption for all channel conditions.

21.1.3.1 Linear Modulation

In the following, we introduce a transmitter and receiver model that is applicable to conventional linear modulation schemes such as M-ary PSK and M-ary QAM, combined with cyclic prefix transmission [2]. In addition, we introduce a different pulse shaping filter in the I and Q branches of the transmitter, so that Offset-QPSK is also supported by our model. O-QPSK is a variant of QPSK digital modulation with a relative delay of half a symbol period between the I and Q branches.

The transmitter and receiver models are shown in Figures 21.1 and 21.2 respectively. In the I branch, the symbol stream is converted to blocks of length B. After that, the functional block T_{CP} inserts a cyclic prefix of length N_{cp}. Finally, the resulting blocks are transmitted through the pulse shaping filter g_{I}. The Q branch performs a similar operation, except that the filter g_{Q} has an additional delay of half a symbol (only for O-QPSK).

Let us define $h_{\mathrm{I}}(t) := g_{\mathrm{I}}^{\mathrm{T}}(t) \times c(t) \times g_{\mathrm{R}}(t)$ and $h_{\mathrm{Q}}(t) := j \times g_{\mathrm{Q}}^{\mathrm{T}}(t) \times c(t) \times g_{\mathrm{R}}(t)$ as the overall channel impulse response encountered by the I and Q data symbols respectively ($c(t)$

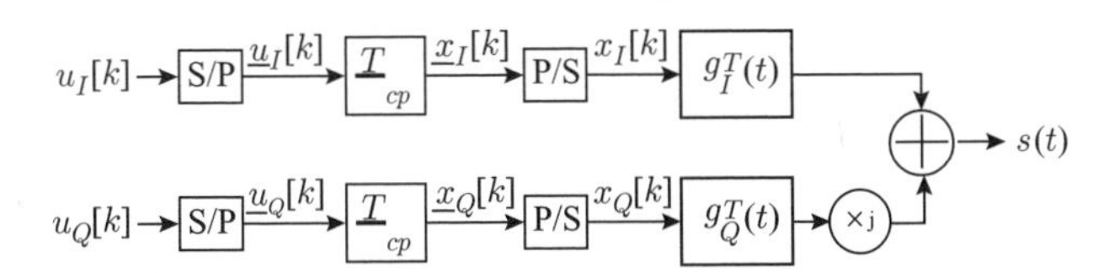

Figure 21.1 Transmitter block diagram for linear modulation with cyclic prefix.

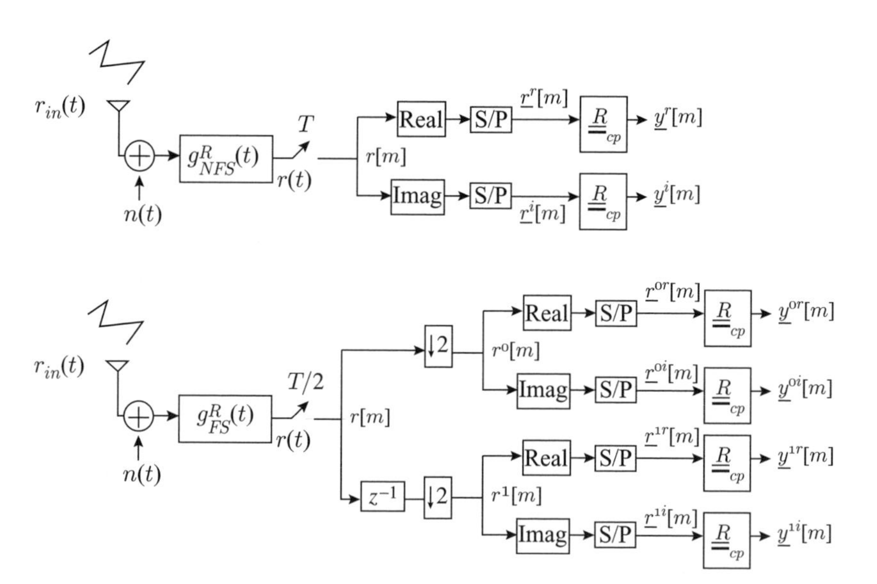

Figure 21.2 Receiver block diagram for linear modulation with cyclic prefix (top: integer sampling; bottom: fractional sampling).

is the channel impulse response and $g_R(t)$ is the receiver filter impulse response). Then, the sampled received signal $r[m]$ is given by:

$$r[m] = \sum_k x_I[k]\, h_I[m-k] + \sum_k x_Q[k]\, h_Q[m-k] + v[m] \tag{21.1}$$

in which $v[m]$ is the discrete-time low-pass filtered noise, $h_I[m]$ and $h_Q[m]$ are the sampled versions of $h_I(t)$ and $h_Q(t)$ respectively. Finally, after separation of the real and imaginary parts and removal of the cyclic prefix, we get the received blocks $\mathbf{y}^r[m]$ and $\mathbf{y}^i[m]$, free from Inter-Block Interference (IBI) when the CP is sufficiently long:

$$\underbrace{\begin{bmatrix} \mathbf{y}^r[m] \\ \mathbf{y}^i[m] \end{bmatrix}}_{\mathbf{y}[m]} = \underbrace{\begin{bmatrix} \dot{\mathbf{H}}^r_I & \dot{\mathbf{H}}^r_Q \\ \dot{\mathbf{H}}^i_I & \dot{\mathbf{H}}^i_Q \end{bmatrix}}_{\dot{\mathbf{H}}} \underbrace{\begin{bmatrix} \mathbf{u}_I[m] \\ \mathbf{u}_Q[m] \end{bmatrix}}_{\mathbf{u}[m]} + \underbrace{\begin{bmatrix} \mathbf{w}^r[m] \\ \mathbf{w}^i[m] \end{bmatrix}}_{\mathbf{w}[m]} \tag{21.2}$$

where $\dot{\mathbf{H}}^r_I$, $\dot{\mathbf{H}}^i_I$, $\dot{\mathbf{H}}^r_Q$ and $\dot{\mathbf{H}}^i_Q$ are circulant matrices and $\dot{\mathbf{H}}$ is the compound channel matrix.

At this point, even as consecutive blocks are IBI free, ISI is still present within each individual block. This ISI is mitigated using a frequency domain equalizer (FDE).

According to [3], the expression of a linear Minimum Mean Square Error (MMSE) detector that multiplies the received signal $y[m]$ to provide an estimation of the transmitted symbols is given by:

$$\mathbf{Z}_{MMSE} = \left[\frac{\sigma_w^2}{\sigma_u^2} \mathbf{I}_2 + \dot{\mathbf{H}}^H \dot{\mathbf{H}} \right]^{-1} \dot{\mathbf{H}}^H \tag{21.3}$$

where σ_u^2 and σ_w^2 represent the variances of the real and imaginary parts of transmitted symbols and of the AWGN respectively. However, the computation of this expression is very complex due to the size of $\dot{\mathbf{H}}$. Fortunately, by exploiting the circulant structure of submatrices $\dot{\mathbf{H}}_I^c$ and $\dot{\mathbf{H}}_Q^c$ composing $\dot{\mathbf{H}}$, the latter can be transformed in a matrix $\mathbf{\Lambda}$ of diagonal submatrices. This is done by using discrete block Fourier Transform operators of size $\mathrm{B} \times \mathrm{B}$.

Afterwards, a careful permutation between the columns and lines of $\mathbf{\Lambda}$ is performed such that the latter matrix is transformed in a block diagonal matrix $\mathbf{\Psi}$, in which the lth block contains the lth subcarrier frequency responses $\lambda_{X,l}^c$ (where c is either r or i and X is either I or Q) of the different channels. Thus, each virtual subcarrier is equalized individually and independently from the others. Note that we use the wording 'virtual subcarrier' since the modulation does not transmit symbols on individual subcarriers as in conventional OFDM. Finally, the expression of the joint FDE-MMSE detector becomes:

$$\mathbf{Z}_{MMSE} = \mathbf{F}_2^H \mathbf{P}^H \left[\frac{\sigma_w^2}{\sigma_u^2} \mathbf{I}_2 + \mathbf{\Psi}^H \mathbf{\Psi} \right]^{-1} \mathbf{\Psi}^H \mathbf{P} \mathbf{F}_2 \tag{21.4}$$

where $\mathbf{F}_2 = \mathbf{F} \otimes \mathbf{I}_2$ is the block Fourier Transform matrix and $\mathbf{P}$ is a permutation matrix. Both are of size $2\mathrm{B} \times 2\mathrm{B}$.

If fractional sampling is used in the receiver, the MMSE equalizer is derived in a similar manner, but uses the polyphase approach to take the fractional sampling into account as in [4]. Note that the sub-blocks of $\mathbf{\Psi}$ are of size 2×2 in the integer sampling case and of size 4×2 in the fractional sampling case. Hence, the MMSE equalizer matrix $\mathbf{Z}_{MMSE}$ has dimension $2\mathrm{B} \times 2\mathrm{B}$ for the integer sampling case and dimension $4\mathrm{B} \times 2\mathrm{B}$ in the fractional sampling case. Thanks to the permutation matrix P, the matrix inverse in 4 is greatly simplified and reduces to B inversions of 2×2 or pseudoinverse of 4×2 matrices.

21.1.3.2 Continuous Phase Modulation

CPM covers a large class of (nonlinear) modulation schemes with a constant amplitude, defined by [5]:

$$s(t, \mathbf{a}) = \sqrt{\frac{2E_S}{T} e^{j\varphi(t,\mathbf{a})}} \tag{21.5}$$

where $s(t, \mathbf{a})$ is the sent complex baseband signal, E_S the energy per symbol, T the symbol duration and $\mathbf{a}$ is a vector containing the sequence of M-ary data symbols $a[n] = \pm 1, \pm 3, \ldots \pm (M-1)$. The transmitted information is contained in the phase:

$$\varphi(t, \mathbf{a}) = 2\pi h \sum_{n=0}^{N-1} a[n] \cdot q(t - nT) \tag{21.6}$$

where h is the modulation index and

$$q(t) = \int_{-\infty}^{t} g(\tau)\, d\tau \tag{21.7}$$

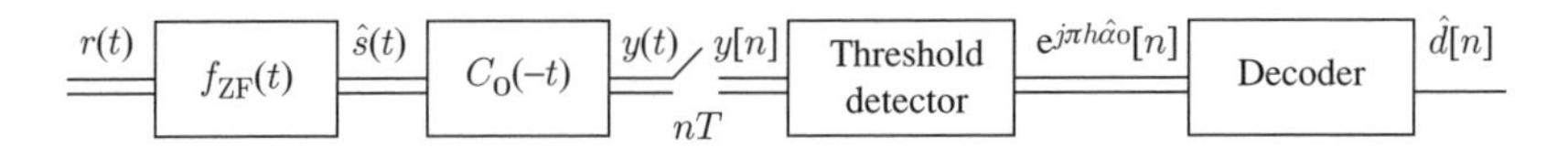

Figure 21.3 Low-complexity CPM receiver based on Laurent decomposition.

It is important to realize that a CPM scheme is defined by four parameters:

- M (number of levels)
- h (modulation index)
- L (pulse filter length, which can be longer than the symbol duration, leading to partial response CPM)
- the pulse filter shape.

Hence, there are many CPM flavours, which allows several trade-offs between spectral efficiency and receiver complexity. For example, Gaussian Minimum Shift Keying (G-MSK) is characterized by a truncated Gaussian pulse filter of length $L = 3$, a modulation index $h = 0.5$ and bi-level modulation ($M = 2$). Furthermore, to simplify the receiver (which, optimally, uses a Viterbi sequence estimator), an approximation due to Laurent [6] is often used, whereby the CPM signal is exactly represented as a linear combination of 2^{L-1} amplitude-modulated pulses. The approximation consists in using only the first term of this linear combination, which contains most of the energy. Based on this, the GMSK receiver consists of a zero-forcing (ZF) linear equalizer, a matched filter $C_0(-t)$ matched to the first Laurent pulse and a simple decoder (Figure 21.3).

Cyclic-prefixed transmission can also be combined with CPM. However, whereas the cyclicity is ensured by copying the last part of a block to the beginning of the block, the phase continuity is lost in this process. Therefore, a few (L-1) extra symbols must be reserved to ensure that the phase continuity is preserved when block transmission is being used. A suboptimal receiver for M-level CPM will consist in a frequency-domain equalizer (to compensate for the channel) followed by a conventional receiver based on the Laurent approximation.

21.1.3.3 Performance of Candidate Air Interfaces

Simulation Set-up: The parameters for the CP-QPSK and GMSK case are shown in Table 21.2.

The channel model is based on the Saleh-Valenzuela (SV) model [7]. The parameters for the SV channel model are shown in Table 21.3.

Table 21.2 Parameters for CP-QPSK (left) and CPM (right).

Filter bandwidth	BW = 1 GHz	Symbol duration	$T = 1$ ns
Sample period	$T = 1$ ns	Pulse shape	Gaussian
Number of bits per symbol	2	Pulse duration	$3 \cdot T$
Number of symbols per block	256	Modulation index	$h = 1/2$
Cyclic prefix length	64	Number of symbol levels	$M = 2$
Roll off transmit filter	0.2	Channel coding	Uncoded

Table 21.3 Parameters for the Saleh-Valenzuela model. Γ is the cluster decay factor and γ is the ray decay factor. The clusters and the rays form Poisson arrival processes that have different, but fixed, rates Λ and λ respectively.

$1/\Lambda$	75 ns
Γ	20 ns
$1/\lambda$	5 ns
γ	9 ns

Finally, the nonidealities were modeled as follows:

- The phase noise was generated based on a PSD cutoff frequency of 1 MHz and a noise floor of −130 dBc/Hz. The integrated phase noise had a value of −16, −20 and −24 dBc respectively, corresponding to phase noise value of −82, −85 and −90 dBc at 1 MHz.
- The ADC had a resolution of 4–6 bits.

Performance in Multipath: First, we show the results of both modulation schemes for the multipath case but without nonidealities (Figure 21.4). For the CP-QPSK case (left), we can see that the integer sampling MMSE scheme (=NFS-MMSE in the figure) is close in performance to the more complex FS-MMSE. This NFS-MMSE shall be used for CP-QPSK in the next results. For CPM, we can observe that the receiver based on the zero-forcing equalizer and the linear approximation is close to the combination of a ZF equalizer and a Viterbi decoder. Hence, this simpler receiver will be used for CPM in the following results.

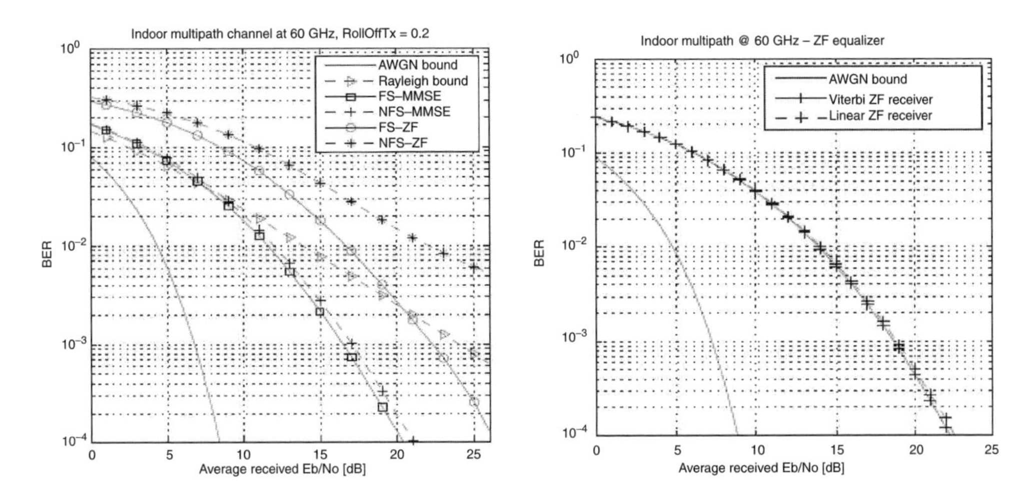

Figure 21.4 Performance of CP-OQPSK (left) and CPM (right) in 60 GHz multipath channel.

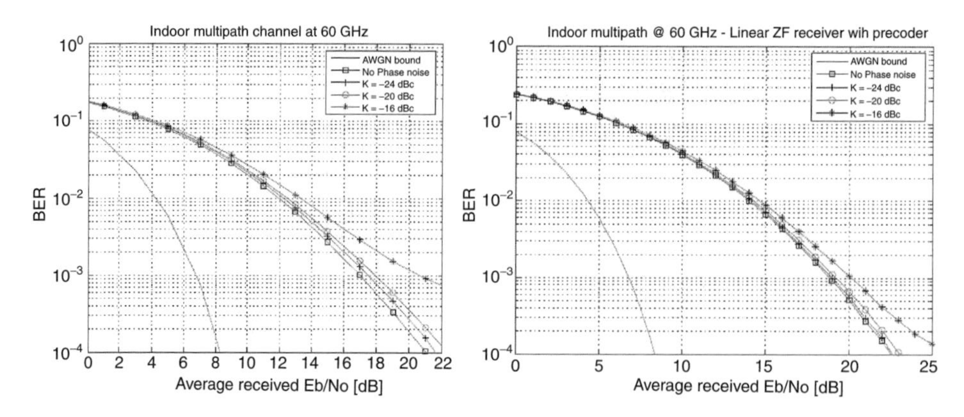

Figure 21.5 Performance of CP-OQPSK (left) and CPM (right) in 60 GHz multipath channel with phase noise added.

Performance with Phase Noise: The performance of CP-QPSK and CPM in a 60 GHz multipath channel with phase noise is shown in Figure 21.5. Both schemes exhibit approximately similar sensitivities to phase noise, CPM being slightly more robust. In any case, both schemes accept a fairly large amount of phase noise, which is mostly due to the fact that the symbol rate is quite large compared to the cutoff frequency of the phase noise PSD. The slightly higher sensitivity of CP-QPSK is attributable to the block transmission mechanism involving blocks of much longer duration than the individual QPSK symbols (in other words, the block rate is more commensurate with the phase noise cutoff frequency).

Performance with ADC Clipping and Quantization: The performance of CP-OQPSK and CPM in a 60 GHz multipath channel with finite ADC resolution is shown in Figure 21.6. Here again, we observe similar performance degradations for both schemes, with a slightly better robustness for CPM. This is mostly due to the PAPR of QPSK, which is slightly higher than

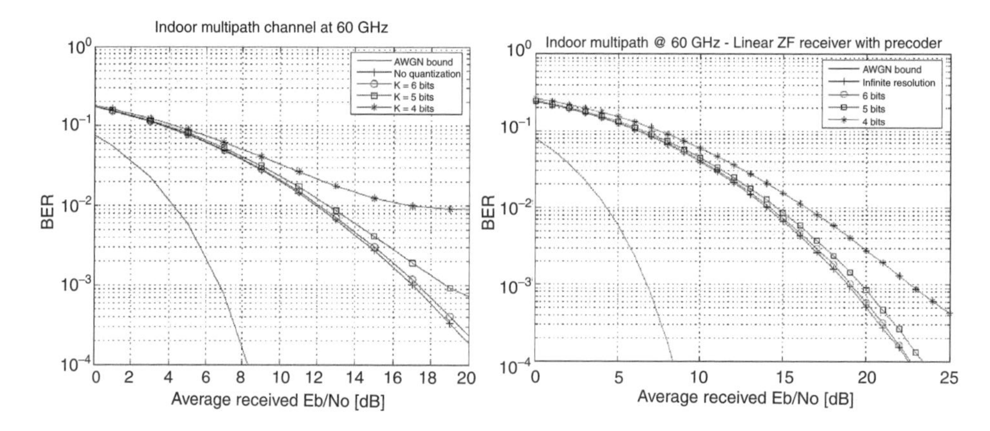

Figure 21.6 Performance of CP-OQPSK (left) and CPM (right) in 60 GHz multipath channel with low ADC resolution.

one, whereas the PAPR of CPM is exactly one. Hence, although multipath propagation induces PAPR degradation, the overall PAPR of CPM over a multipath channel is slightly better than the one of QPSK. It should be noted that these results were obtained with the assumption of a perfect AGC that automatically adjusts the receiver gain so that the signal fits in the best possible manner within the ADC dynamic range.

21.2 Microelectronics Implementation of the Transceiver

Microelectronics can be used to implement both the analog and the digital electronics of mass-market wireless communication systems. The goal is to obtain with microelectronics a solution that meets the specifications at low cost, with a small form factor and with low power consumption.

21.2.1 Available Microelectronics Technologies

Different microelectronics technologies are in use nowadays: CMOS, BiCMOS, which are both based on silicon wafers, and III–V-based technologies.

Among the above technologies, CMOS is the cheapest one. CMOS technology is subject to scaling: according to the law of Moore, CMOS technology is shrinking at a speed that allows a doubling of the density of digital functions every 18 months. To obey this law, silicon technology development is put on a roadmap, the so-called International Technology Roadmap for Semiconductors (ITRS) [4]. The decrease in the size of a transistor also yields a higher intrinsic speed in the transistor. This speed is expressed as the transit frequency f_T, which is the frequency above which the transistor does not have any current gain anymore. Figure 21.7 shows the increasing trend of the transit frequency with CMOS downscaling.

Of course, downscaling does not come for free. A technology becomes increasingly more complex, and hence more expensive, as it is downscaled while meeting the ITRS specifications. The increase of speed and density with scaling is not only more expensive, it also yields some deficiencies in electrical performance. With downscaling, transistors in the off state start to leak significantly. This is especially a problem for digital electronics, and it is not considered further here. Next, as sizes in a transistor structure come closer to the size of atoms, variations at atomic scale during processing, which are difficult to control, yield increasing variations on the transistor's performance. As a result, there is a large uncertainty on this performance. This variability will increase with further downscaling. Another consequence of scaling is the decrease of the maximum power supply voltage. Whereas for digital circuits this leads to a decrease in the power consumption, a low-power supply poses difficulties for analog circuit design. Indeed, in analog circuits most of the MOS transistors need to operate in saturation, which requires a minimum DC bias voltage over the transistor. Stacking different transistors in saturation between the power supply rail and the ground, while maintaining sufficient headroom for signal swings, is an increasing problem in analog circuit design.

The BiCMOS technology is an extension of CMOS that offers a fast bipolar transistor in addition to the MOS transistors. BiCMOS comes at the expense of a few more masks, leading to a cost increase of a small percentage. Typically, the maximum cutoff frequency of the bipolar transistor, which has also increased over the years through technological advances in bipolar transistor architectures, is higher than that of the MOS transistors. Also, for a given f_T, the maximum power supply voltage that a bipolar transistor can stand is typically higher than for a MOS transistor. For these reasons, the first silicon implementations of a full mm-wave transmitter-receiver have been realized in BiCMOS [8]. One could consider a silicon bipolar technology without MOS transistors. However, this technology does not offer the possibility of making complex digital circuits, as these consume a considerable amount of power, even when they are not active.

Technologies based on III–V materials, such as GaAs and InSb, offer transistors with a higher cutoff frequency and/or a higher breakdown voltage than silicon-based transistors. Nevertheless, these technologies do not lend themselves to high-yield complex digital electronics. On the other hand, III–V components are extremely useful for low-noise and/or high-power circuits in the microwave and millimeter-wave domain.

Given the different microelectronics technologies listed above, there are two questions to solve for the implementation of the electronics of mm-wave communication systems: While the technology for the implementation of the digital electronics is certainly CMOS, will other technologies be used for the implementation of the analog electronics? In that case, multiple chips would be required. If CMOS were used for all electronics, a single-chip would be the solution. Secondly, will there be a microelectronics technology that enables us to put all electronics, from the antennas to the digital electronics, on one chip at an affordable price? These two questions are treated in the next two sections.

21.2.2 Microelectronics Technology for mm-wave Applications

To solve the question of the technology choice for future mm-wave transceivers, one should keep in mind that for economical reasons hardware should be reusable for different communication standards. Moreover, the quality of the communication link and the energy consumption of the hardware should be adaptable. These requirements lead to hardware that can be programmed. This programmability is not only required in the digital baseband platform; the radio part, which is the electronics between the antennas and the converters between analog and digital, will also have to be programmable. This is the concept of software-defined radio. Software-defined radios (SDRs) are already being developed in CMOS for wireless communication standards below 10 GHz [9]. The use of CMOS is almost inevitable here: the advanced digital control of many functional blocks in an SDR, which can have feedback loops that cross the analog-digital boundaries, requires so many interconnections between the radio blocks and the digital control that a radio IC that is separate from a digital IC would be unpractical and expensive.

In a similar way to the low-GHz wireless applications, there will be an economic drive for mm-wave applications to reuse the radio blocks for different standards. In a next step, one could try to combine the mm-wave wireless communication standards with low-GHz standards. For the same reasons as above, the preferred solution for such multi-standard radio will be a single CMOS chip.

To design a mm-wave radio in CMOS, the peak cutoff frequency of a MOS transistor should be several times higher than the operating frequency. An often quoted ratio is at least a factor four. This means that for an SDR that could operate with a standard for 60 GHz communication, a peak cutoff frequency is required of at least 300 GHz, which more or less corresponds to the 45 nm CMOS generation. From the ITRS specifications, the power supply voltage of this generation is 1 V. This value will remain the same for the CMOS generations until the end of the ITRS roadmap in 2020 (http://www.itrs.net). This means that for the radio part of a communication system, no extra design complications will arise due to the lower supply voltage. To tackle the effect of higher variability that is expected with further downscaling, the radio circuits will need more digital tuning.

The ratio of f_T versus signal or carrier frequency does not tell everything about the feasibility of designing a complete radio for a standard that uses that carrier frequency. Another important aspect is the power that a device can generate. Here, a CMOS device has a weaker performance than Si bipolar or BiCMOS, and certainly than III–V components. At the time of writing, the output power of CMOS power amplifiers at 60 GHz is limited to a few dBm with efficiency up to 25% [10]. The efficiency or the output power can be increased in two ways: first, when the maximum f_T of a transistor increases while the maximum power supply voltage remains constant (which is the scheduled ITRS scenario), the performance of a PA will improve, as the parasitic capacitances of the transistors decrease [11]; second, the quality of the passive components (capacitors, inductors, transmission lines) can be improved. Here it is not certain whether future digital CMOS will yield higher-quality components with further downscaling. The reason is that metal lines will become thinner, leading to a higher series resistance of the passive components. A higher quality factor inevitably requires dedicated thick metal layers. This is already in use in current commercial RF technologies, at the expense of a higher processing cost compared to a fully digital technology. For the evolution of the quality factor of inductors for power amplifiers at 5 GHz in an RF technology, ITRS predicts a slight increase from 14 to 18 between 2006 and 2020. This evolution will also slightly increase the quality factor at mm-wave frequencies. Figure 21.7 shows the cutoff frequency f_T as a function of the channel length (mask length) of a MOS transistor.

If power amplifier performance does not increase drastically with scaling, the radio architecture and the air interface could take into account this limited performance. The requirement on the output power can be decreased by using multiple antennas (beamforming). Indeed, the power amplifier can then be distributed among the transmit antennas and each power amplifier will only have to deliver a fraction of the total transmit power. Next, the air interface should have a low peak-to-average ratio. In this way, a power amplifier can always operate at maximum efficiency.

21.2.3 Components in a Radio Transceiver

The next question to solve is where the chip boundaries will be: which functions and components will be put on the chip? First there is the question of where to put the antennas, on-chip or off-chip. To solve this question, two boundary conditions must be taken into account. First, in order to obtain a reasonable link budget in the mm-wave range, it is very likely that beamforming will be used in many standards, requiring several antenna elements. Next, the wavelength in free space in the mm-wave range is a few millimeters; at 60 GHz this is 5 mm. On a carrier substrate, this wavelength is a factor $\sqrt{\varepsilon_r}$ smaller, ε_r being the relative dielectric permittivity. For silicon this yields a wavelength of 3 mm at 60 GHz. Since the technology of choice for the active circuits, both analog and digital, is CMOS, one should question the feasibility of antennas on silicon. Given the losses in the silicon substrate and the high area cost of future CMOS, it turns out not to be a good idea to put an antenna array on-chip. Instead, other technologies such as MCM technology can be used.

Next to the antennas are the receive part and the transmit part of the radio. Radio components that are currently not integrated in CMOS are RF bandpass filters, crystals (needed to make a frequency or timing reference), antenna switches and high-power PAs.

As already mentioned in the previous section, the use of high-power PAs can be avoided by a proper choice of the architecture and the air interface. Many bandpass filters can be avoided by a proper choice of the radio architecture. For example, a homodyne or zero-IF architecture [12] avoids IF bandpass filtering, but requires higher linearity in the RF part of the receiver and in the down-conversion mixers because the filtering is achieved at baseband. For the realization of the remaining – passive – components with a small form factor at the expense of some loss in performance, MEMS components (*m*icro*e*lectro*m*echanical *s*ystems) that are integrated on top of the classical layers of IC technology could become an option in the future. Experimental circuits with MEMS above ICs have already been demonstrated [13]. Nevertheless, MEMS components nowadays still suffer from reliability, stability and reproducibility.

It is expected that in the coming years much research will be devoted to reducing the number of analog circuits in a radio. In this way, chip area, power consumption and design time should be reduced. Although connecting the converters directly to the LNA (in the receive part) and to the PA (transmit side) for mm-wave applications is not expected to be feasible in the foreseeable future, we will see that the analog-to-digital and digital-to-analog converters will shift more in the direction of the antenna(s). Some dedicated solutions have already been published for phased-array systems at 24 GHz [14], where the only analog circuits between the LNAs, PAs on one hand, and the converters on the other hand, are a phase-locked loop and some low-frequency buffers. However, for such dedicated solutions the question arises whether they can be easily reused for other standards, according to the SDR idea.

21.3 RF Packaging and Antenna Design

The packaging of the 60 GHz RF transceiver, the interconnection of the RF transceiver and the antenna, and the antenna design and realization at 60 GHz are crucial to obtaining a low-cost, high-performance system.

The antenna interface realizes the transition from the CMOS IC to the interconnect substrate, the routing and necessary filtering on the interconnect substrate and the transition from the interconnect substrate to the antenna. This should be realized with low insertion losses and high integration density, with limited impact on the RF-IC (e.g., chip detuning) and without creating in-band parasitic resonances which would result in a nonfunctional module implementation. For the IC-to-module interface, flip-chip is one alternative; the embedding of (thinned) Si ICs into the interconnect substrate is alternative possibility which could, among others, improve performance while at the same time decreasing overall module thickness.

The antenna of a 60 GHz system consists of a multi-element antenna array in order to realize beamforming. The antenna should be realized with sufficient bandwidth and high radiation efficiency, and this at low cost and small volume. Material choice, layer thickness, antenna type, array performance, radiation pattern, radiation efficiency, polarization but also manufacturability are important parameters to be optimized or taken into account.

In the following, we will first review two architectures which may be used for the realization of the antenna array and discuss the relation between substrate properties and antenna bandwidth and efficiency for patch antennas. Then we will discuss several candidate technologies for realizing the antenna interface.

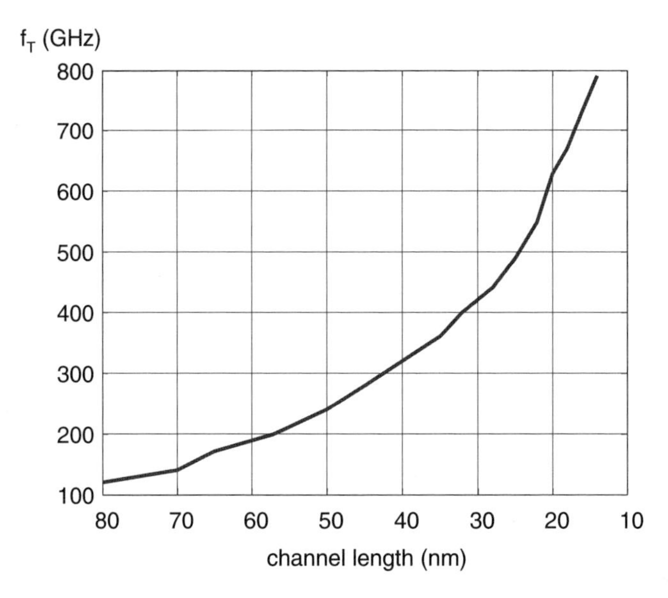

Figure 21.7 Cutoff frequency f_T as a function of the channel length (mask length) of a MOS transistor according to the ITRS roadmap. Clearly f_T increases with downscaling.

21.3.1 Antenna Array Integration

21.3.1.1 Array Architecture

In Figure 21.8 two basic ways of constructing an array are shown. In Figure 21.8(a) the array is assembled with circuits on boards that are mounted perpendicular to the array face. This assembly is called 'brick' construction [15–17]. The area subarrays of Figure 21.8(b) are

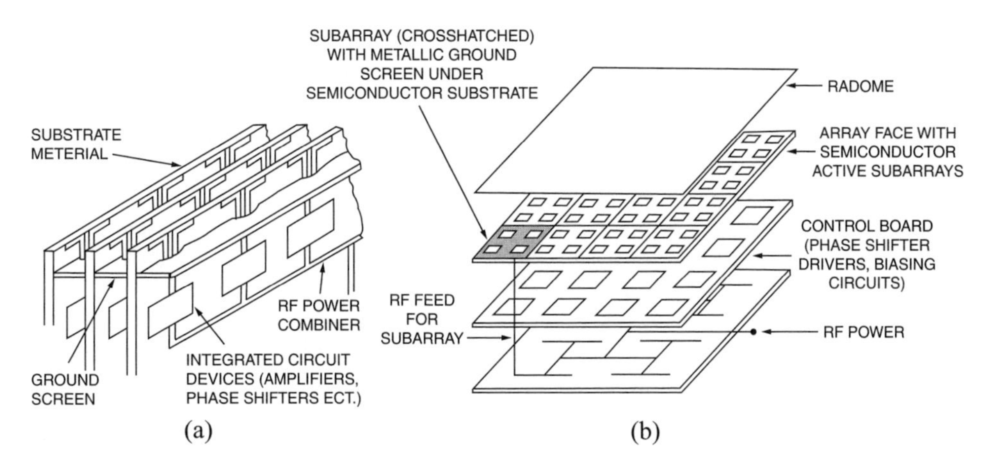

Figure 21.8 Antenna array configurations: (a) column subarrays with 'brick' construction; (b) area subarrays with 'tile' construction.

assembled into a multiple-layer array in what has been called 'tile' construction [15–17]. The signals for tile subarrays are provided by a network that is parallel to the array face, as shown in the figure. In both cases the subarrays could be monolithic active integrated circuits as shown, or else more conventional technology. The terms 'brick' and 'tile' relate to the way the array is assembled, not the organization of the antenna elements in the array. One could assemble an array of column subarrays using the tile construct if the planar power dividers addressed columns of the array, or one could assemble an area subarray by inserting the subarray as a 'brick' from behind the aperture.

'Brick' construction applies when the array can have greater depth: the increased volume allows more room for circuits or thermal management. Within each brick, the circuits can be fabricated by monolithic or hybrid integrated circuits: a 'brick' construct allows for the replacement of failing devices and can thus tolerate lower-yield processes. A major advantage of the 'brick' construct is its compatibility with dipole and flared types of elements, which have significant bandwidth compared to the flat printed elements used in the 'tile' construct. Arrays with large scanning range can be realized when electrically thin substrates are used.

'Tile' construction has several advantages, of which the primary one is that it is thin, with a relatively small volume. The thermal design should however be planned well because of the small volume of a 'tile' architecture. The cost of a 'tile' structure can be significantly lower than that of a 'brick' build-up if the production processes have a sufficiently high yield: 'tiles' can be made by wafer stacking or other layered processes (LTCC, PCB stacking). This construction has however some limitations: it requires relatively narrow-band patch and printed dipole elements, the scanning range of arrays is limited by the choice of substrate material, and finally a low-loss feeding network is more difficult to realize due to the limited volume available.

21.3.1.2 Substrate Materials for Antennas

The largest difference between a 'brick' and a 'tile' concept is the antenna element that can be integrated into both concepts. To investigate the impact of the limitations of a planar antenna element in a 'tile' topology a bit further, we investigate the performance of a square patch on different substrate materials. In Figure 21.9 the antenna efficiency (a) and bandwidth (b) are shown as functions of the thickness and dielectric constant of the antenna substrate. The design graphs are calculated using a cavity model [18] for a square microstrip antenna on a substrate with no dielectric losses (loss tangent = 0). The antenna and ground plane are made by copper layers of at least a few skin-depths thick. In Figure 21.9(a) contours of constant efficiency (in %) are given as functions of the substrate thickness and dielectric constant. The efficiency increases with decreasing dielectric constant. For a fixed dielectric constant, the substrate thickness can be chosen to reach a maximum efficiency. The efficiency of thin substrates is limited by conductive loss, while the efficiency of thick substrates is limited by surface wave loss. Figure 21.9(b) gives contours of constant bandwidth (usable band expressed as % of center frequency). A lower dielectric constant is better for a higher bandwidth. An optimum substrate thickness exists for a maximum bandwidth when using a given dielectric constant. This optimum thickness is however not the same as that required for the antenna efficiency. The choice of the substrate thickness is thus a compromise between efficiency and bandwidth.

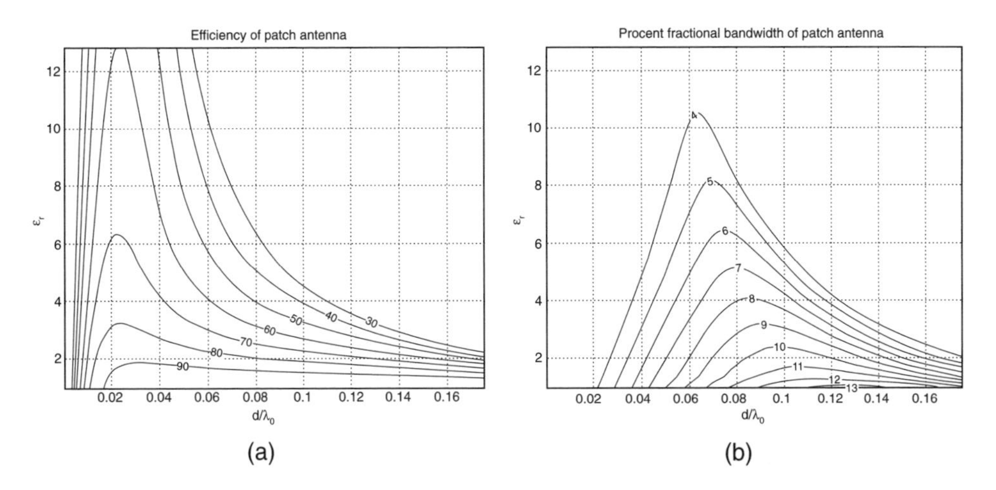

Figure 21.9 Antenna performance for square patch antenna: (a) antenna efficiency; (b) bandwidth as a function of the thickness and dielectric constant of the antenna substrate.

Figure 21.10 shows the power distribution of a planar antenna on two substrate materials. High dielectric materials such as high-resistivity silicon (Figure 21.10(a)) leak a significant amount of the power away at the surface between substrate and free space. This loss is called surface wave loss (gray line in Figure 21.10). Surface wave loss increases with substrate thickness. Planar antenna arrays with substantial surface wave loss also suffer from a reduced scan range. This phenomenon is known as scan blindness [17, 19]. The dielectric losses (black dashed line in Figure 21.10) of the substrate can become important at 60 GHz. A substrate material with a loss tangent lower than 10^{-3} should be used to avoid considerable dielectric losses.

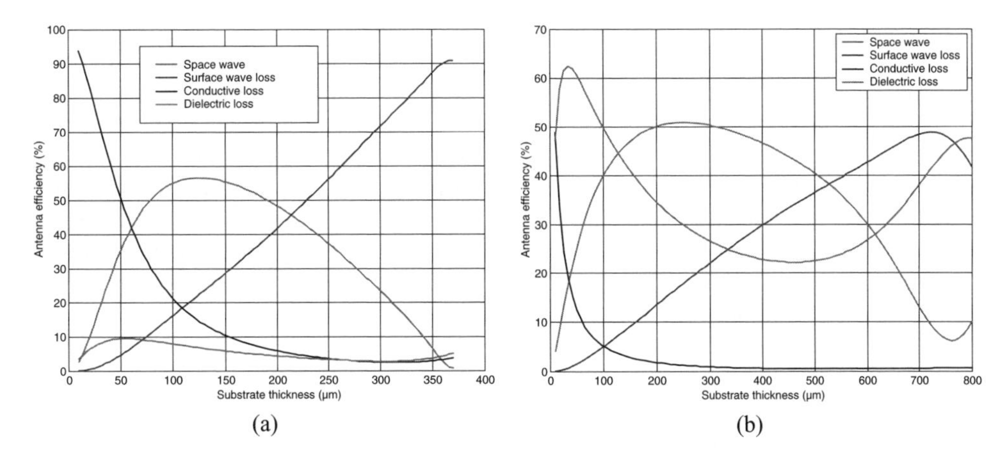

Figure 21.10 Power budget for 60 GHz patch antenna on: (a) high-resistivity silicon (epsr = 11.9, tan $\delta = 0.001$); (b) silicone (epsr = 3.1, tand = 0.026).

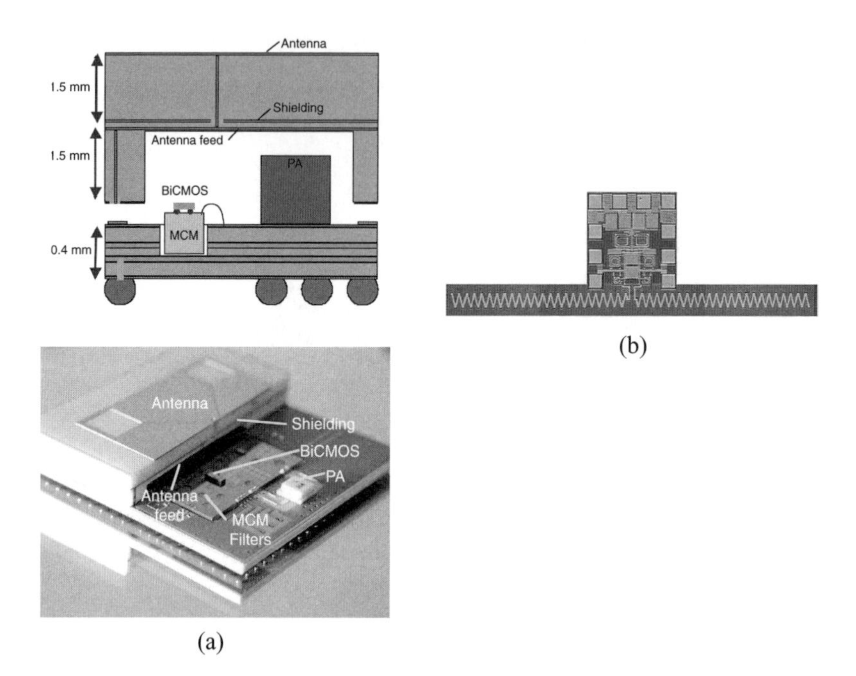

Figure 21.11 Antenna element integration: (a) Example of a SiP integrated antenna at 5.25 GHz using hybrid thin-film (MCM)-board (RO4003C) integration [22]; (b) example of SoC integration of a 15 GHz zigzag dipole antenna on 20 δ cm silicon [23].

21.3.2 SiP versus SoC Integration of Antennas

In Figure 21.11 examples are shown of a System-in-Package (SiP) integrated antenna and a System-on-Chip (SoC) integrated antenna.

The SiP methodology offers many advantages over a SoC integration of antennas:

- A SiP approach offers the best set of materials and processes for the cost-effective integration of an antenna with reasonable performance. Since the electrical size of an antenna has a direct influence on its performance, one should use an inexpensive technology with dimensions that are comparable with the required array size at 60 GHz (centimeters). Furthermore, the technology should have good high-frequency properties such as a low dielectric constant and low loss. The freedom in material choice available for the SiP designer allows them to make the best choice in terms of cost and performance for their application. For SoC however, one is forced to work with rather expensive semiconductor substrates with poor dielectric properties for integration of the antenna: semiconductor substrates have a high dielectric constant and have either high losses or a high cost. This leads to rather narrowband and/or inefficient antennas with a limited scan range. SoC antennas are best used in a 'brick' architecture since the requirements for the antenna substrate are more relaxed in this topology.
- Coupling between antennas and RF circuits should be avoided by a good design methodology. In SiP one can almost completely decouple circuits and antennas by changing the 3D package topology: the separation between the antenna and critical circuit parts can be increased, and a shielding structure can even be foreseen, with no strong effect on system performance

or package size. In SoC however, the system is essentially built on a thin (a few μm) active layer on top of a semiconductor substrate. For an antenna this can be considered as a planar structure. Because of the substrate planarity, it is almost impossible to decouple the circuits from the antenna. Substrate and surface coupling are also much more important for semiconductor substrates because of their high dielectric constant and the potential build-up of substrate charges.

Electromagnetic (EM) coupling structures can however be implemented in the back-end layers or, alternatively, in the postprocessed layers on top of the chip. This could reduce the complexity of the feeding by replacing a number of via connections used for the antenna with EM coupling.

21.3.3 Interconnection Substrates

Building mm-wave communications systems puts forward specific requirements on the interconnection technology. Among many potential candidates, three technologies are currently widely being explored: organic substrate technologies, thin-film technology and LTCC (low-temperature cofired ceramic).

Important requirements for the interconnect technology are the ability to integrate mm-wave active die without adding important contributions of parasitics (e.g., bond wires, flip-chip mounting or die embedding) and the availability of low-loss substrate materials (e.g., glass, HiRes silicon, Teflon, etc.).

21.3.3.1 Thin-film Technology

Multilayer thin-film technology (MCM-D), as developed over the past years [22], is well suited for high-frequency applications, particularly due to its high-precision definition and the possibility of integrating resistive and capacitive layers. The main features of IMECs thin-film technology are the use of electroplated copper for the interconnection lines, use of photo-sensitive benzocyclobutene (BCB) for the dielectric layers and use of Ni/Au for the final contact metallization. Up to four metal layers are used. The minimum copper line width and spacing are 5 μm. Furthermore, a TaN resistor (25 Ω/square), Ta2O5 capacitor (0.75 nF/mm^2) layer is integrated; lower-value capacitors (5.6 pF/mm^2) may be realized using the different BCB-layers. The frequencies of interest vary from 1 to 100 GHz. As a substrate material, low-loss glass (700 μm tick) or 100 μm Hires–Silicon substrates can be used (see Figure 21.12). The latter solution is particularly interesting for mm-wave module integration purposes. A metal finish with Sn-solder and/or Cu/Ni/Au is available for the seamless integration of, for example, flip-chip assembly of RF active die with strongly reduced parasitics. Figure 21.13 shows an example of a 50 GHz bandpass filter as integrated in the thin-film platform. Figure 21.24 illustrated a cross-section of flip-chip mounted and embedded RF devices.

An extensive microwave design library was developed for this thin-film technology in the Agilent ADS microwave design suite. This enables the microwave designer to efficiently and accurately design microwave circuits (Figure 21.13) including those with active devices.

One step further is the possibility to integrate other functional devices such as RF MEMS devices onto the substrates, and even embed (see Section 21.3.3.2) very thin active die into the substrate (Figure 21.14).

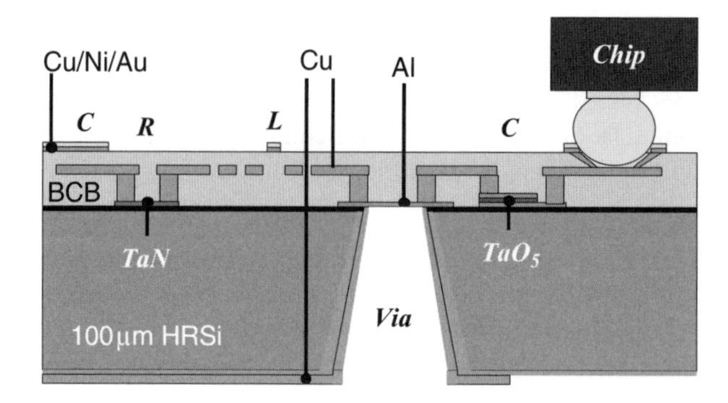

Figure 21.12 Cross-section of the thin-film technology built up on high-resistivity silicon substrate with integrated high Q passives on top and RF through wafer vias.

21.3.3.2 Organic Substrates

Another attractive solution for the integration of mm-wave modules and systems is the use of organic substrates, both rigid and flexible. Due to their low cost they are widely used in numerous portable consumer products. However, the step up towards being used in mm-wave systems still needs considerable research efforts. Up till now, specific organic boards have been used for low-density microstrip PCB-style system integration (e.g., Duroid). As an example, the embedding of active die into organic substrates could enable a low-cost solution to the mm-wave interconnect problem (Figure 21.15). UTCP (Ultra-thin chip packaging) is a technology for embedding ultra-thin chips in polyimide substrates. Silicon chips thinned down to 20–30 μm are packaged in between two spin-on polyimide layers, resulting in a total thickness of only 50–60 μm. Chip, PI and metal layers are so thin that the whole package is actually bendable. The embedded chips can be used as a package (e.g., the package can be solder assembled on interconnection substrates (PCB or flex) like a common component). The ultra-thin chip package is also suitable for embedding in rigid or flex boards, replacing, for example, the bare die. The use of a thin chip package here has two main advantages. First, the alignment constraints for the embedded package are not as severe as for the embedded die: a fan-out metallization provides contacts with more relaxed pitches. Second, the 'Known Good Die' (KGD) issue is solved as the packaged chip can be tested before embedding [23] (http://www.shift-project.org). Figure 21.15 depicts an organic PI substrate.

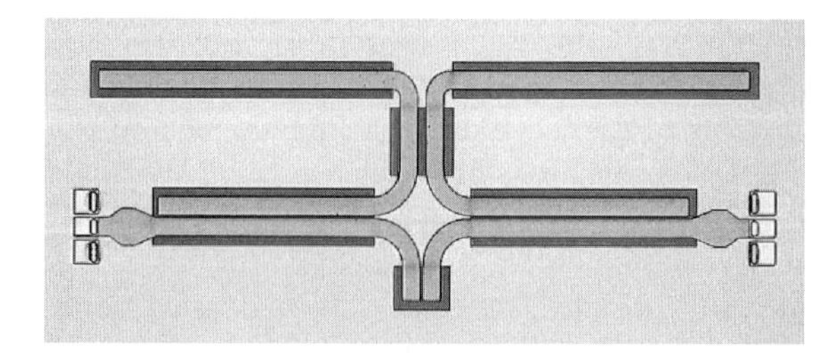

Figure 21.13 Example of $0.7 \times 1.4\,\text{mm}^2$-wave 50 GHz bandpass filter as integrated in the thin-film platform.

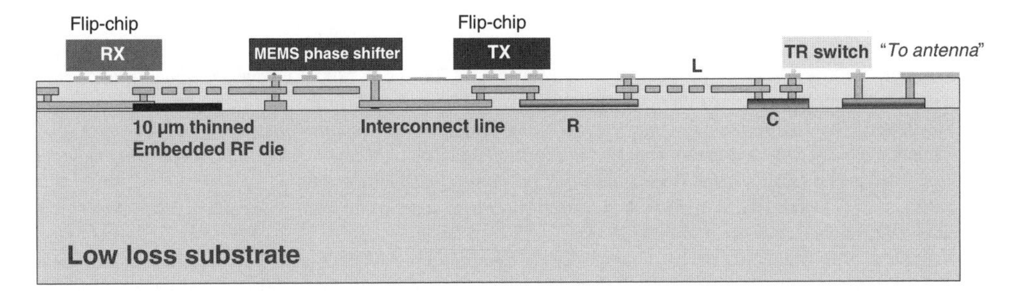

Figure 21.14 Schematic cross-section of the integration of various flip-chip mounted and embedded RF devices to form a mm-wave front-end.

21.3.3.3 Ceramic Substrates

Among candidate technologies for mm-wave applications, LTTC (low-temperature cofired ceramic) is also an important alternative. This technology is basically a 3D built up of different dielectrics (the ceramic material) and metal layers. Dimensional control of the realized structures, the fact that all designs need full 3D EM simulation and optimization, and selection of suitable mm-wave dielectrics are currently the main drawbacks. On the other hand, this technology has evolved during the past decade to a well-known and mature integration technology for consumer products [24].

21.3.3.4 Conclusion

Various interconnect technologies exist today or are under development, each with its strengths and weaknesses, varying from very high-performance technologies to low-cost system integration solutions.

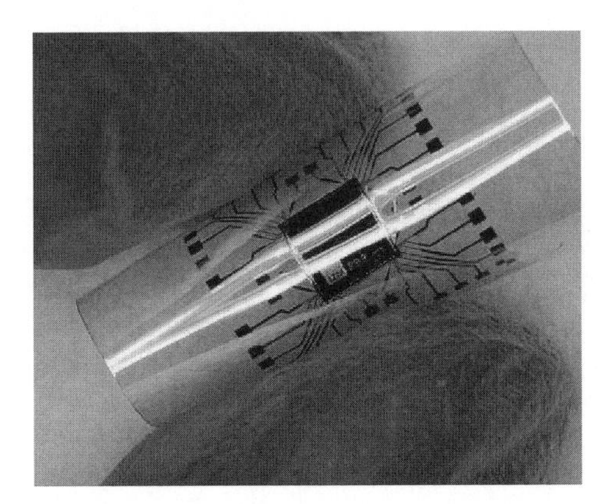

Figure 21.15 Organic PI substrate with embedded 20 μm thick active die.

References

1. Falconer, D., Ariyavisitakul, S.L., Benyamin-Seeyar, A. and Eidson, B. (2002) Frequency domain equalization for single-carrier broadband wireless systems. *IEEE Communications Magazine*, **40** (4), 58–66.
2. Sundberg, C.-E. (1986) Continuous phase modulation. *IEEE Communications Magazine*, **24**, 25–38.
3. Laurent, P. (1986) Exact and approximate construction of digital phase modulations by superposition of amplitude modulated pulses (AMP). *IEEE Transactions on Communications*, **34**, 150–60.
4. Saleh, A. and Valenzuela, R. (1987) A statistical model for indoor multipath propagation. *IEEE Journal on Selected Areas in Communications*, **5**, 128–37.
5. Nsenga, J., Van Thillo, W., Horlin, F. *et al.* (2007) Comparison of OQPSK and CPM for communications at 60 GHz with a non-ideal front-end. *EURASIP Journal on Wireless Communications and Networking*, **2007** (1), 51.
6. Klein, A., Kaleh, G.K. and Baier, P.W. (1996) Zero-forcing and minimum mean-square-error equalization for multiuser detection in code-division multiple access channels. *IEEE Transactions on Vehicular Technology*, **45** (2), 276–87.
7. Horlin, F. and Vandendorpe, L. (July 2002) A comparison between chip fractional and nonfractional sampling for a direct sequence CDMA receiver. *IEEE Transactions on Signal Processing*, **50**, 1713–23.
8. Floyd, B. *et al.* (2006) A silicon 60 GHz receiver and transmitter chipset for broadband communications. Proceedings International Solid-State Circuits Conference, pp. 184–5.
9. Craninckx, J. *et al.* (2007) A fully reconfigurable software-defined radio transceiver in 0.13 μm CMOS. Proceedings International Solid-State Circuits Conference, pp. 346–7.
10. Heydari, B. *et al.* (2007) Low-power mm-wave components up to 104 GHz in 90 nm CMOS. Proceedings International Solid-State Circuits Conference, pp. 200–1.
11. Bennett, H.S. *et al.* (2005) Device and technology evolution for Si-based RF integrated circuits. *IEEE Transactions on Electron Devices*, **52** (7), 1235–58.
12. Razavi, B. (1998) *RF Microelectronics*. Prentice Hall.
13. Ostman, K.B. *et al.* (2006) Novel VCO architecture using series above-IC FBAR and parallel LC resonance. *IEEE Journal of Solid-State Circuits*, **41** (10), 2248–56.
14. Krishnaswamy, H. and Hashemi, H. (2007) A fully integrated 24 GHz 4-channel phased-array transceiver in 0.13 μm CMOS based on a variable-phase ring oscillator and PLL architecture. Proceedings International Solid-State Circuits Conference, pp. 124–5.
15. Kinzel, J., Edward, B.J. and Rees, D.E. (1987) V-band, space-based phased arrays. *Microwave Journal*, **30** (1), 89–103.
16. Mailloux, R.J. (1992) Antenna array architecture. *Proceedings of the IEEE*, **80** (1), 163–72.
17. Mailloux, R.J. (1982) Phased array theory and technology. *Proceedings of the IEEE*, **70** (3), 246–91.
18. Carver, K.R. and Mink, J.W. (1981) Microstrip antenna technology. *IEEE Trans. Antennas Propagat.*, **AP-29** (1), 2–24.
19. Pozar, D.M. (1984) Scan blindness in infinite phased arrays of printed dipoles. *IEEE Trans. Antennas Propagat.*, **AP-32** (6), 602–10.
20. Brebels, S. *et al.* (2004) SoP integration and codesign of antennas. *IEEE Trans. Advanced Packaging*, **27** (2), 341–51.
21. Floyd, B.A. and Chih-Ming Hung, K.K.O. (2002) Intra-chip wireless interconnect for clock distribution implemented with integrated antennas, receivers and transmitters. *IEEE Journal of Solid-State Circuits*, **37** (5), 543–52.
22. Carchon, G., Vaesen, K., Pieters, P. *et al.* (2001) Multi-layer thin-film MCM-D for the integration of high-performance wireless front-end systems. *Microwave Journal*, **44** (2), 96–110 (ref 1).
23. Christiaens, W., Vandevelde, B., Brebels, S. and Vanfleteren, J. (2007) *Polyimide Based Embedding Technology for RF Structures and Active Components*. Proc. SMTA 12th Pan Pacific Microelectronics Symposium (Pan Pac 2007), Maui, Hawaii, pp. 22–9 (Ref 3).
24. Kim, H.-C. *et al.* (2006) *LTCC Technology for 60 GHz Applications*. ISMP, The 5th International Symposium on Microelectronics and Packaging, Seoul, Korea (Ref 5).

22

Antenna and Assembly Techniques for Cost-effective Millimeter-wave Radios

Timo Karttaavi[1], Markku Kiviranta[3], Antti Lamminen[3], Aarne Mämmelä[3], Jussi Säily[3], Antti Vimpari[2] and Pertti Järvensivu[3]

[1]*Nokia Corporation, P.O. Box 407, FI-00045 NOKIA GROUP*
Email: timo.karttaavi@nokia.com
[2]*Refecor Oy, Teknologiantie, FI-90570 Oulu*
Email: antti.vimpari@refecor.com
[3]*VTT, Finland*

22.1 Adaptive Antenna Array Systems Design

22.1.1 Introduction

Adaptive antennas are commonly referred to as smart antennas. A common feature of all adaptive antennas is that their radiated beam patterns can be adapted to different environments in order to improve reception and signal-to-noise ratio (SNR). For example, the antenna beam pattern can have increased gain towards mobile users and have nulls in the direction of interferers like neighbouring cells. This will result in increased range and coverage and allow for smaller and lighter mobile devices [1, 2].

Adaptation mechanisms typically require amplitude and/or phase control of individual antenna elements. The complexity of the system depends on the number of antenna elements and the related controls. The simplest form of adaptation is beamsteering, that is electronic control of the main beam direction, which can be accomplished by phase control of each element. Beamforming is a more complex procedure which can produce more complex beam patterns including several useful lobes and nulls. A review of commonly used beamforming methods is presented in the next section. Typically, the adaptive array beams are generated

Short-Range Wireless Communications Rolf Kraemer and Marcos D. Katz

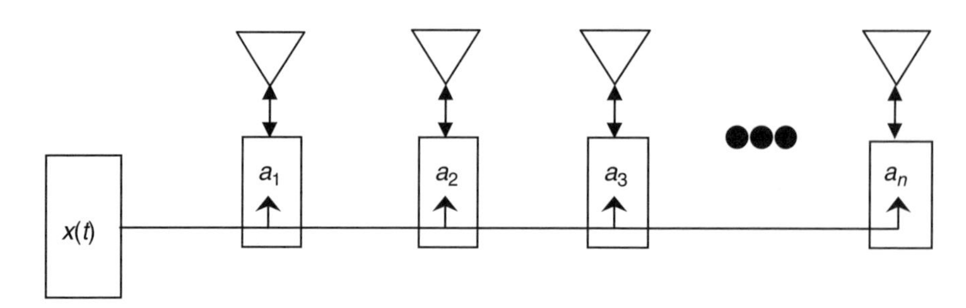

Figure 22.1 Beam pattern controlled by antenna weights.

through digital signal processing (DSP). The so-called antenna weights $a_1 \ldots a_n$ are calculated for each element, and they can be either real or complex numbers. Figure 22.1 illustrates the basic concept of beamformation by antenna weights. The signal to be transmitted is $x(t)$. The antenna weights are recalculated by each adaptation round in the DSP.

Antenna diversity and multiple-input multiple-output (MIMO) techniques are also related to adaptive antennas. The different diversity methods include time, space and polarization diversities. In the MIMO method, several antennas can be transmitting and receiving at the same time and at the same frequency channel. The channels are identified and separated in the receiver with coding. The parallel physical channels in MIMO exist through multipath propagation and space diversity between the antennas. Polarization diversity can also be used. The MIMO method works best in environments which contain many scatterers, such as indoor offices and corridors. A thorough analysis of the benefits of MIMO antenna systems can be found in [3].

22.1.2 Review of Beamforming Antenna Topologies

A recent study on beamforming methods, carried out in the European Commission Sixth Framework Programme Antenna Center of Excellence (ACE), is presented in [4]. The report takes into account different operating cases, for example base station antennas, radars (commercial, civil, military), spaceborne antennas and user terminal array antennas. Beamforming can be accomplished in many ways: analog beamforming, digital beamforming, optical beamforming and reflectarrays.

Analog beamforming is the most mature method, and it has been used in military-grade phased arrays for decades. The method can produce accurate beamforming through calibration and multiple beams. Based on their complexity, the beamforming networks can be divided into single-beam and multiple-beam ones. Passive arrays have only phase shifters and/or switches, while active arrays also have a multitude of power amplifiers and/or low-noise amplifiers connected to the antenna elements. Also, the analog beamforming networks can be based on switchable fixed parallel RF paths if controllable phase shifters with good enough characteristics like power handling or linearity are not available. Limitations of the method are the inherently narrow bandwidth from 10 to 20% due to the phase shifters, and the input matching requirement of all devices [4]. Substantial power consumption of the active arrays may also be problematic in some applications, and a high number of up/down-converters require accurate local oscillator (LO) distribution. Phase shifters can be troublesome in wide-bandwidth applications since their dispersion causes beam-squint, meaning variation of the beam shape with frequency. A typical system using analog beamforming in an active configuration is shown in

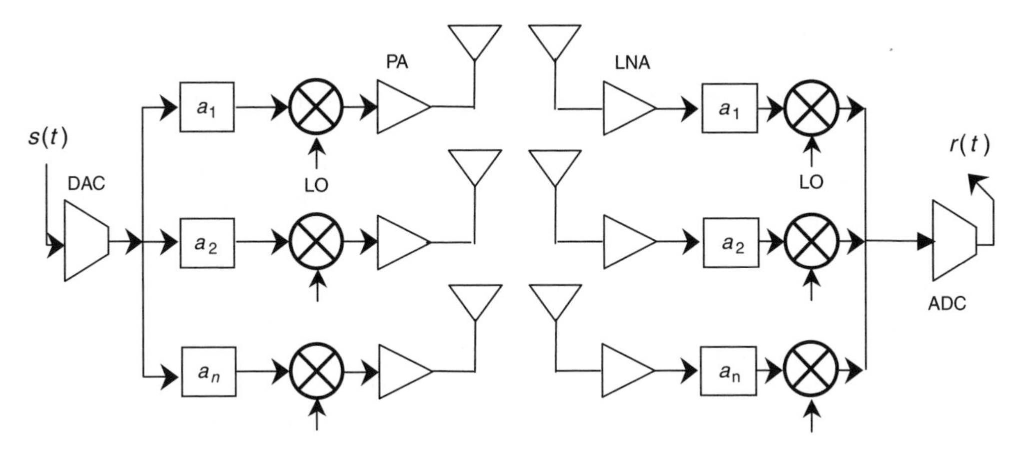

Figure 22.2 Active phased array based on analog beamforming.

Figure 22.2. The antenna weights can be applied either in the baseband (like in the transmitter on the left-hand side) or at the final frequency (the receiver on the right-hand side).

In digital beamforming, the received signals from the antenna array elements are down-converted, digitized and processed in the digital baseband domain. Similarly, the transmitted beams are formed through weighting and summing in the digital signal processor (DSP) before conversion into analog signals, amplification and up-conversion to the final frequency. The advantages of this method are: a potentially wider bandwidth solution compared to analog beamforming; ease of modification of the beamforming coefficients; less complexity increase in expanding the array size when compared to analog beamforming; simple combination of other DSP algorithms with beamforming; adaptive beamforming is easier; and phase shifters and RF switches are not needed. The disadvantages include: a high number of up/down-converters which require accurate LO signals; a high number of high-performance fast ADCs/DACs are needed; and bandwidth is limited by processing power. The power consumption of fast wideband converters is relatively high. A typical adaptive antenna system using digital beamforming is shown in Figure 22.3.

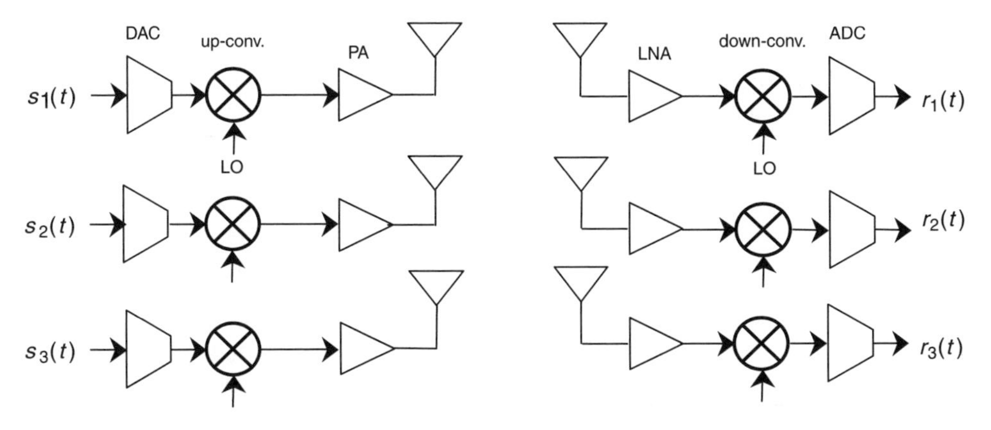

Figure 22.3 Active phased array based on digital beamforming.

Optical beamforming (microwave photonics) uses fiber-optical technology to distribute and weight the element signals [4]. Either optical phase shifters or true time delay (TTD) elements may be used in the feed network. The phase shifting method suffers from the same beam-squint effect as the analog beamforming method, again caused by dispersion. Various TTD architectures are available in microwave photonics and they all have negligible dispersion. A problem with current optical controls is their relative immaturity compared to electronic control devices. This results in systems composed of discrete optical devices which are bulky, expensive and mechanically not stable enough.

A reflectarray is a surface, typically planar, which is illuminated by a feed horn. The surface contains a multitude of antenna elements, which capture the incident radiation, apply the beam weights by modifying the phase and/or amplitude, and radiate it back. The antenna elements may have stubs of different lengths to facilitate beam shaping. Reflectarrays can be either passive or reconfigurable. Passive arrays have fixed weights and thus fixed beam patterns. Reconfigurable arrays have means to control the phasing of the elements with, for example, MEMS switch networks. Beam scanning can be done efficiently with reconfigurable reflectarrays. MEMS-based reflectarrays are being studied around the world but the technology has not yet matured to the point of commercialization.

Beamsteering of the main beam of an antenna array is initiated by separately phasing the individual elements. A full beamforming solution with the capability to shape the beam would also require amplitude controls for each element. Let us first consider the 1-dimensional case; that is, a linear array of radiating elements. It can easily be derived with geometry that, for a beam tilt of θ, the required progressive phasing factor between the elements of an equally-spaced linear array is:

$$\phi = 360^\circ \frac{d \sin\theta}{\lambda}, \tag{22.1}$$

where d is the element separation and λ the wavelength. In order to tilt the beam to a certain direction, the phase difference of adjacent elements must be equal to ϕ. Table 22.1 shows the 243 required progressive phase shifts for typical 0.5 λ arrays. Figure 22.4 further illustrates the emitted beamsteered plane wave direction in one dimension. The same principle can be applied to 2-dimensional planar arrays.

Table 22.1 Progressive phase shifts as a function of the desired beam tilt from normal direction.

θ (deg)	ϕ (deg)
0	0
5	15.7
10	31.3
15	46.6
20	61.6
25	76.1
30	90

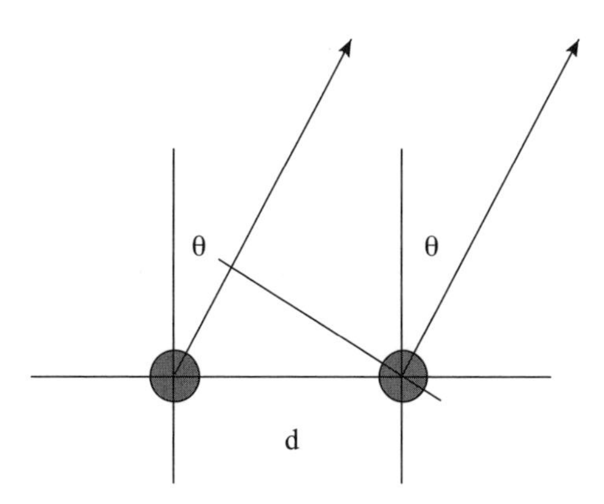

Figure 22.4 Beamsteering in one dimension.

22.1.3 Case Study of 60 GHz Beamsteerable Array on LTCC

Several types of patch antennas are suitable for the 60 GHz frequency band. For instance, either a microstrip-line-fed aperture-coupled patch (ACMPA) or a slot-coupled coplanar waveguide-fed patch (SCMPA) can be used as an element for the beamsteerable array on LTCC. A linear array of four ACMPA elements was designed, fabricated and tested for the 60 GHz frequency band at VTT. Also, an array with delay lines was processed to demonstrate the fixed beamsteering.

22.1.3.1 Design

The geometry of the ACMPA element is shown in Figure 22.5. The ACMPA consists of a microstrip feed line, a matching stub, an aperture in the ground plane and a radiating patch. The matching stub is used to compensate for the aperture's reactance; that is, to match it to the 50 Ω characteristic impedance of the feed line. The relative permittivity of Ferro A6-S LTCC

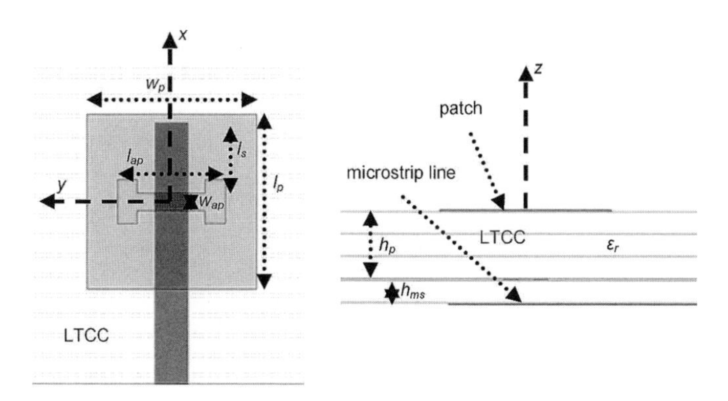

Figure 22.5 Geometry of the ACMPA element on Ferro A6-S LTCC substrate: $w_p = 0.785$ mm, $l_p = 0.79$ mm, $l_{ap} = 0.5$ mm, $w_{ap} = 0.08$ mm, $l_s = 0.315$ mm, $h_p = 0.3$ mm and $h_{ms} = 0.1$ mm.

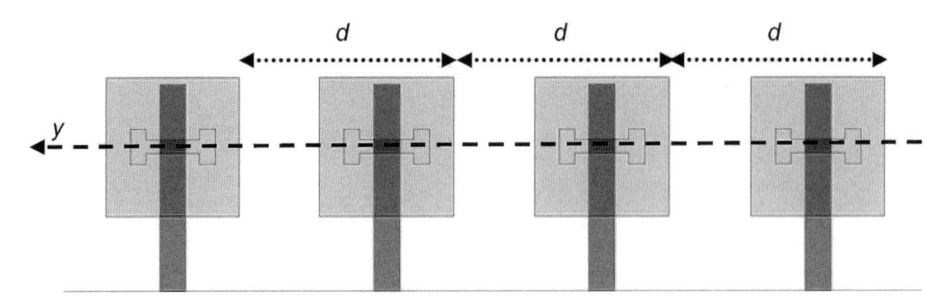

Figure 22.6 A schematic of a four-element linear array of ACMPA elements.

tape material is $\varepsilon_r = 5.99$ and the dielectric loss tangent $\tan(\delta) = 0.0015$. The conductivity of gold $\sigma_{\mathrm{Au}} = 4.1 \times 10^7$ S/m was used in simulations because the actual conductivity of the gold paste was not known. The fired thickness of the Ferro A6-S tape layer is 0.1 mm. The size of the patch substrate is 9.7 mm × 9.7 mm.

The linear array consists of four ACMPA elements placed along the y-axis (Figure 22.6). An element spacing of $d = 2.5$ mm ($0.5\,\lambda_0$) was used. Simulations were conducted using the HFSS^{TM} electromagnetic simulation software by Ansoft. Radiated fields of an array were calculated using far fields of a single element and the array factor. A quarter-wave matched T-junction microstrip-line power divider embedded inside the LTCC substrate was designed to feed the linear array with a V-connector by Anritsu. The power divider consists of three quarter-wave matched T-junctions, straight microstrip lines and 90° corners (see Figure 22.7). The T-junction has 50 Ω line impedances at each port, and the impedance of the quarter-wave matching transformer is 35.4 Ω [5]. Each corner is mitered to obtain optimum design. Also, the discontinuities of the quarter-wave transformers are compensated for by tapering the microstrip lines.

Delay lines were designed to feed array elements with a progressive phase shift. The idea was to demonstrate the beamsteering of the array. The target phase shift of 90° is implemented by properly bending the microstrip line. Phase shifts of 180° and 270° are achieved by placing

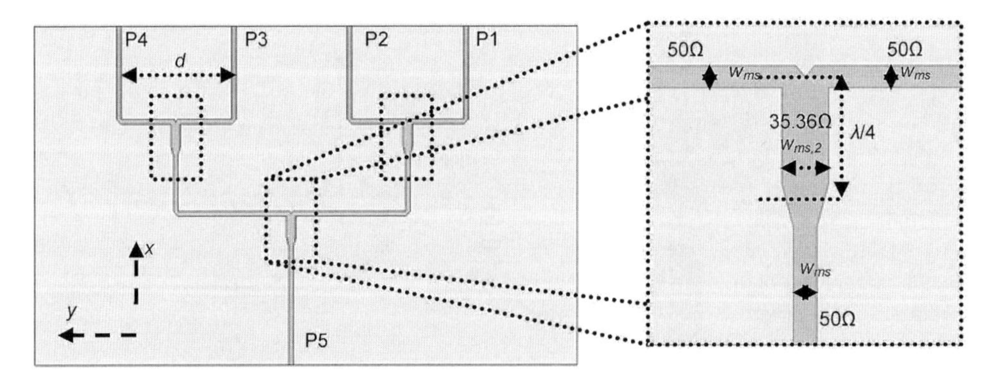

Figure 22.7 An one-to-four microstrip-line feed network based on quarter-wave matched T-junctions: $w_{\mathrm{ms}} = 0.1$ mm and $w_{\mathrm{ms},2} = 0.2$ mm.

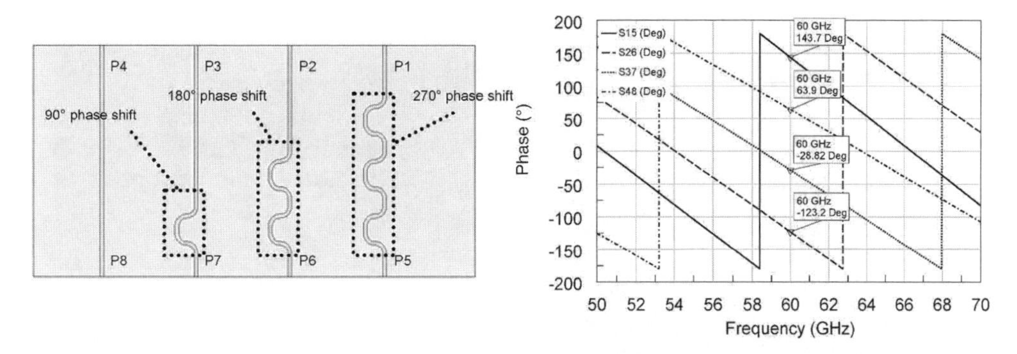

Figure 22.8 Left: the geometry of the delay lines for the fixed beamsteering of an array. Right: HFSS simulation results.

two and three bends, respectively, one after the other (Figure 22.8). The optimal size of the bent line is iterated with IE3D simulation software by Zeland. The final designs of feed lines are also simulated by HFSS. Simulation results are seen in Figure 22.8. Calculated values for the phase shifts are 92.7°, 94.4° and 93.1° between the neighbouring lines at 60 GHz. Insertion losses are: S48 = −0.44 dB, S37 = −0.60 dB, S26 = −0.94 dB and S15 = −1.28 dB. These differences in the feed amplitudes were not included in the array simulations.

The impedance characteristics of the test structures can be measured with an on-wafer probe station. For radiation pattern testing, CPW V-connectors by Anritsu were available. Since the ACMPA uses microstrip lines as feed, a transition from the CPW to a microstrip line was designed to feed the test structures. Designed transition is based on a CPW-to-MS electromagnetic transition [6]. The characteristic impedances of the CPW feed and an embedded microstrip line are 50 Ω. An additional LTCC layer below the microstrip line is used in order to have adequate mechanical persistence for the CPW feed.

22.1.3.2 RF Testing

Impedance matching of the single patch element was investigated with a Cascade on-wafer probe station. The GSG probes with 0.15 mm pitch between signal and ground were used (Figure 22.9). Antennas were positioned on top of a 1 cm thick Rohacell foam sheet ($\varepsilon_r < 1.15$) with a Cuming Microwave C-RAM GDSS absorber slab on the bottom. Since the far-field region starts below a distance of 1 mm, the Rohacell foam and absorber sheets were adequate to simulate far-field conditions and prevent multiple reflections.

The simulated and measured return loss and input impedance of an ACMPA are presented in Figure 22.10. Measured return loss is better than −10 dB across a frequency band from 56.3 to 61.2 GHz with a centre frequency of 58 GHz. The coupling aperture in the ground plane could be larger to obtain even better input matching.

Radiation patterns of the test structures were measured in an anechoic chamber at VTT. The test equipment and photographs of an ACMPA connected to the V-connector and antenna positioner are shown in Figure 22.11. RF absorbers were used to prevent reflections from the objects near the test structure. Results in the E-plane were noisy due to radiation from the transition and the V-connector. However, the H-plane pattern is more interesting in linear

Figure 22.9 Left: S-parameter measurements with a probe station – test setup. Right: GSG probes.

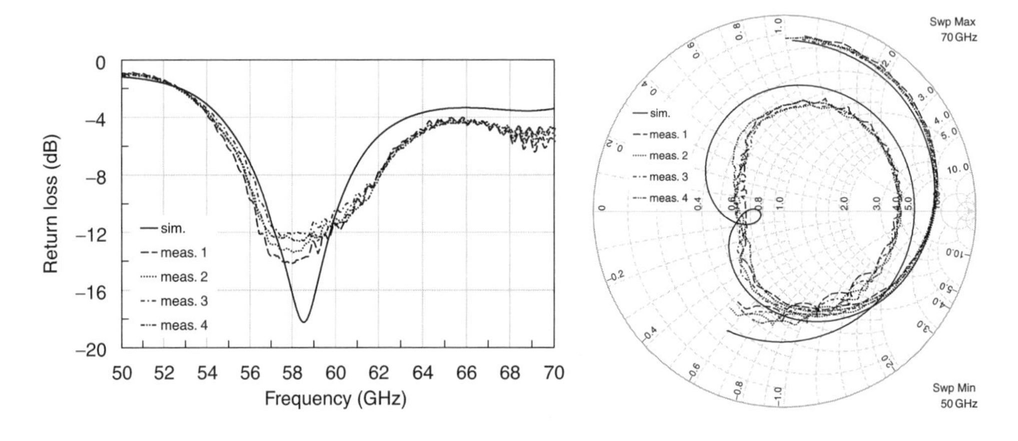

Figure 22.10 Simulated (HFSS) and measured impedance characteristics of an ACMPA.

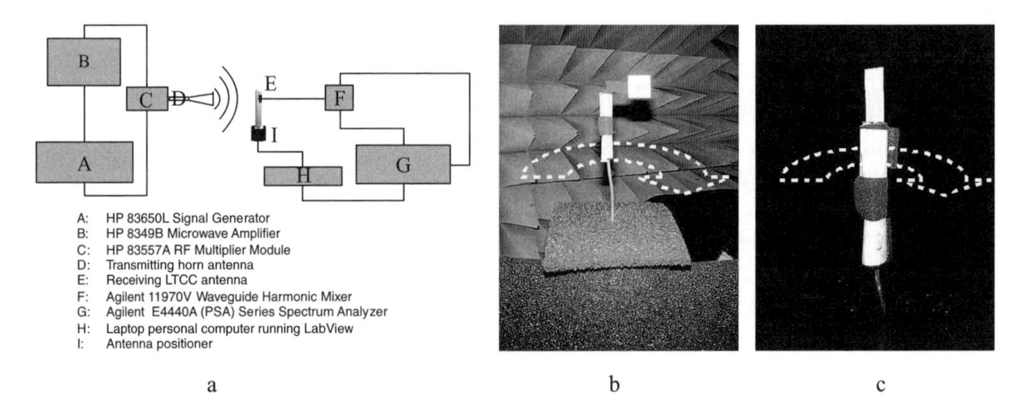

Figure 22.11 Radiation pattern measurements in anechoic chamber: (a) test equipment; (b) E-plane measurement of an ACMPA; (c) H-plane measurement of an ACMPA. The antenna-undertest is rotated in the horizontal plane.

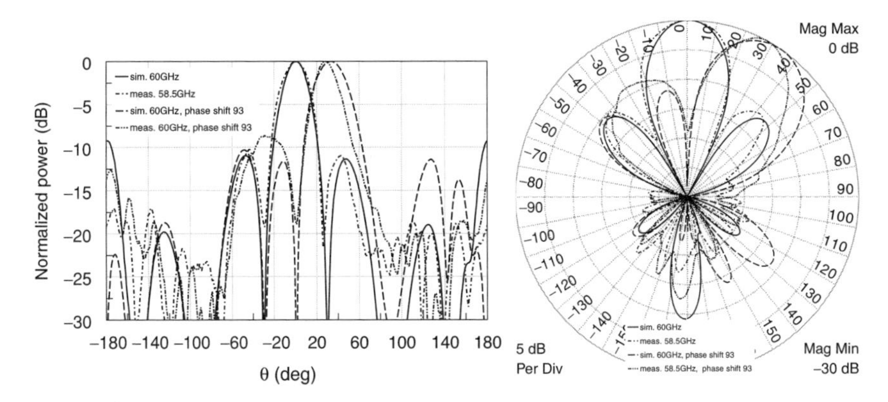

Figure 22.12 Simulated and measured H-plane patterns of the ACMPA array with and without progressive phase shift. The element spacing is 2.5 mm.

array configurations. Simulated and measured H-plane patterns of the arrays are presented in Figure 22.12. It is seen that simulated and measured patterns are very similar. With a progressive 93° phase shift the main beam is directed towards $\theta = 32^{\circ}$, as predicted by Equation (22.1). Small differences in the shape of the simulated and measured patterns with a beamsteered array are caused by the feed amplitude differences between elements in the realized array.

22.1.4 Conclusions, with Predictions

The basic properties of adaptive antennas and beamforming have been reviewed. A case study of a beamsteerable 60 GHz antenna array on LTCC was presented. Adaptive antennas at base station sites, and also in mobile devices, are required to facilitate increased data rates and better coverage in a congested frequency spectrum. We predict that LTCC packaging of antennas and CMOS RF circuitry play a key role in enabling cost-effective smart antennas at millimeter waves.

22.2 Assembly Techniques

22.2.1 Low-temperature Cofired Ceramic (LTCC) Antenna Array

22.2.1.1 Introduction

So far, the high costs of the radio technology for the licence-free band around 60 GHz have prevented many commercial applications from emerging. Both multilayer LTCC packaging technology and integrated semiconductor circuits are widely used at the microwave frequencies in making cost-effective radio systems. Recently, there has been much popular interest in developing these for 60 GHz as well. Studies have been conducted as to whether millimeter-wave antennas or antenna arrays can be integrated with passive components and millimeter-wave integrated circuits (MMIC) into the same LTCC module in order to have complex radio systems in a miniaturized, cost-effective package.

The commonly used Ferro A6-S LTCC material has a dielectric constant of 5.97 and loss tangent of 0.0015 at 60 GHz [5]. The very small loss tangent makes this material a great option for the integration of millimeter-wave components. Also, there are very good-quality silver and gold thick-film conductor materials available for this LTCC system. The LTCC manufacturing technology has a limited physical realization accuracy, which has to be taken into account with millimeter-wave designs. The dimensional errors caused by the alignment tolerance between the LTCC tape layers and the patterning accuracy of the screen-printed conductors can be in the tens of microns. In this field, there has been continuous progress in improving the situation.

22.2.1.2 LTCC Technology in Millimeter-wave Applications

LTCC technology provides a multilayer ceramic process capable of processing high-density electronics wiring boards at a reasonably low cost. LTCC processing starts with punching vias into glass/ceramic sheets with a nominal thickness of about 50–250 μm using mechanical tooling. This is followed by conventional via metallization and Ag- or Au-conductor printing. After that, the sheets, typically 4–10 layers, are laminated to form multilayer panels, and sintered at about 900 °C. After sintering, the panels are cut into pieces in order to separate individual circuits.

The LTCC substrate has several benefits over organic ones. LTCC has good thermal stability, low dielectric loss from DC to over 100 GHz, the possibility to precision-machine cavities and grooves into the substrate, and compatibility with hermetic sealing. In addition, the fair match of the thermal expansion coefficient of LTCC to silicon (5.8×10^{-6} 1/K vs. 2.5×10^{-6} 1/K) decreases packaging-induced thermo-mechanical stresses to devices.

Passive components integrated into the multilayer substrate increase packaging density, reduce price and increase reliability. Standard LTCC technology, in its present form, allows the low-cost integration of low-loss inductors, small capacitors (up to 10 pF in practice) and miniaturized transmission-line resonators with Q values better than 100 at 2 GHz. Future developments will advance high-permittivity dielectric materials, which will enable higher capacitance values and smaller resonator dimensions. Actually, there is a growing interest in new LTCC systems for millimeter-wave applications, due to the development of low-loss dielectric materials, accurate conductors and the possibility to process 3D structures. Conventional screen-printing of conductors is normally used with the resolution of minimum line width and spacing of about 100 μm. However, better resolution will be necessary for many designs at millimeter-wave frequencies. Recently there has been some progress on that, fortunately. So-called trampoline printing screens have enabled printed conductors with line widths and spacing resolutions of as small as 50 μm.

22.2.1.3 Active Circuit Integration

The traditional packaging method for high-performance V-band (50–75 GHz) millimeter-wave circuits involves elaborate machining of metal blocks to be used as carriers or enclosures and, in addition, transitioning to rectangular waveguide connections. For high-volume consumer applications, this approach is prohibitively expensive. The mechanical construction is easily the dominant cost factor. The System-on-Chip (SoC) approach tries to solve this problem.

Here, all the millimeter-wave functions can be integrated onto a single chip and no external connections are needed at the signal frequency. This, of course, works perfectly only if the antennas are also integrated on the chip. But on-chip antennas have limitations regarding size and performance. This approach probably works best at very short communication distances, where the requirements for antenna performance are not too severe.

LTCC as an integration platform can be seen as a System-in-Package (SiP) type of solution which offers a compromise between the traditional approach and SoC. The reasonably high dielectric quality and metallization accuracy give the possibility of fabricating lower loss interconnections and better passive structures than on-chip. This holds true especially for silicon circuitry, where the substrate is very lossy. A hybrid solution also gives flexibility in choosing the optimal active circuit technology for each function. For instance, Si CMOS could be used for the receiver and SiGe HBT or GaAs HEMT for the transmitter. Individual chips are mounted on LTCC using wire bonding or, preferably, flip-chip technology. In some cases, chips can be enclosed in cavities formed within the multilayer structure, which then serve as first-level packaging. The addition of antennas on the LTCC board gives the opportunity to implement more complex radiating structures, with larger area and higher gain than on-chip, and avoids the cumbersome transition to, for example, a rectangular waveguide. This makes medium-range communication systems possible, but for longer ranges a larger external reflector or lens antenna is still probably the better choice.

22.2.1.4 Integrated 60 GHz Antennas in LTCC

On the millimeter-wave frequencies, reflector, lens, leaky-wave, microstrip and slot antennas are widely used. Microstrip and slot antennas are the most suitable for integration in a planar substrate and hence can be fabricated by applying the LTCC technology. Microstrip antennas usually have a large ground plane in the substrate laminate, and the actual antenna elements are on the topmost conductor layer. Sometimes it is desirable to stack the antenna elements on two or three layers in order to widen the impedance bandwidth. It is reasonable to use element stacking for microstrip patch antennas. Microstrip-line dipoles are practically always used as nonstacked antenna elements.

Patch antennas are the most widely-used integrated antenna structures because of their good suitability to the multilayer planar environment. The dimensions of patches remain feasible up to the millimeter-wave frequencies for the current LTCC manufacturing technology. A patch antenna is usually a rectangle-, square- or circle-shaped metal conductor on the surface of the dielectric substrate. A patch can be fed through a coupling aperture or via probe. In practice, this means that the feeding of the patch requires a proper transition from the feeding transmission line to the driven patch element in order to have an appropriate impedance matching. A via or an aperture is just a part of that transition.

There are a few analysis techniques for microstrip patch antennas [7]. Transmission-line models are the simplest, but they are applicable for rectangular patch shapes only. The patch is modelled as two narrow parallel radiating slots, which represent its radiating edges. This analysis technique suffers from numerous disadvantages. Just to mention a couple, it demands empirically-determined parameters and it is not adaptable to inclusion of the feed.

The disadvantages of the transmission-line model are eliminated in the modal expansion analysis technique, whereby the patch is viewed as a thin TM_z-mode cavity with magnetic walls [7]. By treating microstrip antennas as cavities, a simple theory based on the cavity model has also been advanced [8]. With this theory, almost all the properties of microstrip antennas of various geometries can be explained and predicted. The field between the patch and the ground plane can be expanded in terms of a series of cavity-resonant modes, along with the resonant frequencies associated with each mode. The effects of radiation and other losses are represented in terms of either an artificially-increased substrate loss tangent or an impedance boundary condition at the walls. This results in a much more accurate formulation for the input impedance, resonant frequency and bandwidth for both rectangular and circular patches. The modal expansion technique may also be used for the analysis of a circular patch, following the same general lines as for the rectangular patch.

The manual implementation of the electromagnetic-field solution for the patch antennas is a tedious process, of course [8]. However, the resonant frequency of a circular microstrip patch antenna of radius a can be calculated with a relatively simple formula:

$$f_{\mathrm{ro}} = \frac{ck'_{10}}{2\pi a\sqrt{\varepsilon_r}}, \qquad (22.2)$$

where $k'_{10} = 1.84118$.

In general, k'_{10} is the complex eigenvalue (that is, complex resonance frequency) of a cavity. The value is dictated by the physical shape of the patch. The formula is only valid for a cavity with a nonradiating zero-admittance wall. In the case of a radiating patch antenna, the wall admittance is complex, so the resonant frequency becomes complex as well. The real resonant frequency f_r is therefore less than f_{r0}. The difference between f_r and f_{r0} can be tens of percents for electrically-thick substrates ($t > 0.05\lambda_{d0}$) [7], which is the case with real-world LTCC substrates on millimeter-wave frequencies. For example, the realized circular patch antenna with radius of 475 μm on the Ferro A6-S LTCC ($\varepsilon_r = 5.97$) had a measured resonant frequency of 59 GHz. For the same geometry, the presented formula gives 75 GHz for the resonant frequency. The difference is about 27%. However, because of the simplicity of the formula, it can be regarded as quite useful for the initial estimation of the patch size. Actually, it would be easy to determine experimentally the error of the calculated resonant frequency for various substrate parameters, if that was necessary. Nowadays there are several very good electromagnetic simulator softwares with full 3-dimensional capability, so after the initial estimation of the physical parameters of the patch antenna, the design can be finished by computer simulation with a relatively small effort.

The electrical dimensions of the patches are usually between the quarter- and half-wave length, as guided in the substrate by the microstrip structure. The exact size of the patch and the position of the feed point must be selected so that the impedance of the patch is matched to the transition between the feed system and the driven patch. The structure of a 60 GHz circle-shaped patch antenna with a microstrip-line feed is presented in Figure 22.13. The microstrip line and the patch are connected with a via of 90 μm in diameter. The size of the void in the ground plane around the via has a diameter of 450 μm and the diameter of the patch is 890 μm. The substrate material is ceramic, with $\varepsilon_r = 5.97$, which corresponds to Ferro A6-S LTCC. The substrate thickness between the microstrip line and the ground plane is 99 μm, and between the patch and the ground plane, 396 μm. Figure 22.5 presents the principle of the usage of a coupling aperture.

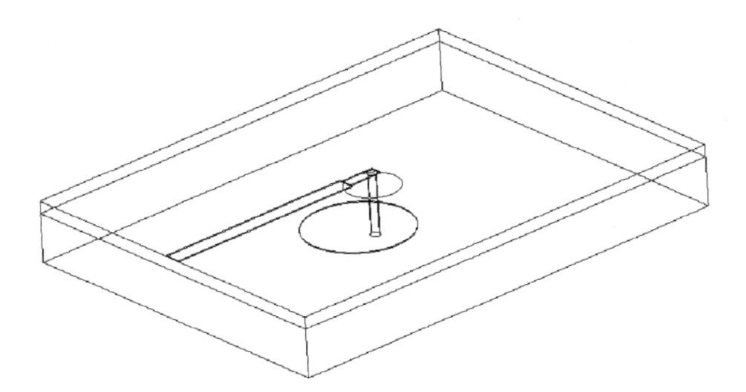

Figure 22.13 Construction of a probe-fed circular patch antenna.

It is also possible to implement a rectangular patch antenna as a directly microstrip line-fed structure, which demands only one conductor layer in addition to the ground plane. Such a structure can be regarded as a dipole of which the width is much greater than the length. The length is approximately half wavelength, electrically, so the antenna is fundamentally a half-wavelength dipole. The feed point is at the centre of one of the wide edges. Making the dipole very wide compared to the guided wavelength in the substrate reduces the feed-point impedance to an appropriate level for the microstrip-line feed, and increases the impedance bandwidth too.

One possibility for realizing the electrical half-wavelength dipole is a relatively narrow microstrip line. The feed-point location and the width of the microstrip line can be selected so that the impedance-matched conditions are achieved. The benefits of this kind of structure are simplicity and ease of design compared to patch antennas. Figure 22.14 presents a microstrip-line dipole designed for 60 GHz. The LTCC substrate has been left out of the figure for better legibility. The length of the dipole is 1.12 mm. The feed point must, of course, be somewhere between the centre and the end of the dipole in order to make the antenna radiate. The feed-point impedance gets higher when the feed-point location is moved towards the end. The cuts next to the feed point are to allow sufficient reactance compensation to generate some serial inductance. In most cases, it is wise to optimize the whole design to as good a return loss and as wide an impedance bandwidth as possible, to ensure the feasibility of the design under the physical realization tolerances of the manufacturing technology used.

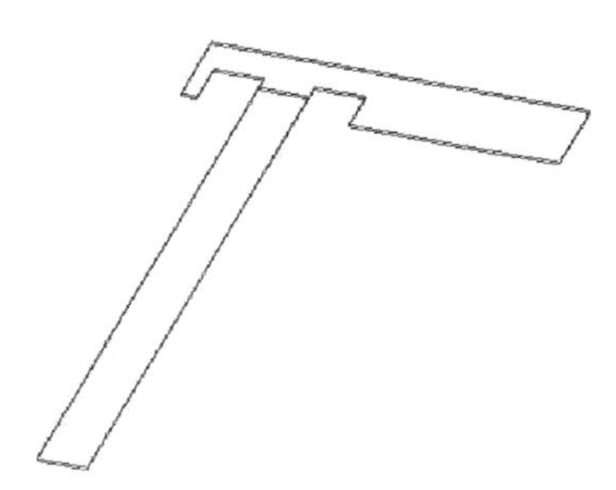

Figure 22.14 Microstrip-line dipole and its feed system.

A slot-antenna realization is a slot on a large solid ground plane, and hence it is not as practical as a microstrip-line antenna. Electrically, it can be regarded as a magnetic dipole, and hence it is a magnetic counterpart for a conventional microstrip-line dipole. The feeding can be realized as a crossing microstrip or strip line on another conductor layer. The feed-point impedance of a slot antenna is inversely proportional to the corresponding microstrip-line dipole.

A conductor-backed patch antenna (a patch with a ground plane) has one main radiation lobe in the hemisphere. The beam width and the maximum achievable gain depend on the exact physical dimensions of the patch and the properties of the substrate. Also, the resistive loss of the used conductor metallization is significant at 60 GHz. At VTT, LTCC-based patch antennas have been designed and tested on the 60 GHz band [6, 9]. The results have been promising for 60 GHz wireless systems. An impedance bandwidth of several GHz has been achieved, which is important for high-speed communication. Antenna gains up to 5.9 dBi at 57 GHz for a single patch antenna have been measured, which is a promising starting point for future antenna-array design. One major source of problems seems to be excitation of the surface waves, which is enhanced by the quite high permittivity of the LTCC. However, the surface waves can be blocked by using electric band gap structures or a so-called soft surface ring around the patch.

22.2.1.5 Conclusions, with Predictions

LTCC seems to be sufficient technology for the realization of integrated antennas for 60 GHz communication systems. The electrical performance of the available LTCC materials is very good, which is essential for millimeter-wave applications. The accuracy of the printed conductors seems to be good enough for the physical realization of the integrated antennas and their feed systems. The accuracy demand is 50 μm of minimum conductor width and spacing. The tolerance of the conductor width and spacing must be in the order of 10 μm. For 60 GHz use, Ferro A6-S and A6-M have for many years been practically the only commercially-available material systems with sufficiently good electrical properties. Hopefully, the emergence of 60 GHz communication systems, and the related business potential, will drive LTCC material manufacturers to develop better materials in the future. The process resolution of the printed conductors is expected to get gradually better in the future, as it has been doing so far. The manufacturing cost level is expected to remain about the same, at least for the near future.

22.3 Review of Effects and Compensation of Nonlinearities Caused by Analog Parts of Radio Modems

22.3.1 Introduction

In fast data transmission, the modulation method is often quadrature amplitude modulation (QAM) or orthogonal frequency division multiplexing (OFDM). The nonlinear analog parts will play a significant role in performance, since the modulation methods are very sensitive. A modulation scheme with a large number of constellation points is sensitive to phase noise. For example, M up to 64 is used in M-ary QAM. The requirements for the amplifier linearity are severe when the modulation scheme is not constant envelope [10]. Therefore, it is crucial to take the nonlinearities into account when comparing different air-interface technologies and when developing new high data-rate products [11].

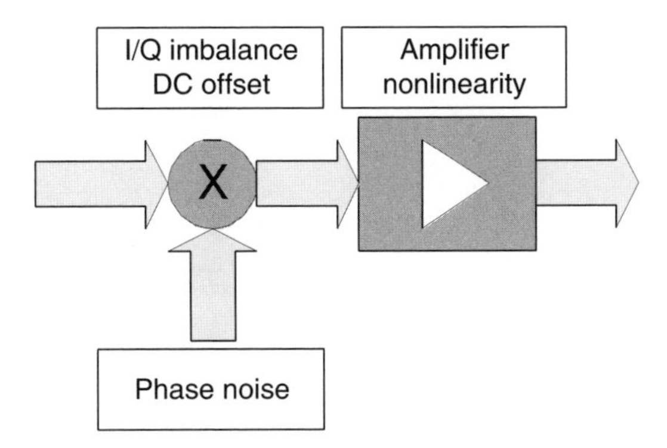

Figure 22.15 Nonideal analog parts.

22.3.2 Sources of Nonlinearities

Both the physical channel itself and the analog parts of the transmitter-receiver are sources of noise and distortions [12]. The ideal mixer can be modeled by a multiplier (see Figure 22.15) and is a linear operation. However, practical mixers are nonlinear devices, and their imperfections include I/Q imbalance and DC offset. Oscillators exhibit phase noise and all amplifiers are nonlinear if driven hard enough.

22.3.3 Effects of Nonlinearities

Nonlinear systems typically create spectral components (harmonic and intermodulation distortions) that are totally absent from the input spectrum. The same phenomenon is also possible in linear time-variant systems. Spectral spreading causes adjacent channel interference (ACI). If the nonlinear amplifier has memory, the system response depends not only on the input amplitude but also on its frequency. Each point of the signal constellation becomes a cluster showing intersymbol interference (ISI). 'Constellation warping' (see Figure 22.16) refers to the situation where the constellation points are not on a rectangular grid as in the original

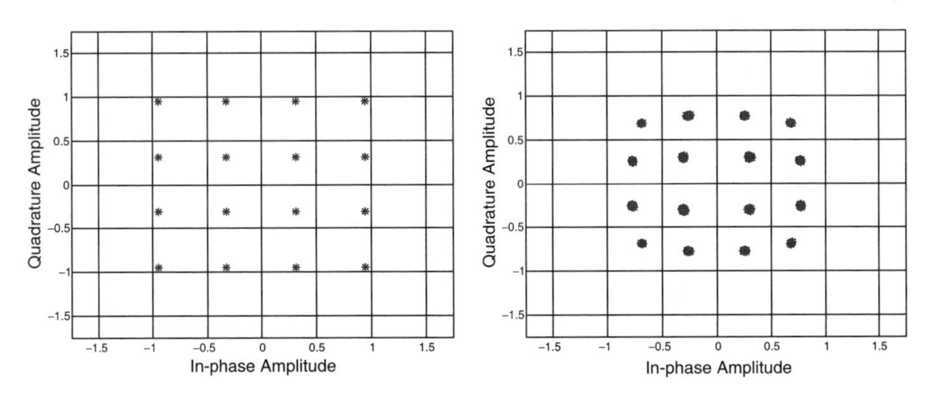

Figure 22.16 Left: ideal 16-QAM constellation. Right: constellation warping with clusters.

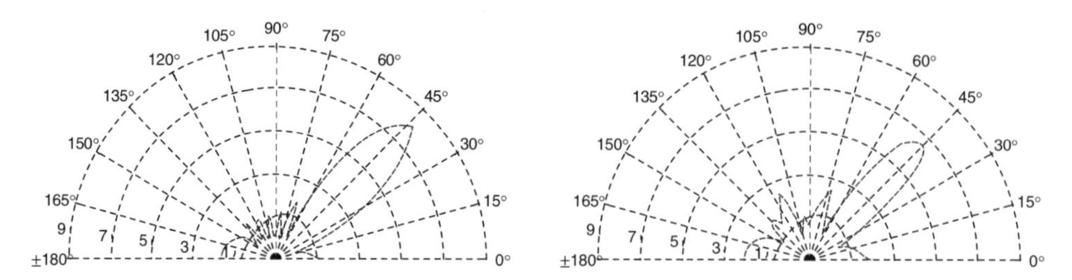

Figure 22.17 Ideal (left) and distorted (right) array factors.

constellation. Other normal changes in the constellation are attenuation, phase offset, phase noise, frequency offset, I/Q imbalance, DC offset and thermal noise [13].

By creating beams and nulls in the antenna array system, we can increase the gain in the direction of wanted signals and decrease the gain in the direction of interference and noise. Nonlinear distortions, however, cause impairments (see Figure 22.17) in beam direction, beam gain, beam width, sidelobe level, null depth and null direction [14].

22.3.4 Countermeasures for Nonlinearities

The general countermeasures for nonlinear distortions can be divided into (1) signal design, (2) linearization in the transmitter and (3) equalization or cancellation in the receiver. In general, we believe that it is not reasonable to compensate transmitter nonlinearities in the receiver if there is a fast-fading channel in between.

The power amplifier can be backed-off to operate in its linear region, but this will result in a poor efficiency. The peak-to-average power ratio (PAPR) can be reduced by using clipping, coding or selective mapping techniques. The probability of different QAM constellation points may not be uniform and therefore the PAPR may be larger than that predicted from the uniform distribution. On the other hand, long OFDM symbols generated with a large fast Fourier transform (FFT) can have high PAPR values, but the peak values occur with very low probability [10].

The constant envelope OFDM signaling technique (see Figure 22.18) combines OFDM and phase modulation where (1) phase modulation creates a constant envelope signal which allows the power amplifier to operate near saturation levels, thus maximizing power efficiency and (2) OFDM increases robustness to multipath fading [11]. In general, we can separate OFDM-CPM (continuous phase modulation) and OFDM-PM (phase modulation). Since a CPM signal can be viewed as both phase and frequency modulation, the term OFDM-FM (frequency

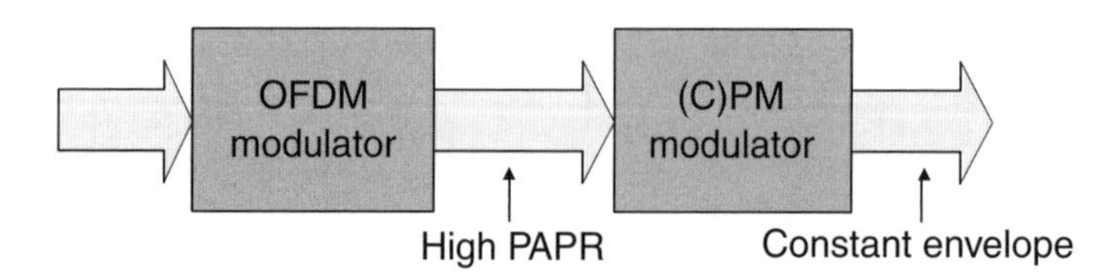

Figure 22.18 Constant envelope OFDM signaling.

modulation) is also used [15]. The idea is that OFDM-FM can be implemented simply and inexpensively by retrofitting existing FM communication systems.

If a PM signal with a rectangular pulse shape and sudden instantaneous phase shifts is first filtered, its envelope will no longer be constant [16] and the subsequent nonlinearity will distort the signal. The CPM signals are better in this respect since the amplitude changes will be small after filtering as there are no sudden changes in the phase. But distortions can exist, and this topic is important since separable nonlinear systems are preferred for complexity reasons. For example, the Wiener model [12] is a cascade of the linear filter and nonlinear memoryless systems.

There are many techniques for linearization. For example, the pre-distorter tries to invert the nonlinear amplifier [17]. Inversion of a nonlinear system with memory may not be possible: not all nonlinear systems possess an inverse, and many systems can be inverted only for a restricted range of input amplitudes. Nonlinear blocks are not in general commutative [18]. We should also use both analog and digital techniques. For example, if we have a separate linearizer for each antenna element in the antenna array systems, the adaptive digital implementation would require down-conversion for each feedback signal. On the other hand, and as a trade-off between the complexity and performance, we can use a common compensator for all antenna elements.

References

1. Kraus, J.D. and Marhefka, R.J. (2002) *Antennas for All Applications*, 3rd edn, McGraw-Hill Inc.
2. El Zooghby, A. (2005) *Smart Antenna Engineering*, Artech House Inc.
3. Suvikunnas, P. (2006) Methods and criteria for performance analysis of multiantenna systems in mobile communications. Doctoral thesis, Radio Laboratory, Helsinki Univ. of Technology, [Online]. Available: http://lib.tkk.fi/Diss/2006/isbn9512282976/.
4. Encinar, J., Piqueras, M.A., Marti, J. *et al.*, Beam-forming methods: Final Report. Tech. Rep. ACE WP 2.4-2, [Online]. Available: http://ace1.antennasvce.org/Dissemination/view/download?id_file=69.
5. Lamminen, A., Säily, J. and Vimpari, A. (2006) *Design and Processing of 60 GHz Antennas on Low Temperature Co-Fired Ceramic (LTCC) Substrates*. Proc. 4th ESA Workshop on Millimetre Wave Technology and applications.
6. Ellis, T.J., Raskin, J.-P., Katehi, L.P.B. and Rebeiz, G.M. (1999) A wideband CPW-to-microstrip transition for millimeter-wave packaging. *IEEE MTT-S Digest*, 629–32.
7. Carver, K.R. and Mink, J.W. (1981) Microstrip antenna technology. *IEEE Trans. Antennas Propagat.*, **29**, 18–20.
8. Lo, Y.T. and Solomon, D. (1979) Theory and experiment on microstrip antennas. *IEEE Trans. Antennas Propagat.*, **27**, 137–45.
9. Vimpari, A., Lamminen, A. and Säily, J. (2006) *Design and Measurements of 60 GHz Probe-Fed Patch Antennas on Low-Temperature Co-Fired Ceramic Substrates*. Proc. 36th European Microwave Conference.
10. van Nee, R. and Prasad, R. (2000) *OFDM for Wireless Multimedia Communications*, Artech House.
11. Kiviranta, M. Mämmelä, A. Cabric, D. *et al.* (2005) *Constant Envelope Multicarrier Modulation: Performance Evaluation in AWGN and Fading Channels*. Proc. MILCOM.
12. Jeruchim, M.C., Balaban, P. and Shanmugan, K.S. (2000) *Simulation of Communication Systems*, 2nd edn, Plenum Press, New York.
13. Karam, G. and Sari, H. (1991) A data predistortion technique with memory for QAM radio systems. *IEEE Trans. Commun.*, **39**, 336–44.
14. Kiviranta, M., Mämmelä, A., Zhang, Y. *et al.* (2005) *A Real-Time Simulation of Impairments in the Analog Parts of the Transmitter-Receiver*. Proc. VTC-Spring.
15. Warner, W.D. and Leung, C. (1993) OFDM/FM frame synchronization for mobile radio data communications. *IEEE Trans. Veh. Technol.*, **42**, 302–13.

16. Ziemer, R.E. and Peterson, R.L. (1985) *Digital Communications and Spread Spectrum Systems*, Macmillan, New York.
17. Benedetto, S. and Biglieri, E. (1999) *Principles of Digital Transmission with Wireless Applications*, Kluwer Academic/Plenum Publishers, New York.
18. Mämmelä, A. (2006) Commutation in linear and nonlinear systems. *Frequenz [Journal of RF-Engineering and Telecommunications]*, **60**, 92–4.

23

Improving Power Amplifier Utilization in Millimeter-wave Wireless Multicarrier Transmission

Peter Zillmann and Gerhard Fettweis

Technische Universität Dresden, Vodafone Chair Mobile Communications Systems, D-01062, Dresden, Germany

23.1 Introduction

Short-range communication for personal area networking and multimedia distribution in the 60 GHz band is expected to be one of the most significant new application areas of wireless communications in the near future. Standardized, low-cost transceivers are required in order to meet the requirements of the consumer market. Consequently, research and development efforts are currently directed towards highly-integrated, all-CMOS transceiver solutions.

The targeted applications, which include distribution of uncompressed HDTV video content, require data rates of 3–10 GBit/s, not including overheads like error correction coding. Keeping in mind that the available bandwidth in the 60 GHz band is in the range of 7 GHz, it is apparent that spectrally-efficient physical-layer transmission is needed, especially when a multiple access is to be supported. In other words, although the available bandwidth seems to be rather huge at first sight, it is actually not in relation to the required data rates.

Another important constraint is the wireless channel: a typical indoor scenario exhibits significant multipath components and thus frequency-selective fading (see, for example, [1]), which requires appropriate equalization methods. Hence, spectrally-efficient multicarrier modulation, implemented as *Orthogonal Frequency Division Multiplexing* (OFDM), is often considered for these applications, although there are physical-layer options requiring less computational complexity.

Multicarrier modulation is also interesting when multimode devices for ultrawideband (e.g., Multiband OFDM [2]), wireless LAN and 60 GHz transmission are considered.

Short-Range Wireless Communications Rolf Kraemer and Marcos D. Katz

However, one of the major drawbacks of multicarrier modulation is the high *Peak-to-Average Power Ratio* (PAPR) of the transmit signal, which is especially critical for the *Power Amplifier* (PA) at the transmitter. Low efficiency and power utilization factors then increase both hardware cost and power consumption [3]. In the context of wireless communications at a carrier frequency of 60 GHz, the achievable average transmit power presents a serious challenge. For a given amplifier, it decreases with increasing PAPR, since a higher *Input Power Backoff* (IBO) is required in order to transmit the signal peaks unaltered. An example illustrates this: the PAPR of typical OFDM signals is in the range of 10–12 dB, depending on the number of active subcarriers. Hence, in order to provide an average output power of 10 dBm, a PA with a peak output power of more than 20 dBm is necessary. Unfortunately, the cost of a PA is essentially determined by its peak output power, not by the average power it is operated at. On top of that, peak PA power comes at a premium, especially at higher carrier frequencies. This can lead to the situation where digital and mixed-signal signal processing is realized in CMOS, while the PA is an external *Gallium-Arsenide* (GaAs) or *Indium-Phosphide* (InP) device – an approach which is hard to justify in price-sensitive consumer market applications.

Information-theoretic investigations show that the reduction of the dynamic range of multicarrier signals results only in a marginal decrease of channel capacity for transmission over an *Additive White Gaussian Noise* AWGN channel [4]. Therefore, it is worthwhile to investigate multicarrier transmission with reduced dynamic range.

Simple reduction of the IBO of the PA creates undesired out-of-band radiation, which is not easily filtered out, since filtering would have to take place on a high-power, low-impedance signal between the amplifier and the antenna. Such a filtering stage leads to insertion losses and consumes a lot of space; it is also subject to manufacturing tolerances and temperature drift.

Therefore, it is desirable to reduce the PAPR of the transmit signal prior to amplification. Like filtering of the amplified signal, analog filtering of the baseband or RF signal suffers from manufacturing tolerances, temperature drift and space limitations. PAPR reduction of the *discrete-time* baseband signal is therefore preferable.

23.2 PAPR Reduction for Discrete-time OFDM Signals

PAPR reduction algorithms for discrete-time OFDM signals can be divided into two basic groups: *data* predistortion and *signal* predistortion. Data predistortion algorithms modify the transmitted bit or symbol sequence such that the PAPR of the resulting OFDM symbols is reduced. Common data predistortion algorithms include *Selected Mapping* (SLM), *Partial Transmit Sequences* (PTS), PAPR reduction *Coding*, *Tone Reservation* and *Constellation Expansion*. An overview of these methods can be found in [5].

A common feature of all data predistortion techniques is their high computational complexity for large numbers of subcarriers or higher-order modulation. The reason is that these

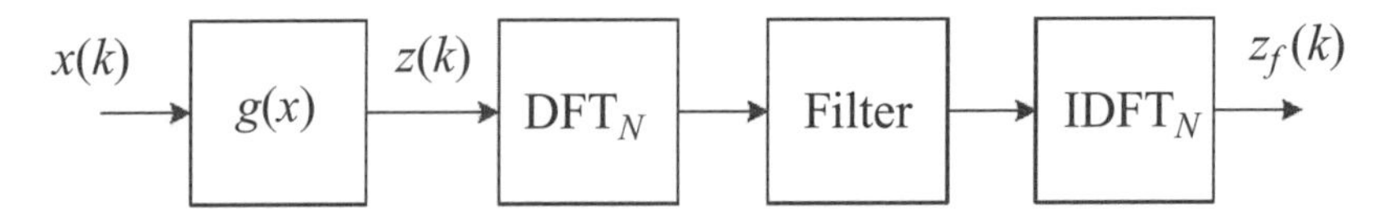

Figure 23.1 Block diagram of CF with frequency-domain filtering.

Table 23.1 Parameters of two OFDM systems.

Parameter	IEEE 802.11a	WIGWAM
DFT/IDFT size, N	64	1024
Used subcarriers, N_u	52	616

approaches require the search of a parameter space, whose size increases with the number of active subcarriers. PAPR reduction coding suffers from a small code rate and difficult design for higher order modulation and high numbers of subcarriers.

The computational complexity of PAPR reduction is a major implementation issue, since it can alleviate the power and cost savings obtained from better utilization of the *Radio Frequency* (RF) PA. Especially in battery-powered equipment, it is not desirable to use a PAPR reduction technique, which comes at the cost of significantly increased signal-processing complexity.

Signal predistortion addresses this problem by simple clipping of the discrete-time baseband signal. Out-of-band power can easily be filtered out in the frequency domain, and the algorithm can be combined with oversampling in order to improve PAPR reduction [6, 7]. In the following, this algorithm will be referred to as *Clipping and Filtering* (CF).

Traditionally, CF was investigated for fully-loaded OFDM systems, where the number of used subcarriers N_u is equivalent to the IDFT-size N of the system. However, in practical systems, guard carriers are used at the edges of the OFDM band and at the centre in order to simplify the implementation of filters and to reduce the impact of RF impairments. These guard carriers are equivalent to oversampling, and it can be shown that already a small number of guard carriers is sufficient to exploit most of the PAPR reduction capability of CF.

Figure 23.1 shows a block diagram of CF without additional oversampling, indicated by the DFT/IDFT size, N. $x(k)$ denotes the discrete-time OFDM transmit signal. $z_f(k)$ is the clipped and filtered discrete-time transmit signal. The frequency-domain filtering stage ensures that clipping noise on the guard carriers is removed, so that the transmitted signal fits into the spectral mask of linear transmission. $g(x)$ is the clipping function; it is usually chosen to be a soft limiter (SL) [7].

Table 23.1 shows the parameters of two OFDM systems which will be used as reference cases in this paper: the IEEE 802.11a system and the WIGWAM[1] system with home/office scenario parameters.

If we define the inherent oversampling factor $a = N/N_u$, we have $a = 1.23$ for the 802.11a system and $a = 1.66$ for the WIGWAM system. Figure 23.2 shows the complementary cumulative distribution function (CCDF) of the PAPR κ of OFDM symbols for IEEE 802.11a and WIGWAM parameters. It was measured at an oversampling rate of 4 in order to capture the peaks of the continuous-time transmit signal. Clipping was done with a soft limiter IBO of $\rho = 0\,\text{dB}$. It can be observed that the PAPR can be reduced by 5–6 dB at an excess probability of 10^{-4}, depending on the number of subcarriers and the inherent oversampling factor. Although unclipped OFDM symbols with WIGWAM parameters have a PAPR of more than 12 dB, the value after clipping and filtering is lower than for 802.11a parameters because of the higher oversampling factor.

[1] For further information, see http://www.wigwam-project.com.

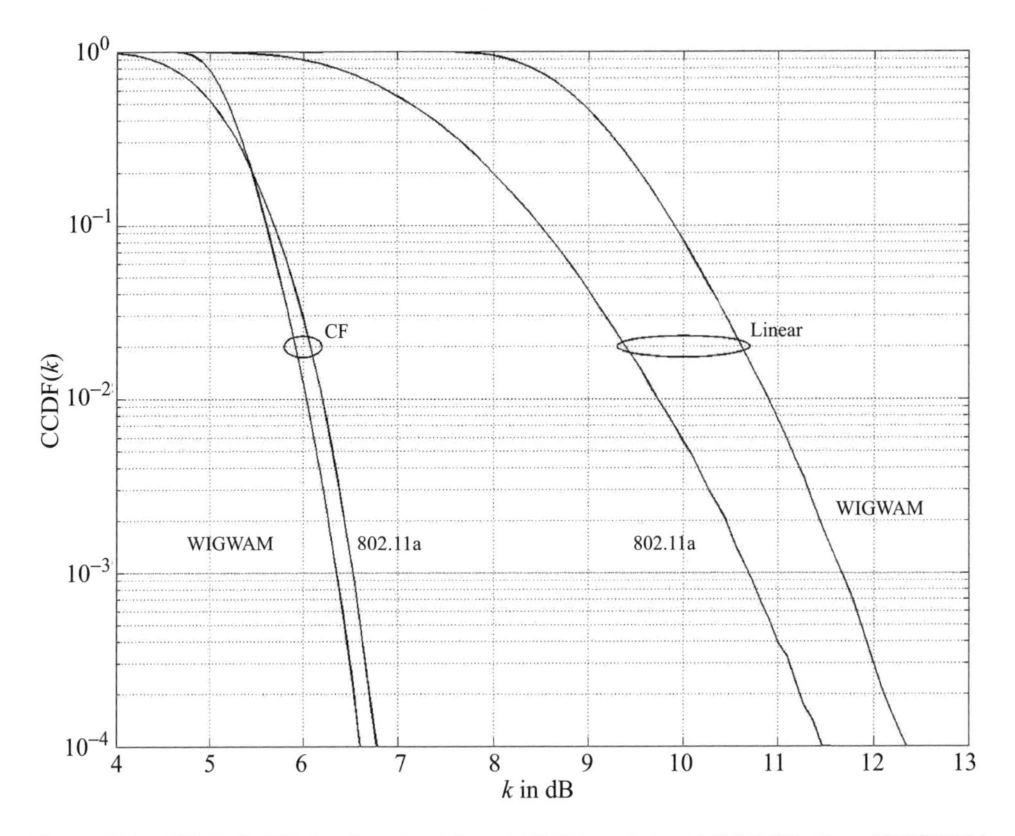

Figure 23.2 CCDF of PAPR for clipped and filtered OFDM symbols with IEEE 802.11a and WIGWAM parameters.

Figure 23.3 shows $\kappa\prime(10^{-4})$, which is the PAPR value that is exceeded by a fraction of no more than 10^{-4} OFDM symbols. It is plotted versus ρ in dB of the SL for the WIGWAM system. The figure also shows that an additional oversampling of factor 4 of $x(k)$ (a DFT of size $4N$ is then also required before filtering) delivers a further improvement of only 0.5 dB.

An IBO of 0 dB is sufficient to capture most of the possible improvement; the lower limit, indicated by the dashed line, is the PAPR after hard limiting of the OFDM signal. That means that all time domain samples have the same magnitude, and the resulting signal is only phase modulated. Digital-to-analog conversion of the complex signal leads to peak regrowth and a PAPR of approximately 5.8 dB.

Of special importance is the computational simplicity of CF: an additional DFT/IDFT pair and clipping have to be implemented. Frequency-domain filtering just consists of discarding signal components falling onto the guard carrier positions after clipping. The computational complexity is independent of the modulation format and the coding scheme, and it is easily characterized by the clipping IBO ρ and the positions of the guard carriers for filtering. CF delivers a strong PAPR reduction of up to 6 dB for typical OFDM system parameters.

Therefore, this PAPR reduction technique is especially well suited to application in battery-powered equipment using multicarrier modulation with a large number of subcarriers and

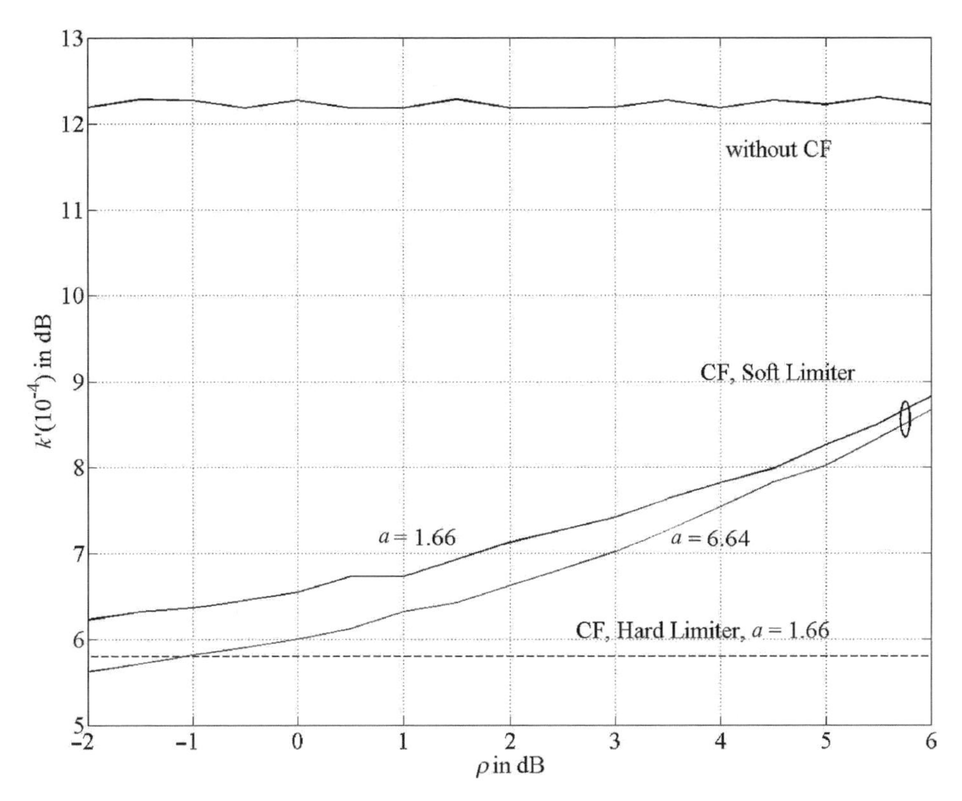

Figure 23.3 PAPR at an excess probability of 10^{-4} for clipped and filtered OFDM symbols and WIGWAM parameters versus IBO of the soft limiter.

high modulation order. It is noteworthy that this approach is different from conventional predistortion (in the sense of linearization) approaches: instead of trying to precompensate for the nonlinearity of an amplifier which is operated at high efficiency, the transmit signal is *deliberately* distorted in the digital baseband, resulting in a low PAPR signal which can be amplified linearly with higher efficiency. For a given average power, the required peak power of the amplifier is reduced by 5–6 dB, which is an important advantage in mm-wave communications.

Naturally, this PAPR reduction comes at the price of in-band distortion noise, which increases the *Bit Error Rate* (BER) of the transmission system if it is not properly equalized at the receiver. The following section shows a powerful approach to soft iterative equalization for clipped and filtered coded OFDM (COFDM) signals.

23.3 Soft Iterative Equalization for Clipped and Filtered COFDM Signals

One basic principle for equalization of nonlinearly distorted OFDM signals was first proposed in [8, 9]. This decision-feedback approach aims at reconstructing the distortion noise and

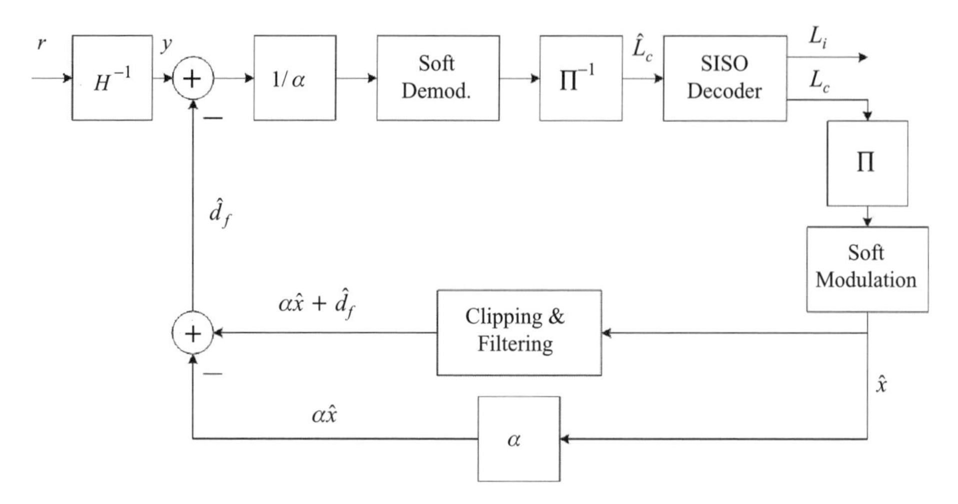

Figure 23.4 Block diagram of soft-decision feedback equalization for clipped and filtered COFDM signals.

subtracting it from the received signal. Initially, the algorithm was investigated with hard-decision symbol feedback for uncoded OFDM transmission. For practical applications, however, it is important to look at the performance for realistic channel scenarios with error-correction coding at relevant SNR values. This analysis was done by the authors in [10]. Due to its manageable computational complexity, the algorithm is well suited to OFDM systems with a large number of subcarriers. Other approaches, like sequential maximum-likelihood equalization and turbo equalization [11, 12], are computationally more expensive.

The authors described soft iterative decision-feedback equalization for coded OFDM signals in detail in [10]; only a brief description shall be given here. It is essential to provide a correction signal which is uncorrelated to the desired signal in order to improve convergence and limit error propagation in the feedback loop. Therefore, it is beneficial to define the distortion noise signal according to the Bussgang decomposition – in principle, any representation, for example as a simple additive error signal, is possible.

Furthermore, the use of soft-symbol decoding and soft feedback turns out to be a huge improvement. Hard-decision decoding and modulation remain attractive when computational complexity at the receiver is an issue.

Figure 23.4 shows the resulting block diagram of soft iterative decision feedback equalization for clipped and filtered COFDM signals. Equalization is done block-wise per OFDM symbol, although one codeword can consist of more than one OFDM symbol. $\boldsymbol{r}$ is the

Table 23.2 Parameters used for performance evaluation.

Parameter	Value
Code	Convolutional code, $r = 1/2$
Generators (octal)	133, 171
Frame size	1 OFDM symbol
Clipping	Soft Limiter, $\rho = 0\,\text{dB}$

received vector of time-domain OFDM samples for one OFDM symbol without cyclic prefix. Equalization of the frequency-selective channel is denoted by H^{-1}, the resulting vector is denoted by $\boldsymbol{y}$. The following scaling factor $1/\alpha$ is a result of the Bussgang decomposition. After soft demodulation and deinterleaving, a *Soft Input-Soft Output* (SISO) decoder delivers Log-Likelihood Ratios (LLRs) $\boldsymbol{L}_i$ of the information bits and $\boldsymbol{L}_c$ of the code bits. The LLRs of the code bits are used for soft reconstruction of the transmit signal, which is denoted by $\widehat{\boldsymbol{x}}$ and then clipped and filtered. The estimate $\widehat{\boldsymbol{d}}_f$ of the filtered distortion noise is then subtracted from $\boldsymbol{y}$, and the next demodulation step delivers improved LLRs. After the desired number of iterations, the information bits are decided according to the sign of the LLRs, $\boldsymbol{L}_i$.

The computational complexity of this method is dominated by SISO decoding, which is done in each iteration. Hard output decoding and, consequently, hard remodulation can reduce the complexity.

The next section shows the performance of the algorithm.

23.4 Performance Results

Table 23.2 summarizes the parameters used for performance evaluation. No standardized specification for an OFDM system at 60 GHz was available when this article was written, but

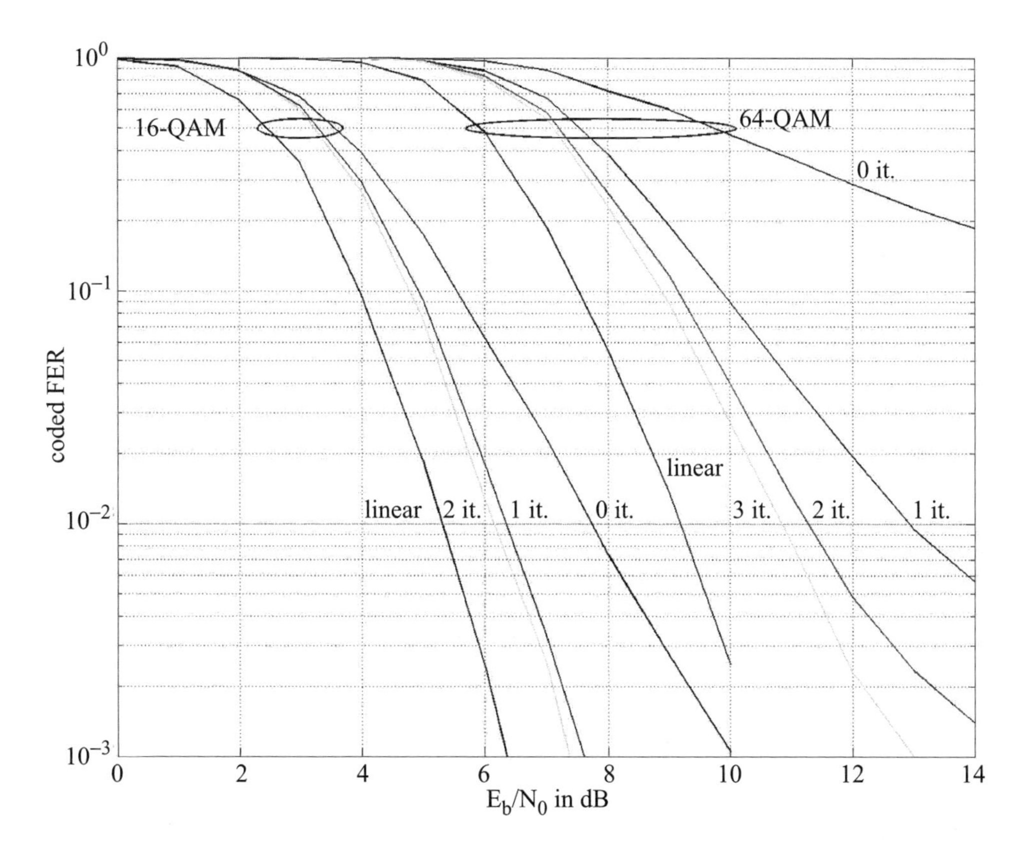

Figure 23.5 Coded FER of an 802.11a system with CF and soft iterative equalization, AWGN channel.

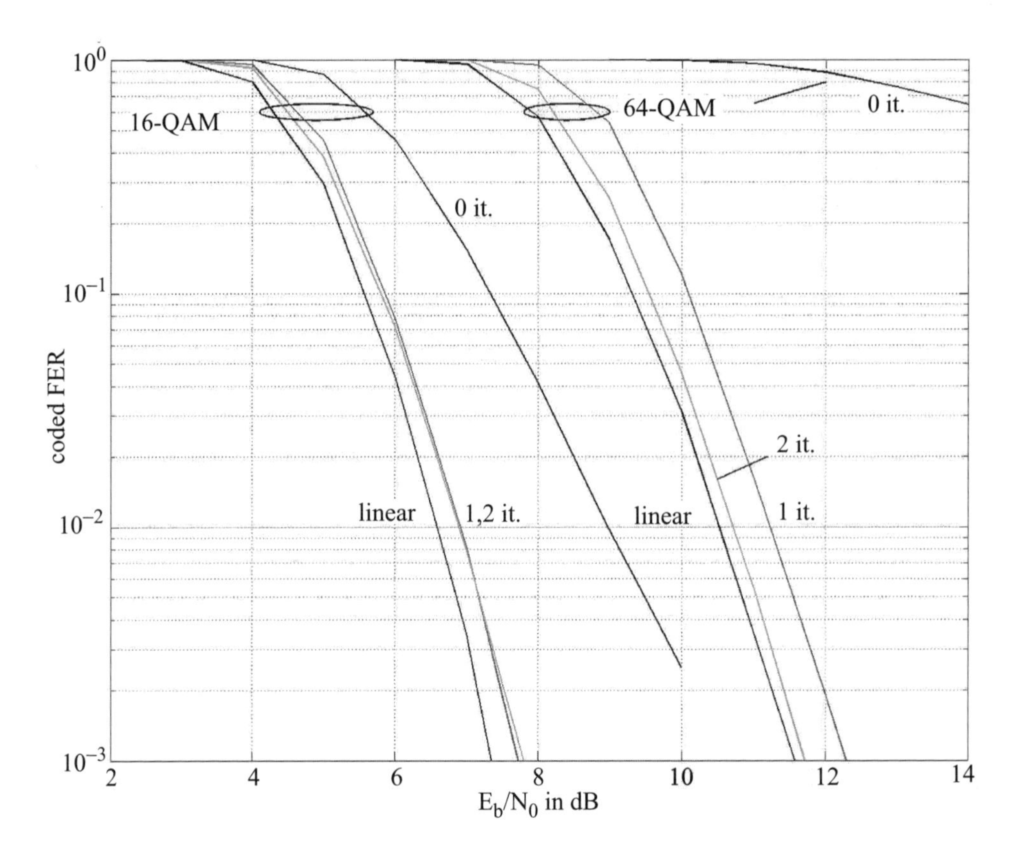

Figure 23.6 Coded FER of a WIGWAM system with CF and soft iterative equalization, AWGN channel.

it can be expected that the parameters of such systems will lie between the IEEE 802.11a and the WIGWAM systems, both in terms of the number of active subcarriers and the inherent oversampling factor. Hence, these two systems deliver useful insights into the behaviour of the equalization process. Since it is a baseband signal processing algorithm, the actual carrier frequency of the OFDM signal does not affect the results. The 802.11a system has a bandwidth of 20 MHz, while the WIGWAM system has a bandwidth of 100 MHz.

The performance was investigated for the AWGN and a frequency-selective channel according to the ETSI Hiperlan A specification. This is a channel model for nonline-of-sight (NLOS) indoor transmission in the 5–6 GHz band. For use with WIGWAM parameters, this channel model was extended to 100 MHz bandwidth. Again, the behaviour of a channel for short-range communications centred around 60 GHz can be expected to lie between these two cases in terms of frequency selectivity.

In all of the following figures, the number of iterations was chosen such that a higher number did not yield further improvements. A coded frame error rate (FER) of 10^{-2} is the reference for comparison.

Figure 23.5 shows the coded frame error rate for 802.11a parameters and the AWGN channel versus the SNR per information bit. For 16-QAM, the FER after two iterations is about 1 dB away from linear transmission; one iteration already delivers most of the improvement. For

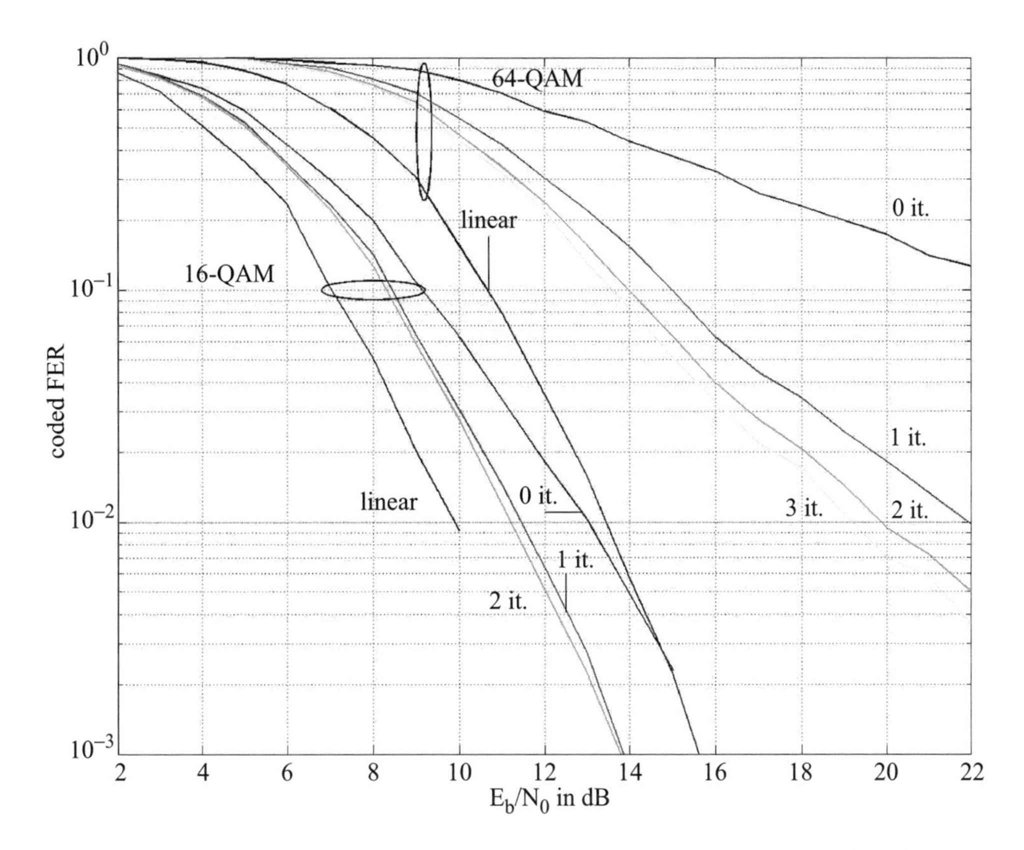

Figure 23.7 Coded FER of an 802.11a system with CF and soft iterative equalization, Hiperlan A channel.

64-QAM, the loss is about 2 dB after three iterations. Figure 23.6 shows the results for the WIGWAM system. For both 16-QAM and 64-QAM, the performance is less than 0.5 dB away from linear transmission after two iterations. The higher number of active subcarriers leads to a better convergence of the algorithm because of the increased codeword length.

The corresponding results for frequency-selective fading are shown in Figures 23.7 and 23.8. While the loss remains at around 1 dB for 802.11a and 16-QAM, it is about 5.5 dB for 64-QAM. Again, a higher number of subcarriers is beneficial: for the WIGWAM system, the performance loss is still below 0.5 dB.

The results indicate that signal reconstruction with SISO decoding and soft feedback is well suited to equalization of clipped and filtered OFDM signals, especially for higher numbers of active subcarriers.

23.5 Conclusions and Future Work

It has been shown that power-amplifier utilization can be significantly improved by clipping and filtering of the discrete-time baseband signal in OFDM systems. This leads to improved efficiency and higher average output power for an amplifier with a given peak power capability,

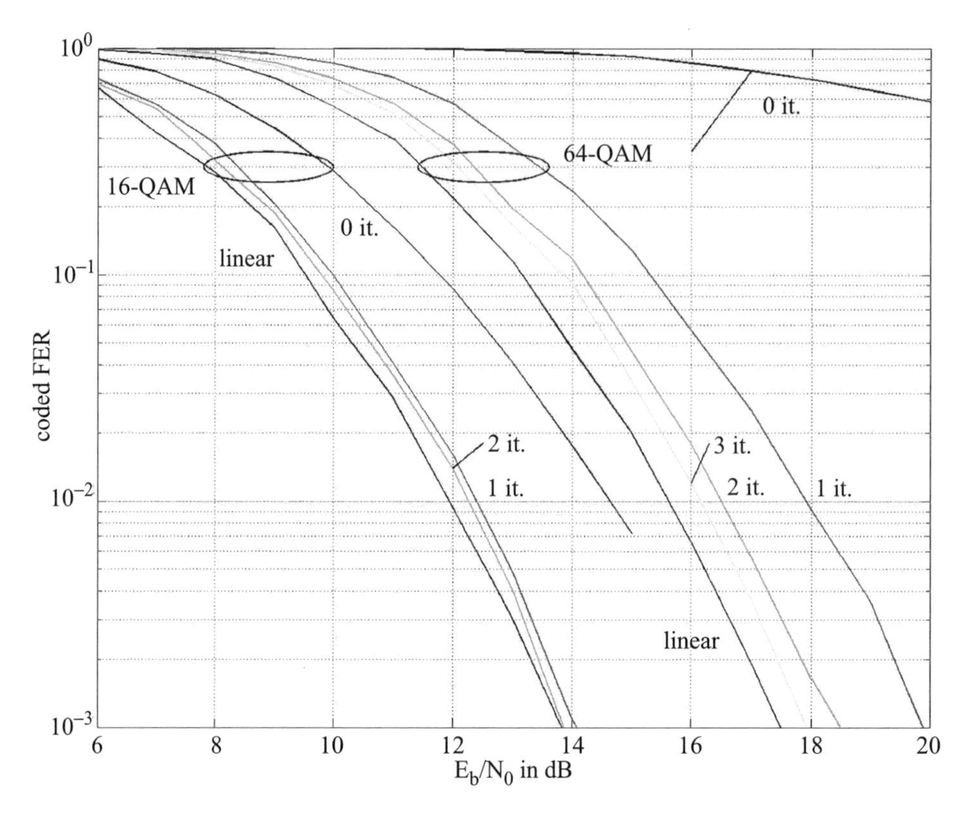

Figure 23.8 Coded FER of a WIGWAM system with CF and soft iterative equalization, Hiperlan A channel.

which is an important advantage for wireless mm-wave communications systems using OFDM modulation. The computational complexity of this PAPR reduction method is low, even for high numbers of subcarriers and higher modulation order. The spectral mask of linear transmission is preserved by frequency-domain filtering.

A matching receive algorithm for equalization of the in-band distortion noise was presented, which restores the coded frame error rate performance to values which are very close to linear COFDM transmission. The computational complexity of this equalization approach is dominated by the SISO decoding operation. Typically, one to three iterations of the algorithm are sufficient for convergence. The presented combination of PAPR reduction and equalization can shift power requirements from the transmitter to the receiver. It is therefore well suited to application in the uplink from a battery-powered device to an access point or stationary receiver.

Future work in this field should be directed towards further complexity reduction. Additionally, turbo equalization methods based on maximum-likelihood detection deserve attention [12]. These approaches can be beneficial for OFDM systems with a smaller number of subcarriers, where decision-feedback algorithms show a less favourable convergence behaviour.

References

1. Lee, W., Kim, K., Kim, J. and Kim, Y. (2006) *Multipath Channel Modeling for 60 GHz Frequency Band*, document IEEE 802.15-06/0038r1.
2. ECMA International – European Association for Standardizing Information and Communication Systems (2005) *High Rate Ultra Wideband PHY and MAC Standard*, 1st edn, Standard ECMA-368.
3. Cripps, S. (1999) *RF Power Amplifiers for Wireless Communications*, Artech House.
4. Zillmann, P. and Fettweis, G. (2005) *On the Capacity of Multicarrier Transmission over Nonlinear Channels*. Proceedings of 61st IEEE Vehicular Technology Conference, Stockholm, Sweden.
5. Han, S.H. and Lee, J.H. (2005) An overview of peak-to-average power reduction techniques for multicarrier transmission. *IEEE Wireless Communications*, **12** (2), 56–65.
6. Ochiai, H. and Imai, H. (2000) Performance of the deliberate clipping with adaptive symbol selection for strictly band-limited OFDM systems. *IEEE Journal on Selected Areas in Communications*, **18** (11), 2270–77.
7. Ochiai, H. and Imai, H. (2002) Performance analysis of deliberately clipped OFDM signals. *IEEE Transactions on Communications*, **50** (1), 89–101.
8. Tellado, J., Hoo, L.M.C. and Cioffi, J. (2003) Maximum-likelihood detection of nonlinearly distorted multicarrier signals by iterative decoding. *IEEE Transactions on Communications*, **51** (2), 218–28.
9. Zillmann, P., Rave, W. and Fettweis, G. (2007) *Soft Detection and Decoding of Clipped and Filtered COFDM Signals*, In Proceedings of 65th IEEE Vehicular Technology Conference (VTC '07), Dublin, Ireland, pp. 23–25.
10. Rave, W., Zillmann, P. and Fettweis, G. (2005) *Iterative Correction and Decoding of OFDM Signals Affected by Clipping*, In Proceedings of the 5th International Workshop on Multi-Carrier Systems and Solutions (MC-SS '05), Oberpfaffenhofen, Germany, pp. 14–16.
11. Zillmann, P., Rave, W. and Fettweis, G. (2007) *Turbo Equalization for Clipped and Filtered COFDM Signals*, In Proceedings of the IEEE International Conference on Communications (ICC '07), Glasgow, UK, pp. 24–27.
12. Hagenauer, J. (2004) *The EXIT Chart – Introduction to Extrinsic Information Transfer in Iterative Processing*, In Proceedings of the 12th European Signal Processing Conference (EUSIPCO '04), Vienna, Austria, pp. 6–10.

Part IV

Emerging Concepts in Short-range Communications

24

Ultra-wideband Radio-over-optical-fibre Technologies[1]

Moshe Ran, Yossef Ben Ezra and Boris I. Lembrikov
HIT–Holon Institute of Technology, Israel

We propose a new concept to address the fundamentally short-range limited ultra-wideband (UWB) wireless communications. The main idea behind the proposed concept is to deliver UWB radio signals over mixed wireless RF and optical channels, where the optical part serves as a super-efficient transparent medium to carry the radio signal of several GHz bandwidths. The new concept is called 'UROOF' (UWB radio-over-optical-fibre).

Future research should therefore aim at developing and enabling the key building blocks for converting UWB signals from electrical to optical domain (E/O conversion) and vice versa (O/E conversion) and at discovering the basic limitations of the overall UROOF channel. Based on UROOF technologies, one can envision a range of new services and applications, such as: range extension of wireless personal area networks (WPAN) by 2–3 orders of magnitude, and new optical/wireless infrastructures capable of delivering broadband multimedia and above 1000 Mb/s traffic to and from subscribers in remote areas. Another application is related to homeland security: collecting data from a large number of sensors and cameras equipped with UWB and transmitting it over existing optical infrastructures using UROOF technologies.

24.1 Introduction

Ultra-wideband (UWB) communications is a fast-emerging technology that offers new opportunities and is expected to have a major impact on the wireless world vision of 4G systems [1, 2]. The important characteristics of UWB signals are their huge bandwidth (3.1–10.6 GHz;

[1]This work is partially supported by the European Commission under project UROOF FP6-2005-IST-5-033615 (see http://www.ist-uroof.org/) and partially appeared in WWRF BoV-III Chapter 7.

Short-Range Wireless Communications Rolf Kraemer and Marcos D. Katz

and more recently, 57–64 GHz) and their very weak intensity, comparable to the level of parasitic emissions in a typical indoor environment (FCC part 15: −41.3 dBm/MHz). The ultimate aim of UWB systems is to utilize broadband unlicensed spectrum (FCC: part 15 : 3.1–10.6 GHz) by emitting noise-like signals. The major UWB advantages are potentially low complexity and low power consumption, which implies that UWB technology is suitable for broadband services in the mass markets of WPAN. Furthermore, UWB combines high data rates with localization and tracking features, and hence opens the door for many other interesting applications, such as accurate tracking and location, safety and homeland security. For these reasons, UWB is considered a complementary communication solution within the future 4G systems.

However, the current high-data-rate UWB systems (e.g., 480 Mb/s; see [3] and future evolving multi-gigabit UWB version IEEE802.15.3c at 60 GHz (http://www.ieee802.org/15/pub/TG3a.html)) are inherently limited to short ranges of less than 10 m. This simply derives from the constraints on allowed emission levels and the fundamental limits of thermal noise and Shannon limits [4]. Larger coverage of high-data-rate UWB, to say 10–10 000 m, is most desired for broadband access technology.

We propose a new concept for converging between the high-data-rate wireless short-range communications based on UWB technologies and the wired optical access technology. The main idea behind the proposed concept is to enable the transmission of UWB radio signals transparently over optical channels. This concept is called UWB radio-over-optical-fibre (UROOF).

In this approach, the UWB RF signals of several GHz are superimposed on the optical CW carrier. This strategy has several advantages: (1) it makes the conversion process transparent to the UWBs modulation method; (2) it permits avoiding the high costs of additional electronic components required for synchronization and other processes; (3) it makes possible integration of all the RF and optical transmitter/receiver components on a single chip. The overall combination of these features results in a cost-effective broadband system that can easily support 1 Gb/s, which is suitable for the residential markets and WPAN.

The UROOF concept is a new paradigm that extends the state-of-the-art radio-over-fibre (RoF) technologies to the short-range communication case. RoF systems in mobile cellular are motivated by the demand for replacing a central high-power antenna with a low-power distributed antennas system (DAS). RoF technologies are successfully deployed for in-building coverage in 2G/3G cellular networks. In this application, many remote access units (RAU) serving as low-cost base stations (BSs) are connected to a single central station (CS).

The converged UROOF proposed in this contribution is a promising approach in the field of short-range communication applications. One of the most important goals of this approach is to overcome the inherently limited range of high-data-rate UWB short-range communications, extending it by 1–3 orders of magnitude compared to the state-of-the-art UWB radio systems in 3.1–10.6 GHz band. Similar to RoF, UROOF allows separation of low-cost BSs from the CS.

The main differences between UROOF and the conventional RoF are:

- The state-of-the-art RoF technologies are used in the wireless access systems, whereas UROOF addresses the challenges of range extended low-cost WPANs.
- In RoF, which targets the 2G/3G cellular systems, the RF signal bandwidth is only a few 10 MHz and its average power is in the range of several 100 mW. This requires high-cost photonic components in the CS and medium-cost components in the BS. UROOF, on the

other hand, targets the PAN market, which is characterized by very low-cost and low-power (10 mW) access points. In UROOF, the optical is used to carry extremely wide RF signals (several GHz).

Other approaches to extending the fundamentally limited short-range nature of the high-data-rate UWB system are either practically not realizable due to link budget considerations or too expensive for the WPAN market. For example, the free-space optical link approach suffers a 41 dB loss over 10 km, roughly the same as 0.375 inch coax cable of 10 km length operating at 10 GHz [5]. In the free-space RF approach, the signal link losses at, say, 4 GHz centre frequency would be at least 125 dB. These should be compared to the extremely low loss level of opticals, which is about 3 dB at a wavelength of 1.55 mm. A legacy approach to demodulate the UWB signal and transmit it digitally as 1 Gbit/s Ethernet data over single mode is too expensive for WPAN applications. Furthermore, this solution is tailored to the specific UWB technology being employed. We are looking for much more generic (e.g., agnostic to the specific modulation technology at the access point) and scalable solutions that can be easily applied to other, less demanding cases (e.g., range extension of wireless local area networks (WLAN)). Other potential solutions using ad-hoc and multihop network topologies to deliver the high data rate between the nodes of WPAN would impose major delay constraints that would prevent the sending of most target WPAN services and applications.

Currently, three main flavours of UWB technologies are proposed for WPAN communication: impulse radio (IR-UWB), direct sequence (DS-UWB) and multiband OFDM (MB-OFDM); see updated survey in [1].

In IR-UWB, information is carried in a set of narrow-duration electromagnetic pulses. The bandwidth is inversely proportional to the pulse width. Unlike conventional wireless communications systems, which are based on carrier modulation, IR-UWB is essentially a baseband ('carrier-free') technology. The centre frequency in IR-UWB is dictated by the zero crossing rate of the pulse waveform. In general, waveforms for IR-UWB are designed to obtain flat frequency response over the bandwidth of the pulse and to avoid a DC component. Various 'monocycle' waveforms have been proposed to meet these characteristics, including Gaussian, Rayleigh, Laplacian and cubic [6]. The time domain representation of a Gaussian monocycle pulse is given by:

$$p_G(t) = A_G \left[1 - \left(\frac{t}{\sigma} - 3.5 \right)^2 \right] \exp \left[-0.5 \left(\frac{t}{\sigma} - 3.5 \right)^2 \right] \tag{24.1}$$

where the monocycle pulse width T_p is chosen to be $T_p = 7\sigma$ in order to ensure that 99.99% of the total energy is captured in the Gaussian monocycle. A_G is the amplitude of the monocycle, chosen in such a way that it has the unit energy. The frequency spectrum of the Gaussian monocycle is given [6] by:

$$S_G(f) = A_G \sqrt{2\pi\sigma^2}\,(2\pi f)^2 \exp\left[-\frac{(2\pi\sigma f)^2}{2} \right] \tag{24.2}$$

Note that the Gaussian monocycle has an even symmetry; this can reduce the mathematical analysis of performance of IR-UWB over various channels. Various data modulation formats

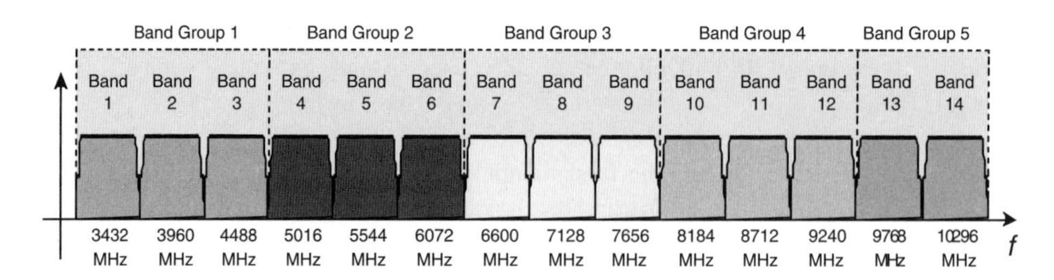

Figure 24.1 MB-OFDM band plan.

may be used with IR-UWB. Among them are pulse amplitude modulation (PAM), pulse position modulation (PPM) and pulse shape modulation (PSM) [7]. Typically, the PPM format is combined with time hopping to spread the RF energy across the frequency band and reduce the spikes in the spectrum of the pulse train.

DS-UWB borrows the concepts of conventional DS spread spectrum (DS-SS). By utilizing a spreading code like pseudo noise (PN) sequence, or ternary spreading codes with prescribed correlation properties [8] and a given chip time T_c, a spreading effect is accomplished. The chip duration plays the same role as the pulse width, T_p, of the monocycle in IR-UWB. However, the DS-UWB is a completely different concept. It has constant envelope modulation format and the information signal is modulated twice: by a spreading code and by a carrier waveform. Unlike IR-UWB, where the interference suppression mechanism is based on time windowing over T_p, the DS-UWB is based on cross-correlation properties of the spreading code and subsequent low-pass filtering at the information bandwidth at the receiver. The DS-UWB waveform proposed in [8] is based upon dual-band BPSK and 4-level biorthogonal modulation (4-BOK) with band limited to baseband data pulse. The mandatory lower band ranges from 3.1 to 4.85 GHz and the optional upper band from 6.2 to 9.7 GHz.

MB-OFDM [9] is based on subdividing the UWB spectrum into 5 band groups and 14 sub-bands of 528 MHz width (see Figure 24.1). Only Band Group 1 is mandatory, while 2–5 are optional. Each 528 MHz channel is composed of 128 subcarriers of the OFDM signal, each QPSK modulated. The OFDM symbol period is 312.5 ns and the time-frequency code switches the signal between the sub-bands of a given group band.

It is the goal of this research to address some of the following questions:

- What are the key requirements for realizing UWB over for each of these applications: range extension, wireless distribution networks, wired (e.g., hybrid fibre coax (HFC)) distribution networks?
- Which UWB technology – IR-UWB, DS-UWB or MB-OFDM – meets best the aforementioned applications of UWB over?
- How can we achieve integration of UWB-over technology into next-generation systems (4G, next-generation HFC, etc.)?
- Which media access control (MAC) protocol and multiple access schemes best fit UWB over?

The rest of this chapter is organized as follows. In Section 24.2 we briefly review UROOF background and state-of-the-art in radio-over-fibre (RoF) technologies. In Section 24.3 we present typical user applications for UROOF technologies. In Section 24.4 we discuss fundamentals of UROOF technologies, and in Section 24.5 we discuss link budget aspects for future UROOF systems. In Section 24.6 we compare several UWB technologies for RoF and show through a list of evaluation criteria and simulations that MB-OFDM technology provides the most promising direction for implementing UROOF. We further investigate the performances of MB-OFDM WiMedia flavour over multimode. In Section 24.7 we develop an all-optical approach for the generation of IR-UWB for UROOF applications. In Section 24.8 we point out the technology trends which must be explored to realize UROOF systems, and give some concluding remarks.

24.2 Motivation and State of the Art

24.2.1 Objectives

In this section we will:

1. Introduce UROOF case scenarios, system and device concepts that can contribute to the 4G vision.
2. Provide detailed performance analyses of typical UROOF applications with different UWB modulation schemes to select the best technique for UWB radio-over-optical distribution, and to determine the optimum RoF distribution configuration.
3. Focus on the challenging problem of low-cost and high-performance conversion of high-data-rate modulated communication signals from optical domain (over single mode and multimode) to UWB radio frequency domain and vice versa.
4. Identify future research areas related to the design and performance of UROOF novel building blocks, including: evaluation of the impact of electro-optical device nonlinearities (pulse chirp/chromatic dispersion) on the UROOF distribution network; device suitability studies for multiple-access architecture VL-DAS; uncooled VCSEL-based design; and optical-UWB link analysis.

24.2.2 State of the Art of Radio-over-Fibre

Radio-over-fibre (RoF) is an analog optical link transmitting modulated RF signals. The development of RoF systems is motivated by the demand for replacing a central high-power antenna with a low-power distributed antennas system (DAS) [10]. RoF systems are usually composed of many base stations (BSs), which are connected to a single central station (CS). Therefore, many efforts have already been devoted to reducing the cost of BSs and moving complexity to the CS [11]. Recently, an RoF-based wireless 'last mile' access network architecture was proposed [12] as a promising alternative to broadband wireless access (BWA) networks. In such a network, the CS performs all switching, routing and operations administration maintenance (OAM). The optical network interconnects a number of simple and compact antenna BSs for wireless distribution. Each BS has no processing function and its main function is to convert the optical signal to wireless and vice versa. Each BS serves as an access point for an area in which many subscriber stations (SS) exist. The proposed architecture assumes a centralized medium access control (MAC) located at the CS responsible for offering a reservation-based, collision-free medium access.

Currently, RoF systems are used for several applications, such as:

- Radio-over-fibre for 2G/3G cellular systems:
  - Radio coverage extension in dense urban environments [13].
  - Capacity distribution and allocation [13].
- RoF for metropolitan area networks (MAN); both wired (Cable TV) and wireless (IEEE 802.16x) broadband access systems.
- Intelligent transport systems (ITS); road-to-vehicle communication systems using, for example the 36/37 GHz carrier frequency [13, 14].
- Other applications, such as use of RoF in radar systems to isolate the radar control station from the radar antenna.

Analog optical links that can support transmission of an entire RF band of a few GHz over a large distance are of great importance. The link losses are very small compared to those of the wireless or wired-coax channel, as mentioned above. The power budget of a UROOF system is determined by the contributions of RF losses, optical losses and conversion losses. The RF–optical conversion losses are dominant in the power budget. However, the total losses can be substantially reduced by using opticals instead of a cable or wireless transmission, due to the extremely low losses per unit length of an optical. For this reason, the development of efficient and cost-effective RF–optical links for UWB technology is a main objective of our research. It is important to understand the factors behind the conversion losses and to develop techniques for reducing them.

We now briefly overview some key building blocks of RoF technologies.

Up-conversion from RF to optical domain can be realized either by direct laser modulation or by external modulation methods [15]. Direct methods have the advantages of simplicity and low cost. Their main disadvantages are relatively limited bandwidth (10 GHz), high chirp, nonlinear and intermodal distortion, and SNR limited by relative intensity noise (RIN). Common external modulation methods are:

- Mach–Zehnder (MZ) interferometer, characterized by limited bandwidths (2–3 GHz), high linearity, low chirp and high bias voltage. In particular, travelling wave (TW) configuration of the MZ modulator permits overcoming the bandwidth limitations.
- Electro-absorption modulator (EAM), characterized by high bit rate and compatibility with the advanced photonic technologies. EAMs based on the quantum-confined Stark effect (QCSE) in quantum wells can exhibit advanced performance.

Down-conversion from optical to RF domain can be implemented by a range of high-speed photodetectors, PIN diodes and avalanche photodiode (APD) photodetectors, and is characterized by simplicity and a typical bit rate of a few 10 Gb/s. A travelling wave PIN based on QWs is used to generate microwave power. Frequency response of a TW photodetector reaches −3 dB at approximately 100 GHz.

Bidirectional conversion based on electro-absorption transceiver (EAT). EAT acts as a receiver for the downlink and as a modulator for the uplink (Figure 24.2). This approach is appropriate mostly for mobile and MAN systems, in which the cost of the BSs is greatly reduced and the complexity is moved to the CS. However, in this case an optical amplifier such as Erbium doped fibre amplifier (EDFA) is used in order to compensate for the link

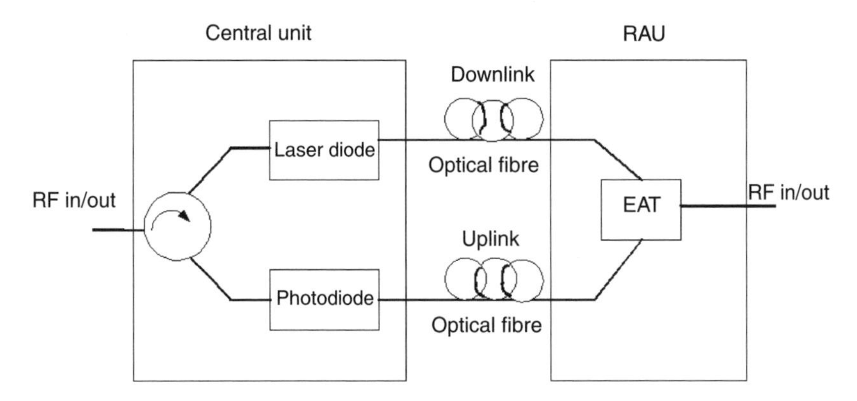

Figure 24.2 A bidirectional RoF BS containing a remote antenna unit (RAU) based on EAT.

losses. Since EAT operates mainly in the 1500 nm range, it is compatible only with single-mode and Fairy-Perot cavity or distributed feedback (DFB) lasers. It can be made to cover other wavelength as well (1550, 1310 nm). However, when it is necessary to amplify the signal with an EDFA it is necessary to choose a 1550 nm wavelength. EDFAs are no longer the only method to provide optical amplification in networks, as in recent years significant progress has been made with semiconductor optical amplifiers (SOA), which can be designed to operate either in the 1310 nm or the 1550 nm band. SOAs have the longer-term potential to be lower-cost devices, so they may be more appropriate to UWB distribution networks with many distribution taps to base stations.

EAT can operate up to 60 GHz and appears to be the most promising candidate for BSs in the future broadband wireless access systems [13, 16]. A promising EAT consists of a multiquantum well (MQW) III–V semiconductor active waveguide.

We will discuss these technologies in detail in Section 24.4.

24.2.3 Potential Impact of UROOF Technologies

UWB over permits us to realize the convergence of different access schemes: optical–wireless, HFC–wireless, short-range personal area network (PAN)–metropolitan area network (MAN). The need for global availability, performance, increase in data rates and network-independent services will determine the required technical characteristics of the converged UROOF system. UROOF will greatly simplify the future multilayered approach of access technologies described in [17]. In this approach, five horizontal layers are introduced: distribution layer, cellular layer, hot spot layer, PAN layer and fixed (wired) layer. UROOF will allow, for example, direct and seamless connection between the distribution layer and the personal area network layer.

UROOF will influence several areas of 4G wireless systems:

- Broadband access technologies (PHY and MAC layers) and interference enhancements in dense multi-user communications.
- Network layer: access across heterogeneous networks, convergence of distribution networks with short range.

- Terminals (technical capabilities versus cost).
- Applications and services.

24.2.3.1 Standards-related Topics

The UROOF concept and its outcomes might be relevant to the following regulation and standards committee bodies: CEPT ECC TG3 and 802.15.4802.15.3c.

24.3 UROOF: User Applications and Basic System Configuration

In this section, several novel applications scenarios are addressed with the proposed UROOF system concepts.

- *Case 1: WPAN range extension.* UWB radio signals are transmitted and received among UROOF nodes over combined wireless–RoF links over distances around 1000 m.
- *Case 2: Very low-cost distributed antenna system (VL-DAS).* Several access points equipped with UROOF transceivers, each located in different pico-net locations, are connected through RoF links to a central station. UWB signals pass over the wireless/RoF channel through the access points.
- *Case 3: Security and homeland applications.* A collection of UROOF nodes (e.g., sensor network) capable of transmitting simultaneously through radio–UWB and RoF-fibre are all connected to a control station via optical, enabling low-latency, location-enabled and secure (by beamforming) communications.
- *Case 4: Capacity expansion of hybrid fibre coax (HFC) systems.*
- *Case 5: Intelligent transportation systems.*

24.3.1 Case 1: WPAN Range Extension

The UWB technology advantages are well known: (i) potentially low complexity and (ii) low power consumption. This implies that UWB technology is suitable for broadband services in the mass markets of wireless personal area networks (WPAN). Inherently, the high data rates, around 480 Mb/s, are available over short ranges of less than 10 m. Several studies point out that the usability range should be around 3 m.

The limited UWB range has applications in home/office-wide communications. The range extension application aims to enable this high data rate in extended areas from 10 to 1000 m. This means that high-data-rate UWB connectivity is provided in an integrated way. This approach is depicted in Figure 24.3; the arrows indicate seamless UWB connectivity.

The UWB range extension application overall performance depends on the number of simultaneous users, the traffic and the modulation type (given the packet length) used. Eventually, the collision probability can be reduced by using the different channels (sub-bands) available in the UWB spectrum (3.1–10.6 GHz). Two different cases of this application can be identified:

- *Case 1.A: In-building UWB extension.* The in-building UWB range extension architecture consists of deploying a single optical fibre in a building structure. Several UROOF nodes act as access points to provide seamless in-house UWB connectivity. Figure 24.3 illustrates this concept; Table 24.1 shows the functionalities of this scenario.

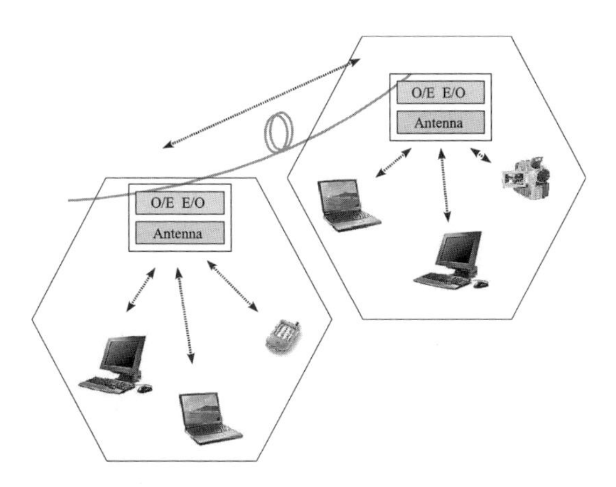

Figure 24.3 UROOF UWB range extension application.

- *Case 1.B: UWB Broadcasting*. This scenario targets the provision of high-definition multimedia services from an operator node or head-end to a number of subscribers employing UWB technology. This application is cost-effective for the operator as no modulation-conversion is required, and it is also interesting for the consumer, as the information arrives to the home as a UWB signal. This application combines the economy and bandwidth advantages of HFC broadcasting with the cost-economy and flexibility of UROOF technology. Figure 24.4 shows this application; Table 24.2 shows the functionalities of this scenario.

24.3.2 Case 2: Very Low-cost Distributed Antenna System (VL-DAS)

When a high number of users are present in a UROOF environment, that is when a large area is covered, the access efficiency of the Case 1 application can be affected heavily.

24.3.2.1 Case 2.A: UWB Access Segmentation

To solve this issue, a cluster-based structure based on a very low-cost distributed antenna system (VL-DAS) can be adopted. This architecture is interesting in terms of security; it avoids direct access from anybody to anyone. It could be leaded by the structure of the building in which the deployment has to be achieved. It is similar to the segmentation done in LAN networks to optimize the network capability. It is depicted in Figure 24.5.

In this architecture, it is clear that all the exchanges on a single floor (intrafloor exchanges) are connected directly on the same optical, but that the interfloor exchanges will go through

Table 24.1 UROOF Case 1.A functionalities.

Min. bit rate required	NO			
Bidirectional capabilities	YES	Node density	HIGH	*1 node every 1–3* m
Location capabilities	NO	Multichannel UWB	NO	
Latency control	NO			
Range extension	SHORT		*100 m span*	

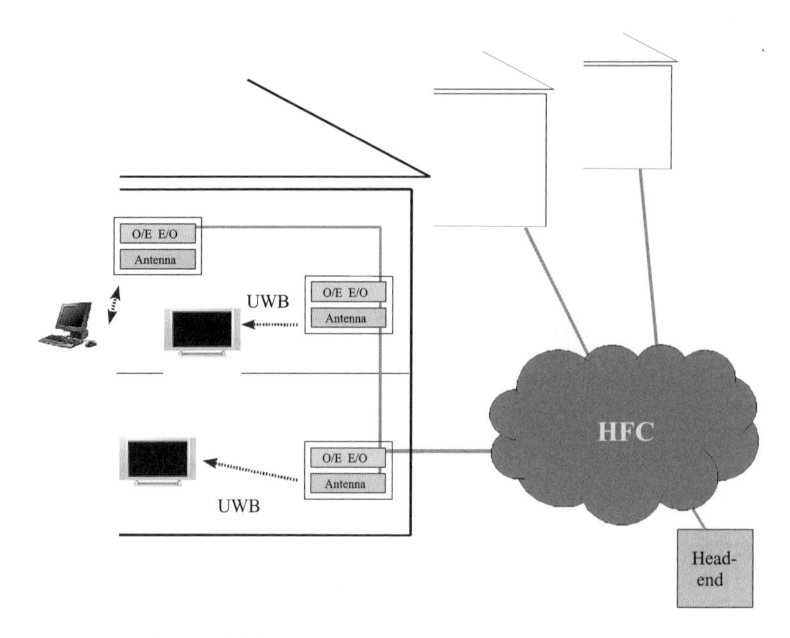

Figure 24.4 UROOF Case 1.B UWB broadcasting.

concentrators. In such a structure, the local traffic of one floor has no effect on the local traffic of the other floors; the traffic is cumulated floor by floor and each concentrator (with router function) will secure the access and act as the real access point (in the definition of the 802.15.3a and 802.15.4a standards). See Table 24.3 for the scenario functionalities.

24.3.2.2 Case 2.B: UWB MIMO Extender Coverage

This architecture is also suitable to provide enhanced smart antenna processing. In Figure 24.6, we consider a structure in which the number of converters is increased in order to enable new antenna processing or new triangulation computation at the concentrator level. In this application, it is considered that the density of converters is sufficiently large to get the signal from one mobile by multiple antennas. The consequence is that the concentrator can achieve a kind of 'macrodiversity', macro meaning that the distance between the antennas is very large in comparison to conventional diversity schemes, thus improving the traditional diversity results.

Table 24.2 UROOF Case 1.B functionalities.

Min. bit rate required	YES	Depends on video quality: min. 384 kb/s (low-Q compressed); max. 55 Mb/s (HD, uncompressed)		
Bidirectional capabilities	NO	Node density	LOW	3 nodes home + head-end
Location capabilities	NO	Multichannel UWB	YES	Depends on the number of users
Latency control	NO			
Range extension	LARGE	*Max. 10 km*		

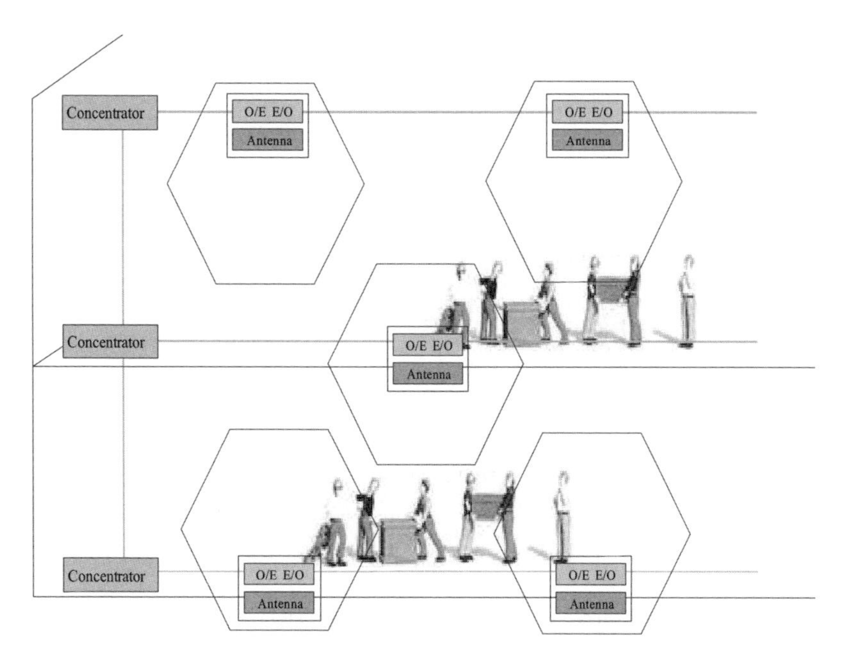

Figure 24.5 UROOF Case 2.A UWB access segmentation.

This application requires precise control of the signal latency. To go further than typical diversity mechanisms, such a structure can be used to implement MIMO algorithms or beamforming techniques. The condition for this is the ability to manage the amplitudes and phases sent to each antenna. It is a complementary way to extend the coverage or to focus the energy in one particular direction, or to avoid interference in certain directions. Table 24.4 summarizes these scenario functionalities.

24.3.3 Case 3: Security and Homeland Applications

24.3.3.1 Case 3.A: Low-latency Communications

Latency control is particularly useful for applications in which time constraints are mandatory, as is the case for homeland security, security in tunnels and so on. A typical scenario is given in Figure 24.7.

Table 24.3 UROOF Case 2.A functionalities.

Min. bit rate required	NO			
Bidirectional capabilities	YES	Node density	MEDIUM	1 node every 20 m
Location capabilities	NO	Multichannel UWB	YES	
Latency control	NO			
Range extension	MEDIUM			

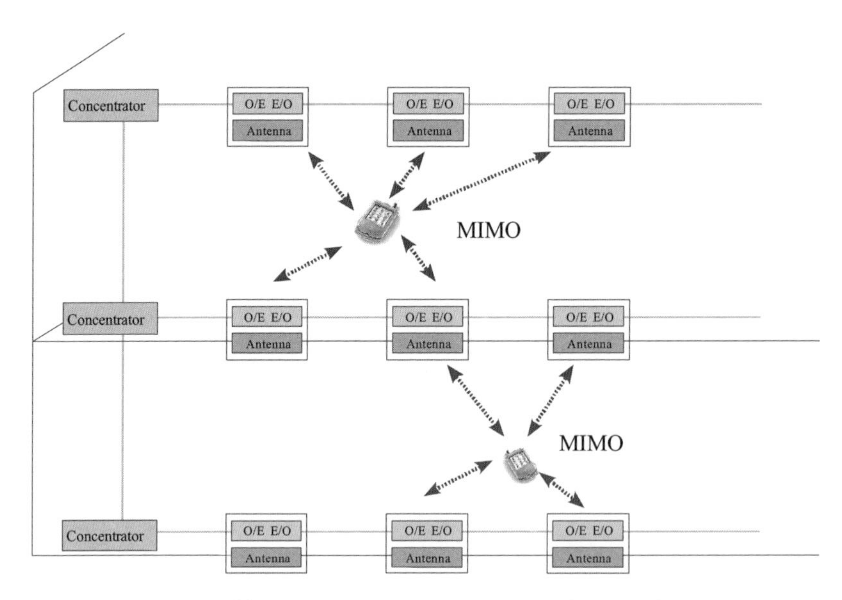

Figure 24.6 UROOF smart antenna MIMO processing.

In the case of long tunnels where security is a major concern, surveillance cameras are deployed in high density. The simultaneous use of a large number of cameras, covering every point in the tunnel and in the parallel corridors, by the security crew puts exceptional strain on the communication infrastructure. This also applies to rescue teams, which can take advantage of the capability to download in real time such video streams. UROOF technology provides an economical way to transfer simultaneously a number of video streams with latency control, if DWDM channels are used.

UROOF technology also enables latency-critical applications, such as traffic control of metro/train convoys. These applications are based on small UWB transceivers attached to the convoys. Traffic control applications require a controlled latency and high availability as mission-critical applications. Table 24.5 summarizes the scenario functionalities.

Table 24.4 UROOF Case 2.B functionalities.

Min. bit rate required	NO			
Bidirectional capabilities	YES	Node density	HIGH	*1 node every 10 m (MIMO requires at least two nodes detecting emission from one UWB terminal of 10 m range).*
Location capabilities	NO	Multichannel UWB	YES	*2 (bidir.)*
Latency control	YES	*Depends on MIMO algorithm performance.*		
Range extension	MEDIUM	*100 m per concentrator*		

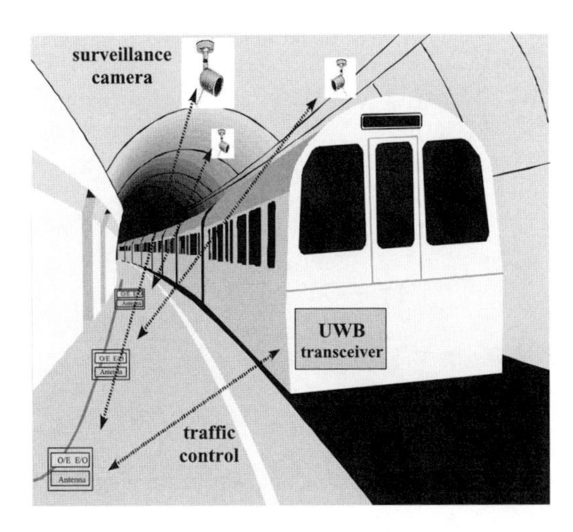

Figure 24.7 UROOF very low latency applications.

24.3.3.2 Case 3.B: Localization-enabled Communications

UROOF technology can also be applied to take advantage of the different signals received by the antenna for triangulation purposes, and the different signals transmitted for beamforming applications. The triangulation is based on the time of arrival (TOA) of the different signals. In this way the distance between a mobile and a set of transceivers is measured sequentially, and its position is evaluated. An accurate estimation can be made as long as the channel is stationary.

If adequate latency control is provided, it is possible to control the time delay (in a coarse mode) between the signals radiated by the antennas. In this case it is possible to include proper beamforming to increase the communication reach and to increase confidentiality. The two above-mentioned applications are depicted in Figure 24.8; Table 24.6 summarizes the scenario functionalities.

24.3.4 Case 4: Capacity Expansion of Hybrid Coax (HFC) Systems

Yet another interesting related application is the use of UWB technology to increase the capacity of existing wired broadband technology, for example HFC distribution systems. This

Table 24.5 UROOF Case 3.A functionalities.

Min. bit rate required	YES	*384 kb/s, surveillance camera video*		
Bidirectional capabilities	YES	Node density	MEDIUM	*1 node every 20 m*
Location capabilities	NO	Multichannel UWB	YES	*Depends on number of cameras*
Latency control	YES	*Specified from traffic control standards.*		
Range extension	LARGE	*10 km*		

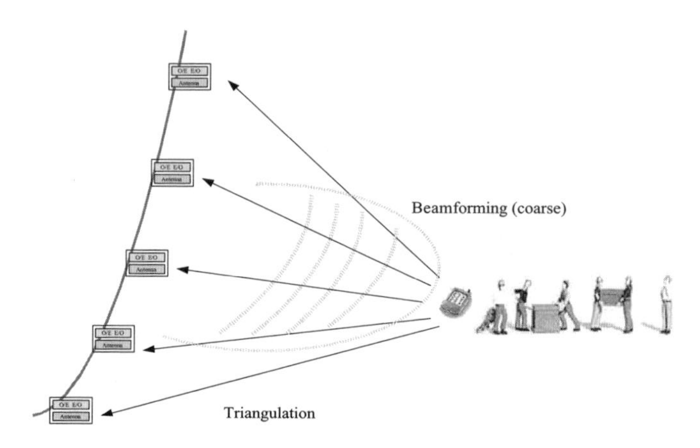

Figure 24.8 Localization-based (triangulation + beamforming) UWB communication.

approach aims to achieve 100–480 Mb/s over existing HFC infrastructure by superimposing a UWB signal into existing data signals. According to the ANSI/EIA-542-1997 US standard, 6 MHz analog video channels are combined by frequency division multiplex (FDM) using frequencies from 55 to 547 MHz. These frequencies (channels 2–78) are the forward analog channels in an HFC network. Forward digital channels occupy a frequency range from 553 MHz (channel 79) to 865 MHz (channel 136). In some extended systems, the range may reach 997 MHz (channel 158). The frequencies between 5 and 40 MHz are allocated in many countries to the digital reverse channel. The two links, downlink (O/E) from head-end to customer and uplink (UL) from customer to head-end, are combined using frequency division duplexing (FDD).

Properly precompensated UWB signals are capable of coexisting with the currently deployed HFC spectrum or even with the actual DC to 3 GHz spectrum of the coax.

24.3.5 Case 5: Intelligent Transportation Systems

Road-to-vehicle and vehicle-to-vehicle are the basic components of evolving technologies which increase the safety of people on the roads and enable a comfortable mobile traffic environment. Recently, a new family of IEEE standards, aimed at providing communications between cars and roadside infrastructure, was approved for trial use. This family, known as the

Table 24.6 UROOF Case 3.B functionalities.

Min. bit rate required	NO			
Bidirectional capabilities	YES	Node density	HIGH	*Depends on triangulation resolution and beam-steering resolution.*
Location capabilities	YES	Multichannel UWB	YES	*2 (bidir.)*
Latency control	YES			*Depends on beamforming angle resolution.*
Range extension	LARGE			*10 km*

IEEE 1609 suit of WAVE (Wireless Access in Vehicular Environments), is based on existing 802.11 short-range communications at 5.9 GHz. Cars with WAVE may be a reality in about 2011.

UROOF technologies can be integrated into the ITS concept and provide means to add broadcasting and anti-collision radar information.

24.4 Fundamentals of UROOF Technologies

24.4.1 Introduction

Optical communication systems use carrier frequencies of about 10^{14} Hz in the visible or near-infrared region of the electromagnetic spectrum. Lightwave systems usually employ optical fibres for information transmission. The well known advantages of the fibre-optic communication systems are, for instance, low losses of transmitted power, compatibility with lasers and semiconductor optical amplifiers (SOAs), the possibility of operation as Erbium doped fibre amplifiers (EDFA) and Raman amplifiers, high bit rate of information transmission (over 40 Gbps), large transmission distances, the possibility of a system capacity increase by using the wavelength division multiplexing (WDM) technique, soliton communication systems, and so on [18]. Fibre-optic communication systems can be classified into three broad categories: point-to-point links, distribution networks and local area networks (LAN) [18]. These categories fall into two main application types:

- Long-haul applications.
- Short-haul applications.

Long-haul telecommunication systems transmit optical signals over large distances ($\geq$100 km). Transmission distances can be substantially increased by up to thousands of kilometers by using optical amplifiers.

Short-haul telecommunication applications are used in intracity and local-loop traffic; that is, over distances of less than 10 km. Multichannel networks with multiple services requiring high bandwidths can be cost-effective in such applications. Only optic communications can meet such wideband distribution requirements [18]. Most UROOF applications are related to the short-haul case.

The information transmitted by optic communication systems is contained in a series of electrical analog or digital signals. UROOF is a novel technology for the transmission of UWB signals by using an optical carrier propagating through an optical fibre. UWB high-speed optical link includes an optical fibre, electrical/optical (E/O) converter and optical/electrical (O/E) converter.

Key problems of optical communication system design are the modulation of an optical signal at the input of the system, and separation of the electrical signal envelope from the optical carrier; that is, detection at the output of the system. The essential advantage of analog optic communications is the possibility to transmit multilevel modulated radio signals over the optical fibre as an envelope of the optical carrier. This approach permits us to keep exactly the same original signal format needed for the wireless at the access portion of the network. After O/E conversion it is possible to use a standard radio receiver for further detection of the multilevel modulated signals.

In the following subsections we discuss the O/E and E/O processes related to the UROOF concept.

24.4.2 Conversion from RF to Optical Domain

In order to convey analog signals over an optical fibre it is necessary to impress the analog signals on the optical carrier via an optical modulation device [19]. Any of the parameters of the optical carrier can be modulated. In practice, intensity modulation of the optical carrier is used at present. The essential parameters characterizing the performance of an analog optical link are the following [19]:

- The gain of the intrinsic link is defined as the power gain between the input to the modulation device and the output of the photodetection device.
- Noise figure (NF) is the ratio of the signal-to-noise ratio (SNR) at the link input to the SNR at the link output.
- Spurious free dynamic range (SFDR) is the SNR for which the distortion terms equal the noise floor.

There are two main methods for the intensity modulation of the optical carrier [15]: direct modulation and external modulation:

24.4.2.1 Direct Modulation

The diode laser intensity is modulated by an analog MW signal. The advantages of this method are its simplicity and its low cost. The rate of the direct modulation is limited to about 10 GHz due to the relaxation resonance of commercial diode lasers.

The best laboratory devices known to date have maximum modulation frequency about 40 GHz [20]. Direct-modulated vertical cavity surface emission lasers (VCSELs) seem to be the most promising candidates for UROOF systems below 10 GHz due to their low cost, efficient coupling with opticals and small number of longitudinal modes. Recently, commercial 850 nm for 17 Gb/s over multimode was demonstrated in OFC/NFOC 2008 by Finisar. In [21] a 300-Gbps 24-channel full duplex 850 nm parallel optical transceiver was introduced, based on an array of 12.5 Gb/s VCSELs.

24.4.2.2 External Modulation

In this case, the laser operates in a continuous wave (CW) regime and the intensity modulation is imposed via a device external to the laser. High efficiency of RF-to-optical conversion requires high slope efficiency, a good matching of impedances and low RF losses. The advantages of external modulation are mainly a bandwidth approaching 100 GHz, high linearity range and spurious free dynamic range (SFDR). Traditionally, external modulation has been implemented by means of bulk nonlinear crystals such as $LiNbO_3$ [15]. A commonly used external modulator is the Mach–Zehnder (MZ) interferometer and modulator. Recently, the focus has been shifted to integrating electro-absorption Mach–Zehnder interferometers (MZIs) or simpler electro-absorption modulators (EAMs) onto the same substrate as the laser diode. Another type of high-speed modulator is based on the quantum-confined Stark effect (QCSE) in multiquantum well (MQW) structures [15]. For instance, an analog EAM based on InGaAsP MQW has been developed which operates at 60 GHz [22].

In this section we will focus on two seemingly promising techniques: (1) direct modulation of VCSELs up to 10 GHz, and (2) EAT that can be operated at higher frequencies, up to 100 GHz.

24.4.2.3 Vertical Surface Emitting Laser (VCSEL) Technologies

VCSELs are characterized as high-performance and cost-effective lasers and have been widely used in on-off digital optical communication systems at bit rates up to 10 Gbps [20]. However, it is not obvious that the same performance can be accomplished for analog signals. The desired characteristics of RF-to-optical converters are quite different. These include RIN, linearity, impedance and modulation efficiency. However, these characteristics are often achieved at the expense of a decreased gain as well as dynamic range, and also an increased noise figure of the optical link.

The common VCSEL structure employs an active region consisting of multiple quantum wells sandwiched between two epitaxially-grown distributed Bragg reflectors. These reflectors are made up of 20–40 alternating quarter-wave layers of semiconductors with different compositions (typically $Al_xGa_{1-x}As$), providing high reflectivity of larger than 99%. Due to their vertical-cavity geometry, VCSELs offer a number of significant advantages over edge-emitting lasers. Since VCSELs have a relatively small active volume, very low threshold currents are attainable of about several μA. Because of the symmetry in the wafer plane, the laser output is suited for high coupling efficiency to optical fibres with relaxed alignment tolerances, as well as easy focusing into a tight spot for optical storage applications. Thus, VCSEL is a low-cost, easily packaged, compact light source.

The quantum nature of photons and electrons, as well as the randomness of the physical process involved in the converting devices, results in residual fluctuations of the laser beam even for the case of no modulation signal. Clearly, the residual intensity fluctuations, that is, noise, at the laser output when no modulation signal is applied will impose a lower limit on the minimum RF signal that can be conveyed by the link. The intrinsic intensity noise of a semiconductor laser is quantified by its relative intensity noise (RIN).

VCSELs are nonlinear devices. Major sources of distortion are the relaxation oscillations and spatial hole burning. These nonlinearities in the VCSELs cause distortion of the analog signal. The third-order intermodulation (IMD3) distortion is very important for the system applications. The IMD3, together with the modulation response and noise, determines the spurious free dynamic range (SFRD). The maximum signal power that gives the maximum dynamic range is determined by the point where noise and intermodulation are equally large. In order to achieve low levels of both RIN and coupling losses, it is necessary to use single-mode VCSELs, which are characterized by low beam divergence and do not have noise due to mode partition.

24.4.2.4 VCSEL Dynamics

The operating characteristics of semiconductor lasers are well described by a set of rate equations that govern the interaction of photons and electrons inside the active region. A

Table 24.7 VCSEL parameters.

Parameter	Description	Typical value
Γ	Confinement factor	0.4
N_0	Electron concentration a transparency	$1.1 \times 10^{18}\ \text{cm}^{-3}$
τ_p	Photon lifetime	3 ps
τ_e	Electron lifetime	1 ns
V	Active region volume	$1.5 \times 10^{-10}\ \text{cm}^3$
α_c	Linewidth enhancement factor	5
β	Fraction of spontaneous emission coupled into a lasing mode	3×10^{-5}
V_g	Group velocity	8.5×10^9 cm/s
α_{Tot}	Total loss coefficient	$65\ \text{cm}^{-1}$
B	Bimolecular recombination	$1 \times 10^{-10}\ \text{cm}^3/\text{s}$
C	Auger recombination coefficient	$3 \times 10^{-29}\ \text{cm}^6/\text{s}$
a	Differential gain	$2.5 \times 10^{-16}\ \text{cm}^2$
ε	The gain compression factor	$5 \times 10^{-17}\ \text{cm}^3$
η_0	The total differential quantum efficiency	0.4

rigorous derivation of these rate equations generally starts from the Maxwell's equations together with a quantum-mechanical approach to the induced polarization. However, the rate equations can also be obtained heuristically by considering the physical phenomena through which the number of photons and the number of electrons are changed with time inside the active region.

The rate equation for the photons is given by the rate of change of photon density, p, defined by the difference between the stimulated emission and spontaneous emission radiation and the cavity losses:

$$\frac{\mathrm{d}p}{\mathrm{d}t} = \left(\Gamma \frac{a(N - N_0)}{1 + \varepsilon p} - \alpha_{T,l}\right) V_g p - \frac{p}{\tau_p} + \frac{\beta \Gamma N}{\tau_e} \tag{24.3}$$

where $\alpha_{T,l} = \alpha_{\text{loss}} + \frac{1}{L} \ln R$ is the total loss coefficient, L is the laser active region length and R is the reflectivity of the mirrors. The rate equation for the carrier density N has the form:

$$\frac{\mathrm{d}N}{\mathrm{d}t} = \frac{I(t)}{qV} - \frac{V_g a(N - N_0)}{1 + \varepsilon p} p - \frac{N}{\tau_e} - BN^2 - CN^3 \tag{24.4}$$

Here $I(t)$ is the current, and carrier loss rate is due to spontaneous, stimulated, bimolecular and Auger recombination processes. All parameters used in Equations (24.3) and (24.4) are described and listed in the Table 24.7.

From a physical standpoint, the amplitude modulation in semiconductor lasers is always accompanied by the phase modulation in semiconductor lasers because of carrier-induced changes in the mode refractive index. Phase modulation can be included through the equation:

$$\frac{\mathrm{d}\phi}{\mathrm{d}t} = \frac{1}{2}\alpha_c \left\{\Gamma V_g a(N - N_0) - \frac{1}{\tau_P}\right\} \tag{24.5}$$

where α_c is the amplitude–phase coupling parameter, commonly called the linewidth enhancement factor as it leads to an enhancement of the spectral width associated with a single longitudinal mode.

The time variations of the optical power and the laser chirp are given by:

$$P(t) = \frac{0.5p(t)V\eta_0 h\nu}{\Gamma\tau_p}; \quad \Delta\nu(t) = \frac{1}{2\pi}\frac{\mathrm{d}\phi}{\mathrm{d}t} \tag{24.6}$$

In (24.6) η_0 is the total differential quantum efficiency. The photon and electron densities within the active region of the laser are assumed uniform and the linewidth enhancement factor and the nonlinear gain compression parameter are taken to be constant for a given laser.

24.4.2.5 Relative Intensity Noise

A common measure of the random fluctuations in lasers is the RIN. In the single mode operation of VCSEL, RIN is almost constant at lower RF frequencies (several MHz), and it peaks at the laser resonance frequency (about several GHz). Rate Equations (24.3)–(24.5) can be used to study laser noise by adding a noise term, known as the Langevin force, to each of them. Equations (24.3)–(24.5) then become:

$$\frac{\mathrm{d}p}{\mathrm{d}t} = \left(\Gamma\frac{a(N-N_0)}{1+\varepsilon p} - \alpha_{T,l}\right)V_g p - \frac{p}{\tau_p} + \frac{\beta\,\Gamma N}{\tau_e} + F_P(t)$$

$$\frac{\mathrm{d}N}{\mathrm{d}t} = \frac{I(t)}{qV} - \frac{V_g a(N-N_0)}{1+\varepsilon p}p - \frac{N}{t_e} - BN^2 - CN^3 + F_N(t)$$

$$\frac{\mathrm{d}\phi}{\mathrm{d}t} = \frac{1}{2}\alpha_c\left\{\Gamma V_g a(N-N_0) - \frac{1}{\tau_P}\right\} + F_\phi(t)$$

where $F_p(t)$, $F_N(t)$, $F_\phi(t)$ are the Langevin forces. They are assumed to be Gaussian random processes with zero mean and to have correlation function of the form:

$$\left\langle F_i(t)F_j(t')\right\rangle = 2D_{ij}\delta(t-t')$$

where $i, j = P, N$ or ϕ, angle brackets denote the ensemble average, and D_{ij} is called the diffusion coefficient (the Markovian approximation). The dominant contribution to laser noise comes from only two diffusion coefficients, $D_{PP} = \frac{\Gamma V_g a_0(N-N_0)}{V}p$ and $D_{\phi\phi} = \frac{\Gamma V_g a_0(N-N_0)}{Vp}$; others should be assumed to be nearly zero. For analog applications, intensity noise is quantified using a signal-to-noise ratio (SNR) which is linked to the relative intensity noise (RIN):

$$SNR = \frac{m^2}{2RIN}$$

where m is the electrical modulation depth and given by $m = \Delta I/(I - I_{th})$. I and I_{th} are the input and the threshold currents, respectively. RIN is defined as:

$$RIN = \frac{\langle \delta P(t)^2 \rangle}{P^2}$$

where $\langle \delta P(t)^2 \rangle$ denotes the mean square power fluctuation and P is average power.

24.4.2.6 Spurious Free Dynamic Range (SFRD)

An important measure of analog link performance is the dynamic range. The dynamic range of the link is defined as the ratio of the largest signal the system can transport to the smallest one. The total lasing power changes linearly in a wide region with operation current. Intrinsic distortions due to the intermodulation in the active region of the laser lead to distortion of the fundamental signal. Due to these nonlinearities, spurious intermodulation products are created. At low modulation depth (small m), intermodulation distortions are still below the noise floor. Increase of the modulation depth leads to distortions rising above the noise floor. These distortions grow faster than the fundamental signal. The largest distortion-free SNR determines the dynamic range and it is reached when the amplitude of the distortion is equal to the noise floor. The SFRD is the dynamic range where the maximum signal is limited by the intermodulation product and is often used as a measure of system performance. Predistortion technique of the fundamental signal can be used to increase the SFRD.

24.4.2.7 Electro-absorption Transceiver (EAT)

General Considerations

The growing demand for UROOF technologies results in the creation of new millimeter-wave devices [22]. They are implemented mostly in the Ka band (26.5–40 GHz). Now the V band (50–75 GHz) is of particular interest, due to the development of ultra-wideband systems with the analogous signal modulation. We will consider optically-fed microwave radio links in a broadband optical network [22]. In the case of so-called external modulation, the optical carrier is emitted by a continuous wave (CW) laser and modulated by a distinct device which operates a conversion from microwave region to optics and provides an efficient control of the optically-transmitted data. The modulators must not give rise to additional noise and must be highly reliable [22]. We consider below the electro-absorption modulator (EAM).

EAM is a compact, efficient and integrable component for optic communications [23]. It utilizes either the Franz–Keldysh effect in a bulk material, or the quantum-confined Stark effect (QCSE) in a multiquantum well (MQW) structure in order to change the absorption of light according to the applied electric field [23]. In EAMs in linear regime, the light input P_{in} and output P_{out} are proportional and the transmitted power depends on an applied voltage or electric field [22]. MQW structures are widely used due to their higher modulation efficiency, but the wavelength dependence is generally higher than for bulk material. It has been shown that EAM based on QCSE in QWs has better figures of merit than similar devices based on the bulk materials [22]. The advantages of EAMs over other types of modulator are as follows [23]:

- A small device dimension, of the order of magnitude of a few hundred micrometers or even less.
- A lower driving voltage, due to the EAMs' highly efficient and nonlinear electro-optical (E/O) transfer function.
- The possibility of monolithic integration with other semiconductor components.

For instance, a compact module containing an MQW EAM integrated with a DFB laser has been realized [24]. A critical design issue of the EAM is the tradeoff between the modulation bandwidth and efficiency. The bandwidth of a lumped-electrode EAM is determined by the resistance-capacitance time constant $\tau = RC$. Consequently, the active waveguide length and width must be short in order to keep the capacitance low and enable high-speed broadband modulation. On the other hand, a short active waveguide length results in reduced modulation efficiency and a higher driving voltage. The optical power handling capability for a short device also sharply decreases, which is especially important for radio frequency photonic links [23].

EAMs with travelling-wave (TW) electrodes were developed in order to exclude the tradeoff between bandwidth and device length. In such a case, the bandwidth is not limited by lumped RC parameters, and the device length can be longer with a higher bandwidth. In the TW design, the driving signal propagates through the active waveguide, which is in fact a microwave transmission line. With a longer device length, the modulation efficiency can be kept high in such a way that a low driving voltage is required. Theoretically, the bandwidth and the useful device length of TW-EAM are limited by the microwave loss and the velocity mismatch between the lightwave and the microwave. These factors determine how long the lightwave can be effectively modulated by the microwave driving signal. The increase of optical scattering loss with device length is also a limiting factor, degrading the insertion loss of TW-EAM. A low characteristic impedance of the active waveguide (25 Ω or even less) originating from the tradeoff between the junction capacitance and optical and microwave losses causes reflections and limits the modulation bandwidth [23]. Low impedance terminations in the range of 12–35 Ω are required in order to optimize bandwidth. For instance, a bandwidth of 43 GHz was measured on a 450 μm device with a 13 Ω termination [23].

EAM using the QCSE in strained InGaAs/AlGaAs MQW waveguide structures has been investigated experimentally. It has been shown that the electrical bandwidth measured in a common 50 Ω system has a cut-off frequency higher than 70 GHz due to an optimum impedance and phase matching. The unique properties of TW-EAMs should be explored and utilized. Consider briefly the unique properties, such as: distributed effect due to TW design, generation of photocurrent and nonlinear E/O transfer function.

Distributed Effect

The most significant implication of the TW design is a finite length inside the TW-EAM [23]. It generates an extra dimension for device optimization, especially when the microwave frequency is large enough that the microwave wavelength is comparable to the device dimension. As a result, both the interaction in time and in space between the lightwave and the microwave should be taken into consideration. This is known as the distributed effect in TW-EAM [23].

Both the copropagation and the counter propagation of the optical and microwave signals can be realized in TW-EAM. The copropagation regime is preferred in terms of output power and shortest optical pulse width. For lower operation frequencies and shorter device lengths, the difference between two configurations becomes insignificant.

Photocurrent Generation and EAT

EAMs inherently generate photocurrents [22, 23]. For this reason, an EAM can be used simultaneously as a modulator and as a photodetector. This exceptional property makes it possible to combine several functionalities within a single TW-EAM. A device thus created is called an electro-absorption transceiver (EAT). EATs have been studied experimentally for operation at 1.3 mm wavelength. The typical device consists of a slightly Si-doped lattice-matched InAlAs top cladding layer and a highly Si-doped lattice-matched InAlAs bottom cladding layer. The active region is formed by 20 n-i-d QWs with a thickness of 7.7 nm each. By implementing 1% tensile strain in the InGaAs QWs and 1% compressive strain in the InAlAs barrier layers, polarization incentive operation is achieved. Microwave operation of up to 70 GHz can be realized by using n-i-n EAMs with hybrid coplanar-microstrip metallization [22].

For applications in ultra-wideband modulation, a strong photocurrent inside the active waveguide should be avoided. Such a photocurrent causes a reduction of effective voltage supplied to the active region and degrades the large signal dynamic extinction ratio.

Nonlinear E/O Transformation

The E/O transfer function of an EAM is determined by the material design, device length, electrode design and waveguide geometries. The polarization dependence may be essentially decreased by using strain-compensated MQWs, as was mentioned above. A typical EAM E/O transfer function dependence on the reverse electric voltage applied consists of three transmission regions:

1. High extinction (<10%)
2. Transition (10–90%)
3. Flat transmission (>90%).

The transition voltage should be small in order to achieve high modulation efficiency and low driving voltage. The slope of the transfer function varies point-to-point because of the nonlinear character of the electro-absorption process. However, it is possible to find certain points with minimized distortion for a high dynamic range RF photonic link. The voltage V_H at which transition begins is typically designed close to 0 at the operating wavelength in order to reduce excess heating. If V_H is shifted towards higher reverse voltage, more flat transmission region can be used without forward-biasing the TW-EAM.

Perspectives

TW optoelectronic devices can meet the contemporary requirements for ultra high-speed operation. In particular, TW photodetectors and modulators are not limited by the usual RC time constants. Microwave properties determine the bandwidth, and the input resistance is given by the characteristic impedance of the waveguide system. TW devices are much more flexible with respect to design parameters and permit monolithic integration in optical MMICs. Simulation and modeling of the devices can be carried out by using equivalent circuits for the optical and electrical domains. The interaction can be described by parametrically controlled elements.

There are still challenges for TW-EAM applications. The impedance mismatch between the active waveguide and driving electronics limits the bandwidth. This shortage can be improved by new materials or by design customizing. The insertion loss of TW-EAMs should be reduced. Waveguide spot-size converters and high-level integrations are both possible solutions.

24.4.3 Conversion from Optical to RF Domain

The optical link speed is mainly determined by the performance of the modulation and demodulation process at the transmitter and the receiver sides. The signal distortion is especially problematic when the signal to be transferred is analog, as a counterpart of digital 3R regeneration (reamplification, reshaping and retiming) does not exist. Therefore, highly linear link is required to prevent signal distortion. As a result, the design constraints in analog links are more stringent. In this section we briefly review RoF techniques that can operate from 1 to 100 GHz, with emphasis on the noise performance and link linearity.

High-speed photodetectors (PDs) are required for use in high-speed optical interconnects, microwave photonic applications, as well as many modern RoF communication systems. Photodetectors are used to convert the optical signal back into electrical form and recover the data transmitted through the lightwave system [18]. A photodetector should have high sensitivity, fast response, low noise, low cost and high reliability. Ultrafast photonic devices based on the microwave optical interactions are expected to be developed and incorporated in future high-speed and high-capacity lightwave systems [25, 26].

24.4.3.1 High-speed Photodetectors

High-speed PDs based on AIIIBV compounds are used in optic communications, optical/wireless systems, detection and conversion of optical signals, and other microwave photonics applications. The high speed and operation efficiency of PDs is essential for an optical-electrical conversion of analog signals in UWB radio-over-optical-fibre (UROOF) technology. To that end, there is an interest in increasing the bandwidth of PDs up to 100 GHz [27].

The main types of high-speed PD are the following:

- PIN PDs
- avalanche PDs
- Schottky PDs
- resonant-cavity-enhanced PDs
- waveguide PDs
- travelling wave PDs
- velocity-matched distributed PDs.

Based on the mode of optical radiation propagation, high-speed PDs are divided into three classes:

- surface-illuminated PDs
- resonant-cavity-enhanced PDs
- edge-coupled PDs.

The quality of high-speed PDs is characterized by the bandwidth-efficiency product [27]. For surface-illuminated PDs this parameter does not exceed 20–30 GHz, being limited by the trade-off between quantum efficiency and bandwidth. In order to increase the quantum efficiency it is necessary to increase PD absorption layer thickness. On the other hand, the increase of the PD thickness results in the decrease of the bandwidth. In resonant-cavity-enhanced PDs, in which the PD is situated in a Fabry-Perot resonator, optical radiation multiply passes through the thin absorption layer and quantum efficiency increases at the resonant wavelength. For this reason, the increase of the device thickness is unnecessary. Hence, the bandwidth may be larger. In the edge-coupled PDs the directions of the optical radiation propagation and charge carrier transport are perpendicular to one another. In such a case it is necessary to increase the device length instead of the thickness of the absorption layer.

Below we briefly discuss the characteristics and peculiarities of these types of PD. We mainly follow the review [27].

24.4.3.2 PIN PDs

A typical PIN PD is based on a $p^+ - InP/i - In_{0.47}Ga_{0.53}As/n^+ - InP$ heterostructure. Wide bandgap p^+- and n^+- InP layers are hardly doped, while the absorption layer $i\text{–}In_{0.47}Ga_{0.53}As$ is not doped and has low background impurity concentration. Under the optical illumination at wavelength $l = 1.1$ to 1.6 mm, the electron–hole pairs are generated in the absorption i-layer and separated by an internal electrical field of p-n junction by means of the drift mechanism in the depletion region and by diffusion in the neutral region. The high-speed operation is determined by the reverse bias large enough for full depletion of the i-layer. The quantum efficiency η of the PIN PD is given by:

$$\eta = \varsigma(1 - R)(1 - \exp(-\alpha d)) \tag{24.7}$$

where ς is the internal quantum efficiency, which is typically close to 1; R is the reflection coefficient of the PD surface; and α, d are the absorption coefficient of the PD surface and the thickness of the absorption i-layer, respectively. The PIN PD bandwidth Δf is limited by the transit time and the RC time of the PIN PD equivalent circuit. It has the form [27]:

$$\Delta f = \left(\frac{1}{(\Delta f_t)^2} + \frac{1}{(\Delta f_{RC})^2}\right)^{-1/2} = \left[\left(\frac{2\pi d}{3.5\overline{v_d}}\right)^2 + \left(2\pi\varepsilon_0\varepsilon_r S\frac{(R_s + R_l)}{d}\right)^2\right]^{-1/2} \tag{24.8}$$

where $\bar{v}_d$ is the average electron and hole drift velocities in the absorption i-layer; ε_0, ε_r are the permittivity of vacuum and of the absorption i-layer, respectively; S is the PIN PD photosensitive area; and R_s, R_l are the series and the load resistances. It can be easily shown that the optimum absorption layer thickness for a maximal bandwidth for the given photosensitive area S is given by [27]:

$$d \approx \sqrt{3.5\bar{v}_d\varepsilon_0\varepsilon_r R_l S} \tag{24.9}$$

The numerical evaluations show that for the PIN PD, a bandwidth larger than several tens GHz requires a small photosensitive area with a diameter of about 10–20 μm. The absorption layer should not exceed 1 μm. The bandwidth can be enhanced by using only fast electrons for charge carrier transport, as is achieved in uni-travelling carrier (UTC) PDs. The transit time limited bandwidth of the UTC UPD can be 200 GHz and above due to the large electron diffusion and drift velocities [27].

24.4.3.3 Avalanche PDs

Avalanche PDs (APDs) are used for increasing photoreceiver sensitivity in optic links. The increase of the APD sensitivity is due to the internal amplification during an avalanche gain of photocarriers. The process of the avalanche gain is characterized by ionization coefficients α_n, α_p for electrons and holes, respectively. The multiplication factor M is given by [27]:

$$M = \frac{\left(\alpha_n - \alpha_p\right)}{\alpha_n - \alpha_p \exp\left[\left(\alpha_n - \alpha_p\right) d_m\right]} \tag{24.10}$$

where d_m is the multiplication layer thickness. Low noise, fast response and high multiplication factor can be realized only if the ionization coefficients α_n, α_p significantly differ from each other. At small reverse bias voltages, APDs operate as usual PIN PDs with a gain of about unity. The amplification in APDs occurs only at comparatively large reverse bias voltage of about 30–40 V. The bandwidth of the high-speed APD is limited by the finite avalanche formation time and does not exceed 10–20 GHz. The APD performance, including the bandwidth increase, can be improved by usage of InAlGaAs/InAlAs superlattice with ionization coefficient ratio $(\alpha_n/\alpha_p) \approx 3 - 4$.

24.4.3.4 PDs Based on Schottky Barrier

The basic electrical characteristics of the Schottky PD are very similar to the characteristics of the PIN PD. The main advantage of the PDs based on Schottky barriers is the smaller capacitance per unit area compared with the PDs based on p-n junction. For high-speed Schottky PD creation, the thin barrier-enhanced $In_{0.52}Al_{0.48}As$ layer having Schottky barrier height 0.8 V with Ti/Au metal contact is used. The main disadvantage of the Schottky PD is the lower quantum efficiency compared to the PIN PDs. The Schottky PD bandwidth is limited by drift time and RC time.

24.4.3.5 Resonant-cavity-enhanced PDs

It is possible to increase the quantum efficiency of the surface-illuminated PDs by means of a resonant-cavity-enhanced (RCE) structure. In the RCE PD the surface-illuminated PD is placed in a Fabry-Perot resonator. The RCE PD quantum efficiency is given by:

$$\eta(\lambda) = \frac{[1 - R(\lambda_1)][1 + R_2(\lambda)\exp[-\alpha(\lambda)d]][1 - \exp[-\alpha(\lambda)d]]}{|1 - \sqrt{R_1(\lambda)R_2(\lambda)}\exp[-\alpha(\lambda)d]\exp(-j\Phi)|^2} \tag{24.11}$$

where:

$$\Phi = 4\pi \frac{n_1 d_1 + n_2 d_2 + nd}{\lambda} + \psi_1(\lambda) + \psi_2(\lambda) \tag{24.12}$$

R_1, ψ_1 and R_2, ψ_2 are the reflection coefficients and phase shifts of the forward and reverse reflectors; n, α, d are the refractive index, absorption coefficient and thickness of the absorption layer; and n_1, d_1 and n_2, d_2 are the refractive index and thickness of layers above and below the absorption layer. The quantum efficiency has a maximum at the wavelength λ_0 that satisfies the condition:

$$4\pi \frac{n_1 d_1 + n_2 d_2 + nd}{\lambda_0} + \psi_1(\lambda_0) + \psi_2(\lambda_0) = 2\pi \tag{24.13}$$

Then we have the maximal efficiency:

$$\eta_{\max} = \frac{[1 - R]\left[1 + R_2 \exp[-\alpha d]\right]\left[1 - \exp[-\alpha d]\right]}{\left(1 - \sqrt{R_1 R_2} \exp(-\alpha d)\right)^2} \tag{24.14}$$

and the half-maximum $\Delta\lambda_{1/2}$:

$$\Delta\lambda_{1/2} = \frac{\lambda_0^2 \left(1 - \sqrt{R_1 R_2} \exp(-\alpha d)\right)}{2\pi(n_1 d_1 + n_2 d_2 + nd)\sqrt{\sqrt{R_1 R_2} \exp(-\alpha d)}} \tag{24.15}$$

Due to multiple propagation of optical radiation at the resonant wavelength through the absorption layer, its thickness can be reduced for receiving necessary bandwidth without any loss of efficiency. In this case, the bandwidth-efficiency product can reach hundreds of GHz [27].

While various photodetectors, such as of PIN, metal–semiconductor–metal (MSM) and APD can be used for high-speed applications, each of these devices has certain drawbacks which makes it rather unsuitable for use in UWB over communications. Lateral PIN and MSM devices are simple to fabricate, but are characterized by low responsivity at wavelengths around 850 nm. APD devices, on the other hand, require relatively high bias voltages and have increased noise.

24.4.3.6 Optically-controlled Microwave Devices

Novel photonic techniques for optical signal generation and processing are required in ultra-wideband systems [25]. Microwave photonic links may serve efficiently in analog interconnects in state-of-the-art wideband microwave functions. Optical signals introduced into microwave devices or controlling them may be used for signal detection [26]. This approach possesses the following advantages:

- The absence of extra electronic circuits required to process the detected signals before application to the microwave device.
- The absence of circuit parasitics limiting the response speed.

- Creation of an optical control port to the microwave device.
- The optical control signal is not influenced by any microwave electromagnetic disturbances.

The optical control of microwave devices has been widely discussed in the last decade (see, for instance, [25, 26, 28, 29]).

Consider briefly the optical control of some typical microwave devices [26].

Amplifiers

The gain of MESFETs and high electron mobility transistors (HEMTs) can be controlled by illuminating the gate region and including an appropriate series resistor in the gate bias in order to produce a change in gate bias caused by the optically-generated current. Gain of up to 20 dB in MESFET amplifiers can be achieved using optical powers of about several μW. HEMT amplifiers exhibit even larger optical sensitivity; around 7–10 times higher.

Oscillators

There are three main forms of oscillator optical control: (i) optical switching; (ii) optical tuning; (iii) optical injection locking. In the case of optical switching, the oscillator optical power is changed depending on the optical control signal intensity. The case of optical tuning is similar to the first one, however the optical intensities used are rather small and the output power change is also small. In the case of optical injection locking, the optical control is optically modulated at a frequency close to the free-running frequency of the oscillator, or one of its harmonics or subharmonics. The modulated optical signal fed into the device active region creates the current at the modulation frequency. These phenomena occur in avalanche diodes, MESFETs and bipolar transistors.

Opto-electronic Mixers

Two regimes can be realized. In the first case, the signal to be converted in frequency is supplied electrically, and the local-oscillator signal is an intensity-modulated optical source. In the opposite case, an electrical local-oscillator signal is used to down-convert an intensity-modulated optical signal. Opto-electronic mixers have been realized using photoconductive devices, diodes, field-effect transistors and bipolar transistors.

24.4.3.7 Optically-controlled Microstrip Converter (OCMC)

A novel concept named OCMC is based on the generation of photo-carriers within semiconductor devices. The substrates possess a strongly manifested contrast between dark conductivity and photoconductivity under illumination by the optical signal, allowing an electrical modulation from the incident light. As a result, in depletion regions a photocurrent occurs, and the built-in potential changes. In other cases, the conductivity of semiconductors increases due to the photoconductive effect. Consequently, optically-controlled semiconductor devices such as MESFETs, HEMTs, PIN diodes and APDs can be integrated as an optically-controlled load with different types of microstrip and coplanar lines.

Such a system, combining all the advantages of the microwave photonic devices mentioned above, can be successfully used as an ultra-wideband OCMC. Indeed, microwaves can be fed into microstrip and coplanar lines realized on GaAs, Si or novel SiGe substrates. The optical signal modulated with an ultra-wideband envelope is introduced into the optically-controlled load that provides the rapid detection of the microwave signal. Different regimes of the signal transmission can be realized, including heterodyning. Microstrip antennas can be also integrated into the system, making the detected ultra-wide signal radiation possible.

The design of OCMC with a 60 GHz bandwidth and minimum losses is a substantial challenge from the point of view of optimization of microstrip and coplanar wave guiding elements, efficient matching of the microwave transmission line with an optically-controlled load, and the choice of high-speed operating PD devices [30].

24.4.4 Optical Microwave Mixing Used for UWB Over Systems

All-optical microwave mixing techniques could be used to up-convert a UWB signal around an RF carrier in the (3.1–10.6) GHz bandwidth. Two configurations can be employed. The first possibility is to simultaneously modulate the optical carrier and realize the RF up-converting. The second is to simultaneously detect the baseband UWB signal and up-convert it around the RF frequency, after the optical transmission.

24.4.4.1 Optical to RF Conversion

All-optical microwave mixing has been demonstrated by direct nonlinear photodetection using CW signals [31, 32]. The simultaneous injection of a microwave signal at the electrical port of the PD and a modulated optical signal permits the mixing of the two signals. The mixing process results from the nonlinearity of the PD current–voltage relationship. Due to the fact that the characteristics exhibit the maximum nonlinearity in the vicinity of 0 V, it is the optimal operation point for efficient mixing. We propose to extend this technique to optical UWB systems by replacing optical the CW signal with an optical UWB one. With this technique, the optical source (e.g., the VCSEL) bandwidth could be even more reduced, to the value of the baseband UWB bandwidth (500 MHz for MB-OFDM signals).

24.4.4.2 RF to Optical Conversion

One way to generate optical–microwave mixing is to insert a nonlinear element in the optical transmission link. Different methods have already been demonstrated. Two input microwave signals at f_{RF} frequency containing data, and a local oscillator signal at f_{LO} frequency, are then converted at mixing frequency $f_{RF} \pm f_{LO}$ at the output of the optical link. Generally, mixing is achieved by the use of optical intensity modulation (direct modulation of laser diodes or external modulation) in different configurations: two cascaded linear modulations [33] or one modulation working in a nonlinear regime [34]. Other methods can be used by employing frequency modulation into intensity modulation conversion achieved by passive optical interferometer [35] or by chromatic dispersion properties [36].

To implement low-cost optical UWB systems, direct modulation of VCSELs in a nonlinear regime could be used to generate optical–microwave mixing. VCSELs are low-cost components but are today limited in modulation bandwidth. Nevertheless, using optical-microwave techniques imposes less stringent requirements on the modulation bandwidth because the VCSEL is modulated by a UWB signal centred around an IF frequency, and by an unmodulated signal at LO frequency, close to the IF one. The mixing process up-converts the signal at IF+LO frequency. More precisely, requirements concerning the VCSELs can be divided by two: 5.5 GHz modulation bandwidth is sufficient to generate signals in the 3.1–10.6 GHz range.

24.4.5 *Integrated UROOF Transceiver (IUT)*

The key element in UROOF architectures is the integrated UROOF transceiver. This transceiver is responsible for translating the UWB signal from the RF domain to the optical domain and vice versa. Two IUT approaches are investigated:

1. The optically-controlled microstrip converter consists of an open-ended microstrip line implemented on a semiconducting material. This device is intrinsically unidirectional, as it produces an O/E conversion. The first IUT approach combines OCMC for the O/E conversion with E/O conversion based on VCSEL.
2. The second approach is the enhanced electro-absortion transceiver (E-EAT). This device is capable of O/E and E/O conversion simultaneously. In UROOF it is implemented in an electro-absorption modulator (EAM) optimized for bidirectional conversion. For optical input signals at a wavelength close to the absorption edge the E-EAT operates as a modulator because the small voltage changes applied are translated to a shift in the position of the absorption edge (i.e. transparency edge) of the device. Optical signals at a shorter wavelength than the absorption edge are absorbed and it effectively operates as a photodiode. The E-EAT is intrinsically bidirectional and is the most suitable option for bidirectional application scenarios. It must be taken into account that bidirectionality implies the use of two WDM channels, or the use of two fibres (uplink and downlink).

The studies presented in this document are based on OCMC/VCSEL. This approach is taken without lack of generality, as the parameters employed can be directly applied if an E-EAT device is used. The OCMC/VCSEL IUT structure is depicted in Figure 24.9.

The integrated UROOF transceiver (IUT) includes the OCMC and the VCSEL systems in order to convert the UWB signal from the optical to the RF domain (OCMC) and vice versa (VCSEL). The IUT includes an amplifier, which prepares received RF and optical signals to be converted and transmitted.

The front-end amplifiers, standard in a typical UWB transceiver, can raise the emission power and the system sensibility considerably, and will compensate the propagation losses during free space or transmission.

24.5 Link Analysis of UROOF Systems

The optical link speed is mainly determined by the performance of the modulation and demodulation process at the transmitter and the receiver sides. The signal distortion is especially problematic in the case where the signal to be transferred is analog, as a counterpart of digital

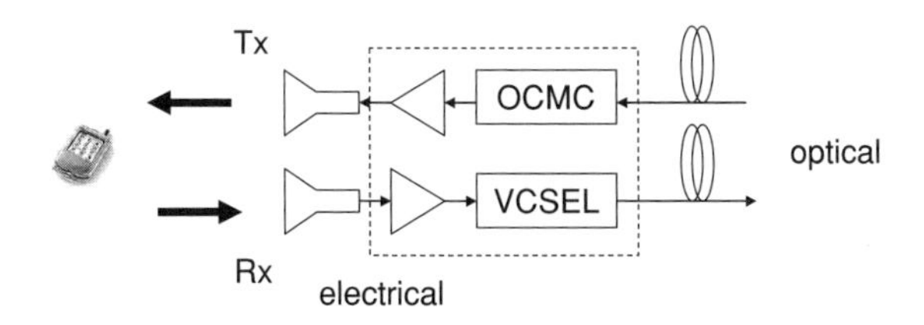

Figure 24.9 IUT structure including amplifier in the front-end.

3R regeneration (reamplification, reshaping and retiming) does not exist. Therefore, highly linear link is required to prevent signal distortion. As a result, the design constraints in analog links are more stringent. We will briefly review a radio frequency (RF) optical analog link design that can operate up to 100 GHz.

24.5.1 Mixed Wireless-wired UROOF Channel

24.5.1.1 Basic UROOF Point-to-point Link

In this section we evaluate the performance of a simple two-integrated-UROOF-transceiver (IUT)-nodes point-to-point link. The study of this simplified scenario is the keystone of the multiple-access scenarios. The simple point-to-point link is depicted in Figure 24.10, where:

P_{Rx} is the received power at the UWB receiver,

P_{Tx} is the RF power delivered to the TX antenna at the UWB transmitter,

P_{IUTRx} is the received power at the IUT receiver antenna,

P_{IUTTx} is the power the IUT is going to transmit to the UWB receiver,

G_{FeRx} stands for the IUT front-end amplifier gain included in the IUT before VCSEL conversion,

G_{FeTx} stands for the IUT front-end amplifier gain included in the IUT after the OCMC conversion and before the IUT transmission,

G_{VCSEL} [dB] stands for the electrical (power) to optical (intensity) conversion gain,

G_{OCMC} [dB] is the OCMC conversion gain from optical power to RF power,

G_{IUTTx} [dBi] and G_{IUTRx} [dBi] are IUT TX and RX antenna gain respectively,

G_{Tx} [dBi] and G_{Rx} [dBi] are the UWB device TX and RX antenna gain respectively,

d is the length of the SSMF,

α is the attenuation at the operation wavelength,

r is the distance the signal travels in free space from the UWB transmitter antenna to the IUT receiver. In order to simplify the calculations, without lack of generality, we will suppose this to be the distance between the IUT transmitter and the UWB receiver antenna.

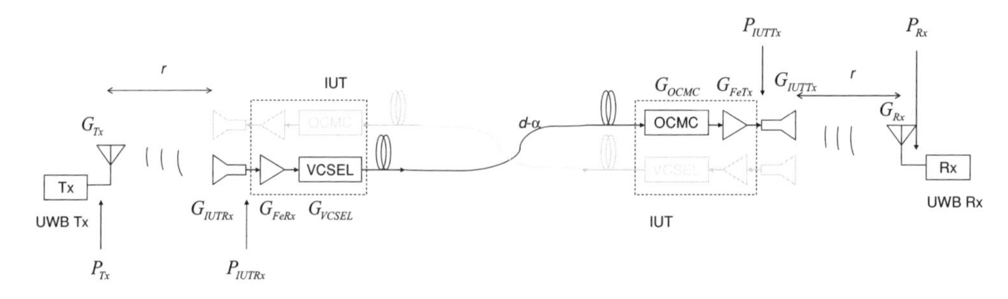

Figure 24.10 Simple point-to-point link.

24.5.1.2 Power Budget

We will summarize the power budget and calculate the expected noise from a system point of view:

1. $\frac{P_{IUTTx}}{P_{IUTRx}}$ stands for the power gain of the UROOF distribution stand-alone, from electrical to electrical point: $G_{e/e} = P_{IUTTx}/P_{IUTRx}$.
2. $\frac{P_{RX}}{P_{TX}}$ stands for the UWB terminal to UWB terminal power budget, including the wireless link.

Let us first calculate:

$$\frac{P_{IUTTx}}{P_{IUTRx}} = \frac{G_{FeRx} \cdot G_{VCSEL} \cdot G_{OCMC} \cdot G_{FeTx}}{d \cdot a} \tag{24.16}$$

Considering free-space propagation over a range r:

$$\frac{P_{Rx}}{P_{Tx}} = \left(\frac{P_{IUTTx}}{P_{IUTRx}}\right)\left(\frac{P_{IUTRx}}{P_{Tx}}\right)\left(\frac{P_{Rx}}{P_{IUTTx}}\right) \tag{24.17}$$

Then the overall power budget for the complete link is given by:

$$\frac{P_{Rx}}{P_{Tx}} = \frac{G_{FeRx} \cdot G_{VCSEL} \cdot G_{OCMC} \cdot G_{IUTRx} \cdot G_{Tx} \cdot L_{F_s}^2 \cdot G_{FeRx} \cdot G_{IUTTx} \cdot G_{Rx}}{d \cdot a} \tag{24.18}$$

where L_{Fs} is the losses due to free-space transmission, given by:

$$L_{Fs} = \left(\frac{l}{4\pi r}\right)^2 \tag{24.19}$$

Typical values are shown in Table 24.8.
Expected power budget:

	dB
$\frac{P_{IUTTx}}{P_{IUTRx}} = \frac{G_{FeRx} \cdot G_{VCSEL} \cdot G_{OCMC} \cdot G_{FeTx}}{d \cdot \alpha}$	49,98
$\frac{P_{IUTRx}}{P_{Tx}} = G_{Tx} \cdot L_{Fx} \cdot G_{IUTRx}$	−47, 9635974
$\frac{P_{Rx}}{P_{IUTTx}} = G_{IUTTx} \cdot L_{Fs} \cdot G_{Rx}$	−47, 9635974
$\frac{P_{Rx}}{P_{Tx}} = \left(\frac{P_{IUTTx}}{P_{IUTRx}}\right)\left(\frac{P_{IUTRx}}{P_{Tx}}\right)\left(\frac{P_{Rx}}{P_{IUTTx}}\right)$	−45, 9471947

Table 24.8 Typical values for power budget calculation.

System parameters	Value	Units
IUT front-end transmitter amplifier gain	20	dB
IUT front-end receiver amplifier gain	20	dB
Electrical (power) to optical (intensity) conversion gain VCSEL	0	dB
OCMC conversion gain from optical power to RF power	10	dB
UWB device TX antenna gain	2	dBi
UWB device RX antenna gain	2	dBi
IUT TX antenna gain	6	dBi
IUT RX antenna gain	6	dBi
SSMF fibre attenuation at the operation wavelength	0.2	dB/km
Length of the SSMF fibre	0.1	km
Distance of the UWB transmitter antenna to the IUT receiver	3	m
Frequency	5.00E+09	Hz

24.5.2 Carrier-to-noise Ratio

Let us consider the degradation of the carrier-to-noise ratio(CNR) due to the signal propagation through the UROOF distribution architecture. Several noise sources contribute to signal degradation in conventional link: thermal noise, laser noise (RIN), photodetector noise and intermodulation distortion due to the nonlinear response of the system. All these are evaluated in this section from a system point of view. The work presented here covers standard devices (VCSEL, APD/PIN photodetector) as introductory calculations.

24.5.2.1 Thermal Noise

The CNR after the IUT Rx antenna is called CNR_{IUTRx}. The CNR_{IUTRx} at the other side of the transmission link is given by:

$$CNR_{IUTTx} = \frac{CNR_{IUTRx}}{f_{FeRx} \cdot f_{VCSEL} \cdot f_{\text{fibre}} \cdot f_{OCMC} \cdot f_{FeTx}} \tag{24.20}$$

where f_{FeRx} and f_{FeTx} are the noise factor of the RX and TX front-end amplifiers respectively; f_{fibre} stands for the equivalent noise factor due to the signal attenuation when travelling down the fibre. In this case it is equivalent to the attenuation along the link; f_{VCSEL} and f_{OCMC} are the noise factor of the laser and detector respectively.

24.5.2.2 Intermodulation Distortion

The UROOF topology supports different UWB analog channels multiplexed in frequency. This is a particular implementation of a subcarrier multiplexing (SCM) system. However, SCM technology is subject to some important system penalties. Because the channel spacing between subcarriers is small, a high nonlinear distortion may be expected. The nonlinearity is reflected in intermodulation distortion – composite second-order (CSO) distortion and composite triple beat (CTO). The UROOF architecture system has very stringent requirements on the noise and

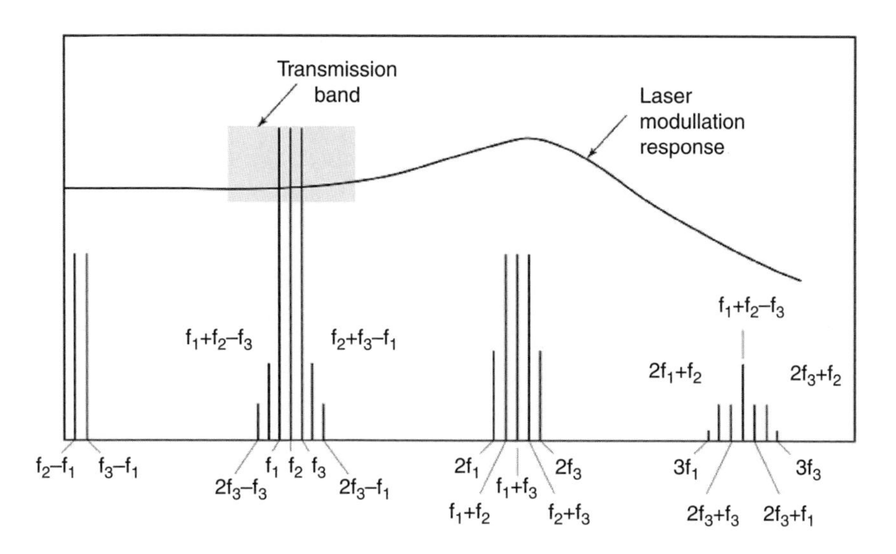

Figure 24.11 Intermodulation products and harmonics generated by a three-tone modulation of a nonlinear device.

the linearity of the system. The linearity of the two key conversion devices (OCMC, VCSEL) is of paramount importance, so it is necessary to calculate the carrier-to-interference ratio (CRI) due to nonlinearity.

In subcarrier multiplexed systems the various electrical subcarriers are combined and used to modulate the optical signal, which is then direct detected. If the devices used in the electrical to optical (E/O) and optical to electrical (O/E) conversion are nonlinear the various subcarriers are mixed to form intermodulation products.

Second- and third-order intermodulation products (IMPs) are generated by every combination of two and three input frequencies, respectively.

The interference resulting from source nonlinearity then depends strongly on the number of channels and the distribution of channel frequencies. Let us consider transmission of three channels with subcarrier frequencies f_1, f_2 and f_3 (Figure 24.11).

In the UROOF application second-order IMPs, $f_i \pm f_j$, will have to be taken into account, since the transmission bandwidth occupies more than one octave. The most troublesome third-order distortion products are those that originate from frequencies $f_i + f_j - f_k$ and $2f_i - f_j$, since they lie within the transmission band, leading to interchannel interference. The interference thus depends strongly on the number of channels, and generally on the allocation of channel frequencies. For an N-channel system with uniform frequency spacing, the number of IMPs, of type $2f_i - f_j$ and $f_i + f_j - f_k$ respectively, coincident with channel r is given by [37]:

$$ {}_rIM^N_{2\bar{1}} = \frac{1}{2}\left\{N - 2 - \frac{1}{2}[1 - (-1)^N(-1)^r]\right\} \tag{24.21} $$

$$ {}_rIM^N_{11\bar{1}} = \frac{r}{2}(N - r + 1) + \frac{1}{4}[(N-3)^2 - 5] - \frac{1}{8}[1 - (-1)^N](-1)^{N+r} \tag{24.22} $$

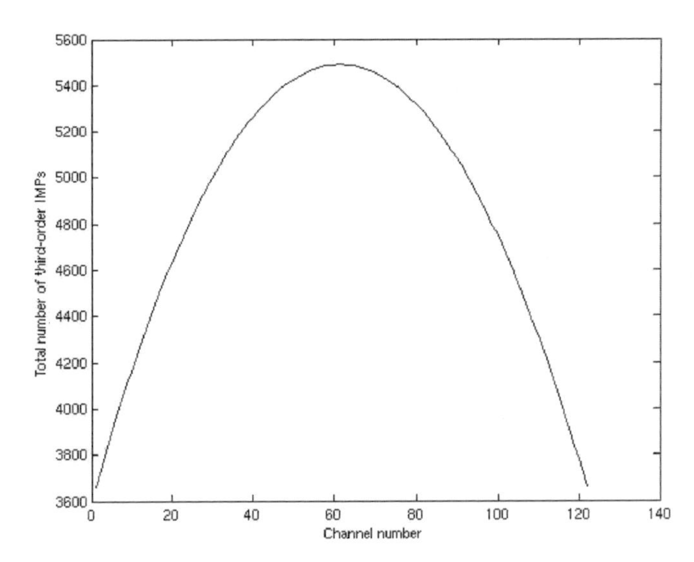

Figure 24.12 Total number of third-order intermodulation products as a function of channel number for a $N = 122$ channel system.

Considering an OFDM-UWB system and considering only one sub-band, the total number of channels $N = 122$. Figure 24.12 shows the total number of third-order IMPs as a function of channel number.

It can be seen that the central channel is the one with the large number of third-order IMPs. For large N, ${}_rIM_{111}^N$ and ${}_rM_{21}^N$ approach the asymptotic values of $3N_2/8$ and $N/2$ respectively, whereas the number of second-order IMPs is proportional to N. In order to quantify the effect of the intermodulation distortion on the system performance we define the carrier-to-intermodulation ratio for the rth channel (CIR) as [38–40]:

$$CIR^{-1} = m^4(D_{111}N^2 + D_{21}N) + m^2(D_{11} + D_2) \tag{24.23}$$

where m is the optical modulation depth and D_{111}, D_{21}, D_{11} and D_2 are the distortion coefficients associated with each type of distortion.

For a device with a static nonlinearity, the distortion coefficient will depend only on the nonlinear characteristic of the device and so is a measure of its linearity performance. However, if a device possesses a dynamic nonlinearity (sometimes called nonlinearity with memory), which is a nonlinearity that is frequency dependent, the distortion coefficients will depend also on the specific channel frequency allocation. The direct modulation of the VCSEL corresponds to the latter case since the laser dynamics are intrinsically nonlinear. This is a result of the interaction between the carriers and photons in the laser cavity. The OCMC should fall in the first category, but further work is required to confirm this.

24.5.3 Laser and Photodetector Noise Baseline

Let us evaluate now the noise expected in a conventional laser/photodiode configuration. This study must be addressed now because this very standard configuration gives the baseline for the complete UROOF architecture. The carrier-to-noise ratios obtained here are considered baseline figures, and they will be checked against the developed devices in order to find their technical advantage.

For a standard APD or PIN photodetector, the determination of the overall system CNR requires us to take into account the relative intensity noise(RIN), receiver shot and thermal noise. The CNR at the receiver may then be expressed as:

$$CNR_r^{-1} = CNR_{TXr}^{-1} + CNR_{RXr}^{-1} \tag{24.24}$$

where CNRTX is the ratio of the carrier power to the noise power generated by the laser diode and CNRRX is the ratio of the carrier power to the noise power generated at the receiver. These ratios are relative to a specific channel r. For simplicity of notation we will drop the subscript r.

The ratios are given by:

$$CNR_{TX}^{-1} = \frac{2B \cdot RIN}{m^2} + CIR^{-1} \tag{24.25}$$

$$CNR_{RX} = \frac{1/2(mgI^2)}{\left(2eIg^2F + \langle I_r^2 \rangle\right) B} \tag{24.26}$$

where *RIN* is relative intensity noise; g is the APD gain; I_r^2 is the receiver noise spectral noise density A^2/Hz; I is the primary DC photocurrent; $F = gx$ is the excess APD noise factor; e is the electron charge; B is the signal bandwidth.

The total CNR is then written as:

$$CNR = \frac{\frac{1}{2}m^2I^2g^2}{\langle I_r^2 \rangle B + g^2I^2RIN \cdot B + 2eBIg^2F + \frac{1}{2}g^2I^2\left(m^6C_1 + m^4C_2\right)} \tag{24.27}$$

with:

$$C_1 = D_{111}N^2 + D_{21}N \tag{24.28}$$

$$C_2 = D_{11}N + D_2 \tag{24.29}$$

CNR may be bipartitely maximized in m and g for a given I, equivalent to balancing the total contributions of signal-independent and signal-dependent noise terms [39]. The maximum CNR for an optimum modulation depth m_{opt} and a given I, obtained by differentiation of Equation (24.27), is:

$$CNR^{-1}(m_{\mathrm{opt}}) = 3m_{\mathrm{opt}}^4\left[D_{111}N^2 + D_{21}N\right] + 2m_{\mathrm{opt}}^2\left[D_{11}N + D_2\right] \tag{24.30}$$

or, conversely, it may be determined by specifying the required *CNR*:

$$m_{\mathrm{opt}}^2 = \frac{-C_2 + \sqrt{C_2^2 + 3C_1/CNR}}{3C_1} \tag{24.31}$$

The optimum APD gain is given by the usual expression:

$$g_{\text{opt}} = \left[\frac{\langle I_r^2 \rangle}{eIx} \right]^{1/(2+x)} \tag{24.32}$$

The maximum CNR for an optimum OMD and APD gain is then:

$$CNR(m_{\text{opt}}, g_{\text{opt}}) = \frac{\frac{1}{2} m_{\text{opt}}^2 I^2 g_{\text{opt}}^2}{K_3 B \left\{ \langle I_r^2 \rangle \left[1 + \frac{2}{x} \right] + g_{\text{opt}}^2 I^2 RIN \right\}} \tag{24.33}$$

where:

$$K_3 = \frac{3m_{\text{opt}}^2 C_1 + C_2}{2m_{\text{opt}}^2 C_1 + C_2} \tag{24.34}$$

The necessary primary DC photocurrent to achieve CNR is readily obtained by combination of Equations (7.32) and (7.33):

$$I_{APD} = \left[\frac{ex}{\langle I_r^2 \rangle} \right]^{\frac{1}{1+x}} \left[\frac{CNR(m_{\text{opt}}, g_{\text{opt}}) K_3 B \langle I_r^2 \rangle (1 + 2/x)}{\frac{1}{2} m_{\text{opt}}^2 - RIN \cdot K_3 B} \right]^{\frac{1+x/2}{1+x}} \tag{24.35}$$

For a PIN receiver, the gain and the excess noise factor are both unity; the required photocurrent to achieve CNR (m_{opt}) is then:

$$I_{\text{PIN}} = \frac{e + \sqrt{e^2 + W \langle I_r^2 \rangle}}{W} \tag{24.36}$$

with:

$$W = \frac{m_{\text{opt}}^2}{2CNR \cdot K_3 B} - RIN \tag{24.37}$$

In practice, we are interested in obtaining the receiver sensitivity for a desired CNR and a specific number of channels with the laser biased at a certain point. These last two parameters, number of channels and laser bias current, will determine the levels of distortion at the laser output and are included in the analysis through the distortion coefficients D. Once the system parameters are specified, the distortion coefficients are determined, and from these the optimum modulation depth for the desired CNR follows from Equation (24.31).

Depending on whether the system uses a PIN or an APD, the receiver sensitivity IPIN (APD) times the photodetector responsivity (R0) is readily obtained from Equations (24.35) and (24.36), respectively.

In cases where only either second- or third-order intermodulation effects are significant, considerable simplification of the previous equations is possible.

If only second-order distortion is significant, the optimum modulation depth (Equation 24.18) may be reduced to:

$$m_{\text{opt}} = [2CNR\,(D_{11}N + D_2)]^{-1/2} \tag{24.38}$$

Similarly, for third-order distortion effects:

$$m_{\text{opt}} = \left[3CNR\left(D_{111}N^2 + D_{21}N\right)\right]^{-1/4} \tag{24.39}$$

Note that, in accordance with the previous definition of the distortion coefficients, these equations are bias- and frequency-dependent: as the frequency subcarrier allocation gets closer to the resonance frequency the distortion coefficients reach maxima, which forces m_{opt} to correspondingly lower values.

24.5.4 Clipping Distortion Implication

RIN and intermodulation distortion establish a limit to the maximum obtainable optical modulation depth or carrier-to-noise ratio.

It has been shown that clipping effects determine a fundamental limit to the total modulation depth [41, 42]. Thus, even if the distortion coefficients are all zero, a maximum CNR still exists. This limit imposed by clipping distortion will be determined based on the model by Saleh, which has been revised in [43] and is shown to agree with simulation results to within 2 dB for values of the effective modulation depth, $\mu = m\sqrt{N/2}$, greater than 0.224. In Saleh's analysis, the nonlinear distortion is calculated by approximating the sum of multiple randomly-phased subcarriers as a Gaussian probability density of the amplitude. The total nonlinear distortion is then assumed to be proportional to the power in the Gaussian tail that falls below zero, and to be distributed uniformly over all channels. The total mean-square value of the clipped portion of $I(t)$ is then [42, 43]:

$$\left\langle I_{\text{clip}}^2 \right\rangle = \frac{1}{\sqrt{2\pi}} \frac{\mu^5}{1+6\mu^2} e^{-1/2\mu^2} \tag{24.40}$$

Additionally, a calculation by Mazo [44] indicates that over a wide range of μ, most of the distortion is thrown out of the transmission band. Thus, the carrier-to-clipping-distortion-noise ratio (CCR) becomes:

$$CCR^{-1} = \Lambda\sqrt{\frac{2}{\pi}\frac{\mu^3}{1+6\mu^2}e^{-1/2\mu^2}} \tag{24.41}$$

where Λ represents the fraction of the clipping distortion power which falls in the transmission band, which for the US-CATV plan (50–500 MHz) takes the value $\Lambda = 1/2$ [44].

Rewriting (24.27) to include clipping noise we obtain:

$$CNR^{-1} = \frac{KN}{\mu^2 I^2 g^2} + \Lambda\sqrt{\frac{2}{\pi}\frac{\mu^3}{1+6\mu^2}}e^{-1/2\mu^2} + 4\mu^4 D_{111} + 2\mu^2 D_{11} \tag{24.42}$$

where K includes all the other noise contributions:

$$K = \langle I_r^2 \rangle B + g^2 I^2 RIN \cdot B + 2eBIg^{2+x} \tag{24.43}$$

The CNR at optimum μ and APD gain, μ_{opt} and g_{opt} respectively, now becomes:

$$CNR^{-1}(\mu_{opt}) = \Lambda \frac{e^{-1/2\mu_{opt}^2}}{\sqrt{2\pi}} \frac{\mu_{opt} + 11\mu_{opt}^3 + 18\mu_{opt}^5}{\left(1 + 6\mu_{opt}^2\right)^2} + 12\mu_{opt}^4 D_{111} + 4\mu_{opt}^2 D_{11} \tag{24.44}$$

and g_{opt} remains unchanged. The relation between the photocurrent I and μ_{opt} is determined from the solution of the following equation:

$$\left[\langle I_r^2 \rangle (1 + 2/x) BN + g_{opt}^2 I^2 RIN \cdot B \cdot N\right] = \mu_{opt}^2 g_{opt}^2 I^2 \times \left[\Lambda \frac{e^{-1/2\mu_{opt}^2}}{\sqrt{2\pi}} \frac{\mu_{opt} + 9\mu_{opt}^3 + 6\mu_{opt}^5}{\left(1 + 6\mu_{opt}^2\right)^2} + 8\mu_{opt}^4 D_{111} + 2\mu_{opt}^2 D_{11}\right] \tag{24.45}$$

It may be advantageous to accept some distortion rather than constrain the total modulation depth to 100% [45]. An adequate measure of the total effective modulation depth is the r.m.s. modulation, defined as $m_{ms} = m\sqrt{N}$ [45].

24.5.5 Latency

The system latency, T_d, is given by the maximum time required by any signal received to propagate by the complete topology; that is, to reach the most far-end IUT. For the latency calculation, it is considered the time delay between the IUT i-node and the k-node. We can calculate the accumulated time delay by:

$$T_d = T_{d\_RX} + T_{d\_VCSEL} + T_p + T_{d\_OCMC} + T_{d\_TX} \tag{24.46}$$

where T_{d_RX} and T_{d_TX} stand for the delay (phase response) of the RX and TX antenna respectively; T_{d_VCSEL} stands for VCSEL of the group velocity delay; T_{d_OCMC} stands for OCMC of the group velocity delay; T_p stands for the propagation time along the path. T_{Rx} and T_{Tx} are the group velocity delay of the IUT front-end amplifier; the sum given by Equation (24.46) must be under this upper limit.

24.5.6 Electrical/Optical Conversion Gain

Let us now calculate the theoretical electrical/optical conversion gain. We evaluate this parameter from a high-level system point of view. This model will also be adequate for an E-EAT implementation. See Figure 24.13.

The electrical power is converted to optical power via the VCSEL efficiency $\xi[W/A]$ according to the following relation: $P_{el} = I^2 R$, and consequently $I = \sqrt{P_{el}/R}$. After the

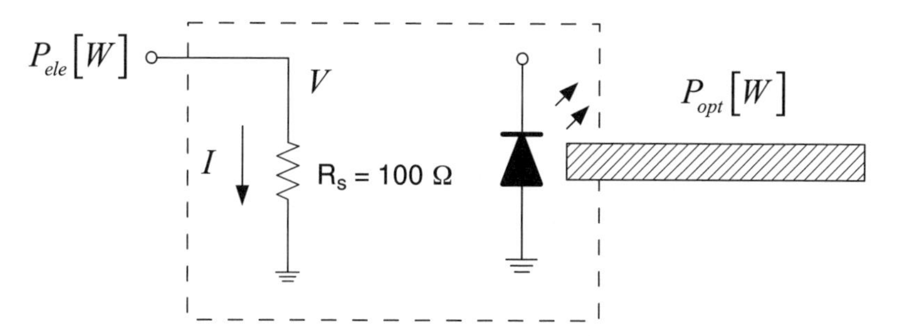

Figure 24.13 System-level generic electrical/optical converter (VCSEL/E-EAT) models.

conversion we have:

$$P_{\text{opt}} = xI = \xi\sqrt{\frac{P_{\text{ele}}}{R}}$$

Expressed in dB, this has the form:

$$P_{\text{opt}}[\text{dBw}] = 10\log(\xi I) = 10\log\left(\xi\sqrt{\frac{P_{\text{ele}}}{R}}\right) = 10\log\left(\frac{\xi}{\sqrt{R}}\right) + \frac{1}{2}10\log(P_{\text{ele}}) \quad (24.47)$$

The $^1/_2$ factor is not a problem, as in the OCMC the opposite effect occurs, compensating this. For actual data we can expect for the VCSEL efficiency, see Table 24.9.

From the efficiency values above, the expected E/O conversion loss can be calculated as in the following Examples:

For $\lambda_{\text{opt}} = 850\,\text{nm}$, $\xi = 0.3$ we have: $P_{\text{opt}}\,[\text{dBw}] = -15.23 + \frac{1}{2}P_{\text{el}}\,[\text{dBw}]$

For $\lambda_{\text{opt}} = 1310\,\text{nm}$, $\xi = 0.28$ we obtain: $P_{\text{opt}}\,[\text{dBw}] = -15.53 + \frac{1}{2}P_{\text{el}}\,[\text{dBw}]$

For $\lambda_{\text{opt}} = 1550\,\text{nm}$, $\xi = 0.1$ we obtain: $P_{\text{opt}}\,[\text{dBw}] = -20 + \frac{1}{2}P_{\text{el}}\,[\text{dBw}]$

24.5.6.1 O/E Conversion Gain

The same analysis can be applied to calculate the optical to electrical conversion gain (Figure 24.14).

Table 24.9 Comparison of 100% conversion efficiency for VCSEL at UROOF wavelengths.

Wavelength (nm)	100% conversion efficiency (W/A)	Actual conversion efficiency (W/A)	G(laser) (dB)
850	1.46	0.3	−13.7
1310	0.947	0.28	−10.58
1550	0.8	0.1	−18.061

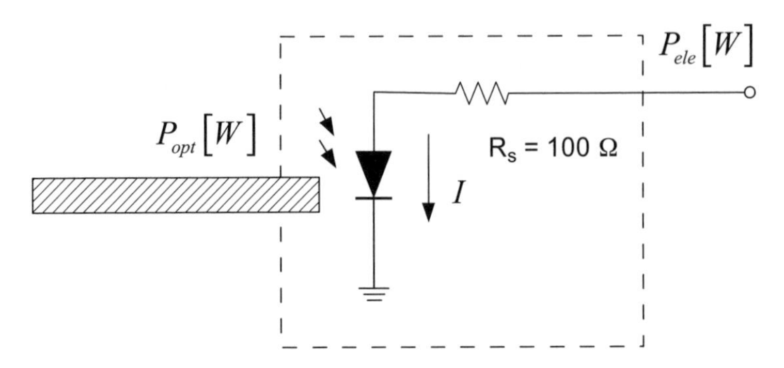

Figure 24.14 System-level generic photoreceiber (APD/PIN/OCMC) model.

The electrical power is converted to optical power via the OCMC efficiency $\eta[A/W]$, as follows:

$$I = \eta P_{\mathrm{opt}} = \sqrt{P_{\mathrm{el}}/R}. \quad \text{Then we get } P_{\mathrm{el}} = R\eta^2 P_{\mathrm{opt}}^2.$$

The same relation expressed in dB has the form:

$$P_{\mathrm{ele}}[\mathrm{dBw}] = 10\log(R\eta^2 P_{\mathrm{opt}}^2) = 10\log(R\eta^2) + 2 \times 10\log(P_{\mathrm{opt}}) \tag{24.48}$$

This confirms that the $^1/_2$ factor in Equation (24.48) is not a problem, because in the OCMC there is a 2 factor that compensates it. Table 24.10 reports actual data we can expect for photodiode efficiency (DSC-R402HR).

For example, for $\lambda_{\mathrm{opt}} = 850\,\mathrm{nm}$, $\eta = 0.25$ we obtain:

$$P_{\mathrm{opt}}\,[\mathrm{dBw}] = 7.96 + \frac{1}{2}P_{\mathrm{el}}\,[\mathrm{dBw}]$$

and for $\lambda_{\mathrm{opt}} = 1310$ and $1550\,\mathrm{nm}$, $\eta = 0.8$:

$$P_{\mathrm{opt}}\,[\mathrm{dBw}] = 18 + \frac{1}{2}P_{\mathrm{el}}\,[\mathrm{dBw}].$$

Table 24.10 Comparison of 100% conversion efficiency for receiver at UROOF wavelengths.

Wavelength (nm)	100% conversion efficiency (A/W)	Actual conversion efficiency (A/W)	G(PIN) (dB)
850	0.685	0.25	−8.75
1310	1.055	0.8	−2.403
1550	1.249	0.8	−3.869

24.5.7 Modelling the Wireless UWB Indoor Channel

Models for propagation through typical WPAN of UWB signals were developed in IEEE802.15.3a [46], based on a modified Saleh–Valenzuela (S–V) model. Like in the original S–V double-exponential decay power delay profile model, multipaths arrive in 'clusters' of rays. Cluster arrival times and rays' arrival times within the cluster are modelled by statistically-independent Poisson processes. A typical power profile of the S–V model is depicted in Figure 24.15.

Here $\alpha^{i}_{k,\ell} \cdots$ gain coefficients of multipath channel.

A recent survey of UWB channel models is provided in [47].

24.5.8 Modelling the Propagation through the Fibre

The dispersive nature of the media is characterized by:

- frequency-dependent refractive index – $n(\nu)$
- absorption coefficient – $\alpha(\nu)$
- phase velocity v and propagation constant $\beta(\nu)$.

Since a pulse light wave is the sum of many monochromatic waves, each component is modified differently by the medium. As a result, the pulse is delayed and broadened ('dispersed in time'). The simplest model for a dispersive [48] assumes a pulsed plane wave propagating along the z axis of a linear, homogeneous and isotropic medium with known $n(\nu)$, $\alpha(\nu)$ and propagation constant $\beta(\nu) = 2\pi \nu n(\nu)/c_0$. The complex wave function is given by:

$$U(z,t) = A(z,t) \exp\left[j\left(2\pi \nu_0 t - \beta_0 z\right)\right]$$

where ν_0 is the central frequency, $\beta_0 = \beta(\nu_0)$ is the central wave number and $A(z,t)$ is the complex envelope of the pulse. We assume that $A(0,t)$ is known and wish to determine $A(z,t)$ at the distance z in the medium.

24.5.8.1 Linear System Description

A transfer function linear system description that relates $A(z,t)$ to $A(0,t)$ is based on the following. Let $A(0,t) = A(0,f)\exp(j2\pi f t)$ be a harmonic function with frequency f so that the wave is monochromatic with frequency $\nu = f + \nu_0$. Then $\mathcal{A}(z,f) = \mathcal{A}(0,f)H(f)$ where the transfer function of the linear system is given by:

$$H(f) = \exp\left\{-\frac{1}{2}\alpha(f+\nu_0)z - j\left[\beta(f+\nu_0)\right]z\right\} \tag{24.49}$$

Hence the complex envelope of the pulse is extracted using the Fourier transform:

$$A(z,t) = F^{-1}\{A(z,f)\} = \int_{-\infty}^{\infty} A(z,f)\exp(j2\pi f t)\,df \tag{24.50}$$

This can be equivalently expressed as the convolution $A(0, t)$ with $h(t) = F^{-1}\{H(f)\}$ as follows: $A(z, t) = \int_{-\infty}^{\infty} A(0, t')h(t - t')dt'$.

24.5.8.2 Slowly Varying Frequency-dependent Approximation

$A(z, t)$ slowly varies with respect to the central optical frequency ν_0 and its Fourier transform $\mathcal{A}(z, f)$ is a narrowband function of f with width $\Delta\nu << \nu_0$. It can be assumed that:

$\alpha(\nu) = \alpha$; that is, the absorption coefficient is not frequency-dependent;
$\beta(\nu)$ varies only slightly with frequency. Hence it can be approximated by three-term Taylor series: $\beta(\nu_0 + f) \approx \beta(\nu_0) + f\frac{d\beta}{d\nu} + \frac{1}{2}f^2\frac{d^2\beta}{d\nu^2}$.

The approximate transfer function can now be presented by:

$$H(f) \approx H_0 \exp(-j2\pi f \tau_d) \exp\left(-j\pi D\nu f^2\right) \tag{24.51}$$

where $H_0 = e^{-\alpha z/2}$, $\tau_d = z/v$.

$\frac{1}{v} = \frac{1}{2\pi}\frac{d\beta}{d\nu}$, where v is the group velocity.
$D_\nu = \frac{1}{2\pi}\frac{d^2\beta}{d\nu^2} = \frac{d}{d\nu}\left(\frac{1}{v}\right)$, where D_ν is the dispersion coefficient.

Notes: (1) when the dispersion is negligible, the third term of the propagation constant (function of f^2) is small and hence the transfer function contains $H(f) = H_0 \exp(-j2\pi f_d) = \exp\left(-\frac{\alpha z}{2}\right)\exp(-j2\pi f_d)$. The first term is an attenuation factor and the second term is a delay $\tau_d = z/v$. In this case the original pulse passed through a linear system (with losses) and is not distorted. Hence, $A(z, t) = e^{-\alpha z/v} A(0, t - \tau_d)$.

(2) For an ideal linear system without losses, $\alpha = 0$ and $\frac{1}{v} = \frac{1}{2\pi} \times \frac{2\pi}{c} = \frac{1}{c}$. Hence $v = c$ and the pulse envelope travels at the free-space light velocity c.

(3) In general cases, the group velocity depends on frequency, so that different frequency components undergo different delays. Hence the quantity $\tau_d = z/v(\nu) = \frac{z}{2\pi(d\beta/d\nu)}$ is frequency-dependent and the pulse is distorted (broadened and attenuated). If a pulse has a spectral width σ_ν (Hz) then after z meters the pulse width can be approximated by $\sigma_\tau = |D_\nu|\,\sigma_\nu z$.

24.6 Analysis of UWB Technologies for UROOF

24.6.1 Comparing UWB Technologies for Radio-over-fibre

The introduction of this chapter defines the three UWB technologies investigated for ultra-wideband radio-over-fibre: MB-OFDM, DS-UWB and IR-UWB. Detailed simulation results, along with a detailed description of the three UWB technologies, were presented in [49]. Although these technologies are well known in literature for their ability to transfer data through RF and optical channels, a combination of wireless and optics as one medium (as suggested by the UROOF project) has never been investigated thoroughly. Therefore, the simulation in [49] served as the first (and one of the most highly weighted) of nine criteria

Table 24.11 Comparison of three UWB technologies over a specified UROOF channel [52].

	Simulation	Performance	BW	TX	Cost	Robustness	Scalability	Dynamic range	Topology
CRITERION	0	1	2	3	4	5	6	7	8
WEIGHT	1	1	0.9	0.9	1	0.8	0.8	0.8	0.7
MB-OFDM	8	9	9	7	9	9	8	7	10
DS-SS	9	9	8	6	8	8	7	5	8
Impulse radio	8	5	8	6	7	7	7	9	7
	Total	out of							
MB-OFDM	**65.7**	79							
DS-SS	60.2	79							
Impulse radio	55.9	79							

for selecting the most suitable of the three technologies for pursuing this goal. The simulation reflected the impairments of each medium on the relevant signal, and the effects of conversion from one medium to another and back again on the quality of the original UWB signal. The simulation results showed each technology's inherent ability to overcome impairments in this difficult case. The clear winner of the first criterion (simulation) is the MB-OFDM. Further benefits of MB-OFDM over multimode fibre have recently been discussed and published in [50, 51].

The criteria were first presented in [49] and used to compare and select the best UWB technology for the combined wireless-wired channel. The results from [52] are given in Table 24.11. High weighting was given to simulation, performance (high data rate) and cost. Evidently, the UROOF concept is aimed at low-cost solutions that will fit the budget of the mass market of WPAN. If it is to be a commercially available technology it will sure help to reduce the price, but trying not to compromise on performance. The most cost-effective technology for the UROOF channel is clearly the MB-OFDM [52]. Apparently, the MB-OFDM is the ONLY available high-data-rate UWB technology on the market. We note that OFDM variants are becoming the preferred technology for a range of wireless local area networks (e.g., 802.11a/g/n) and wireless metropolitan area networks (802.16e/m, 3GLTE, etc.); this has certainly helped reduce the price and increase the availability of MB-OFDM for UROOF applications.

Although MB-OFDM had a quite similar performance in simulation to DS-UWB in terms of the operating distance from the UROOF access node(AN), the other criteria showed MB-OFDM to be the best UWB flavour for UROOF technology.

24.6.2 MB-OFDM Over Multimode Fibre

This section addresses the range extension of the UWB communication, taking both wireless and optical channels into account, using the concept of MB-OFDM UWB over fibre. Previous works have studied the feasibility of pulsed and OFDM-based UWB transmission on optical channels only [53–55]. The optical link consists of a Vertical Cavity Surface Emitting Laser Diode (VCSEL-HFE4192-58 LC TOSA), multimode fibre (MMF) and PIN photodetector (HFD3180-203). Using the above mentioned low-cost optical devices, the possibility of error-free transmission of MB-OFDM UWB signals is demonstrated. Following [50], this section

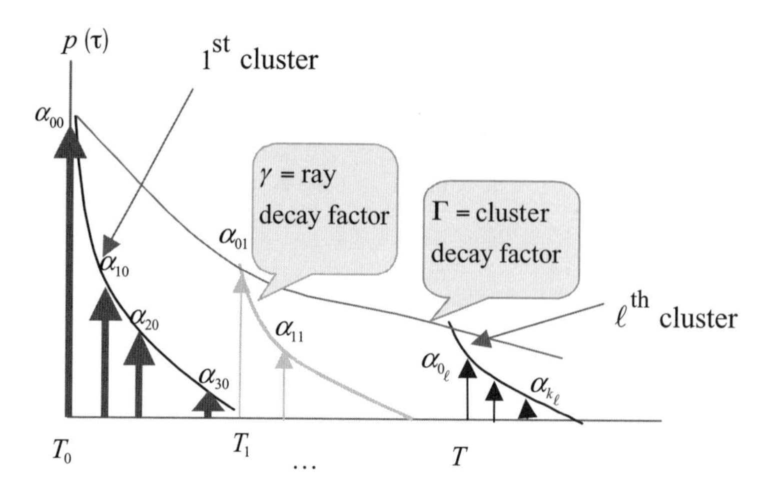

Figure 24.15 Power delay profile of S-V model.

provides performance at different bit rates and different lengths of wireless and optical links by means of Packet Error Rate (PER), Link Quality Indicator (LQI) and Received Signal Strength Indicator (RSSI). It is shown both experimentally and theoretically that the range extension using MMF is limited by 400\,m (excluding the wireless channel) due to the modal dispersion effects in MMF.

Figure 24.16 shows the UROOF experimental setup. The system comprised a pair of MB-OFDM evaluation boards (WisAir), a VCSEL for the up-conversion of the UWB signal, MMF and a PIN photodetector. UWB signals transmitted by UWB-Tx (Wisair-FX2) were detected by the UWB antenna (2 dBi gain). In order to compensate for the attenuation of the UWB signal in the wireless channel, we used a 40 dB, low-noise, highly-linear RF amplifier (Hittite- HMC326MS8G). The VCSEL was directly modulated by the amplified UWB signal (modulation index: 0.34–0.36). The up-converted optical signal propagated through the MMF and was detected by the PIN diode, then onward transmitted through the wireless channel to the UWB-Rx (Wisair-FX2).

The Wisair evaluation board generated MB-OFDM signals in sub-bands of 528 MHz in Band Group 1 from 3.168 to 4.752 GHz with 80 μW average RF power according to FCC regulations [56]. In the set-up, VCSEL emission was at 850 nm, with 4.8 GHz bandwidth and an average optical power of approximately −3 dBm. Standard MMF fibre was used, with 50 μm diameter, 0.24 numerical aperture and attenuation of 0.5 dB/km at the operational wavelength. The PIN photodetector had 0.6 A/W responsivity at 850 nm and 12 GHz bandwidth.

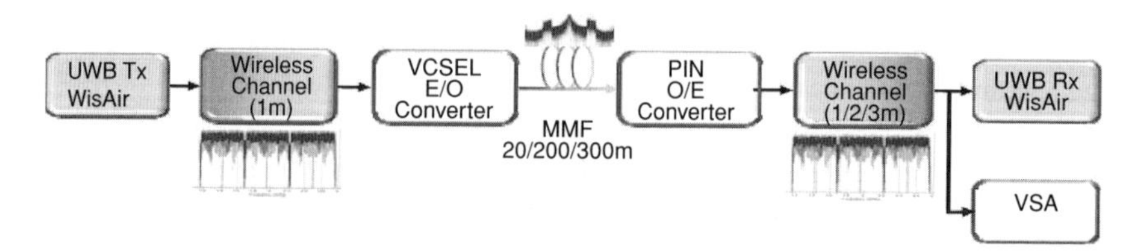

Figure 24.16 Experimental set-up for UWB radio-over-optical (UROOF) system.

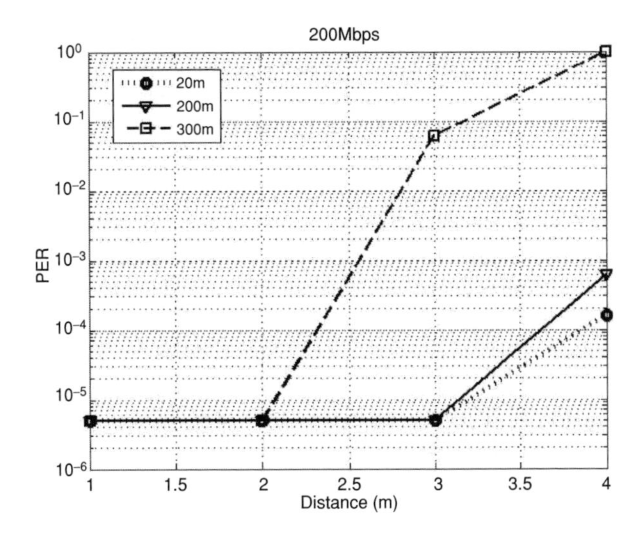

Figure 24.17 PER versus wireless link at 200 Mb/s.

Various experiments were performed with a fixed 1 m distance at the transmitting end of the wireless link, varying the wireless channel at the receiving side from 1 to 4 m for various MMF fibre lengths. UROOF system performance was measured with a Vector Signal Analyzer (Agilent-89600 VSA). For different fibre lengths (20, 200 and 300 m), the packet error rate (PER) versus varying length of the wireless channel at the UWB-Rx side for 200 and 480 Mb/s is shown in Figures 24.17 and 24.18. According to TG3a

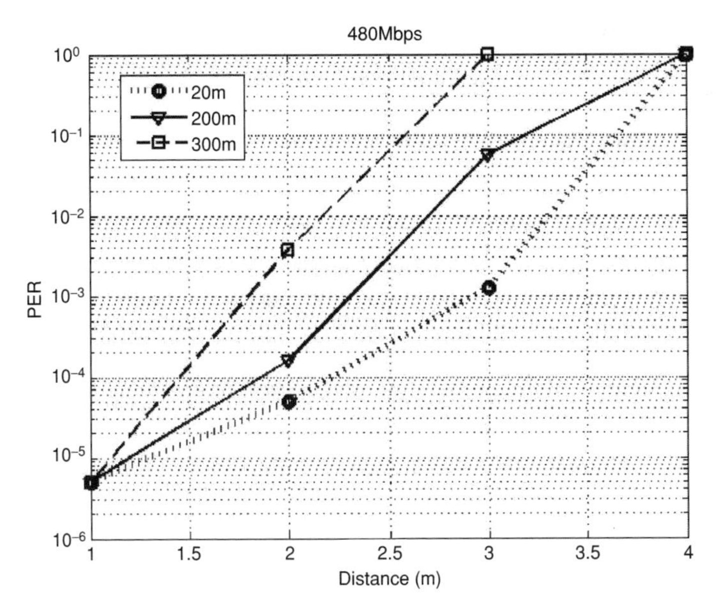

Figure 24.18 PER versus wireless link at 480 Mb/s.

(http://www.ieee802.org/15/pub/TG3a.html) technical requirements, the maximum PER allowed is 8% for a 1024 octet frame body. At 200 Mb/s data rate on 20 and 200 m of MMF, the system showed good performance in terms of PER for the entire test range (1–4 m) of the wireless link (Figure 24.17). However, with 300 m MMF, the PER was significantly degraded for wireless links greater than 2 m (Figure 24.18). Also, at 480 Mps operation rate the system performance was strongly affected by the signal distortion in the MMF fibre, degrading the PER for the wireless links longer than 2 m.

Major limitations of the UWB-over-fibre system are noise and nonlinearities induced by the optical link, such as third-order intermodulation distortion induced by the nonlinear nature of the lasers and relative intensity noise.

The impulse response of the MMF fibre is given by $h(t) = \sum_{i=1}^{M} a_i(t - \tau_i)$, where M is the number of modes in MMF, a_i is the relative power coefficients and τ_i is the time delay corresponding to each mode. The transfer function of MMF can be calculated numerically and given by the following expression [57]:

$$|H(j\omega|^2 = \sum_{i=1}^{M} \left(a_i^4 + 2a_i^2 \sum_{k=i+1}^{M} a_k \cos w(\tau_k - \tau_i) \right) \tag{24.52}$$

Figure 24.19 shows the MMF transfer function at different fibre lengths. The calculations are done for 50 μm diameter MMF with numerical aperture NA = 0.24 at 850 nm wavelength. As shown in Figure 24.20, the transfer function is flat in the required bandwidth for 300–400 m MMF and strongly decays after that. This deterioration affects the transmitted power and spectrum of the UWB signal, as shown in Figures 24.21 and 24.22. Figure 24.21 presents the SSRI of the received signal. The SSRI decreases with increasing fibre length but remains

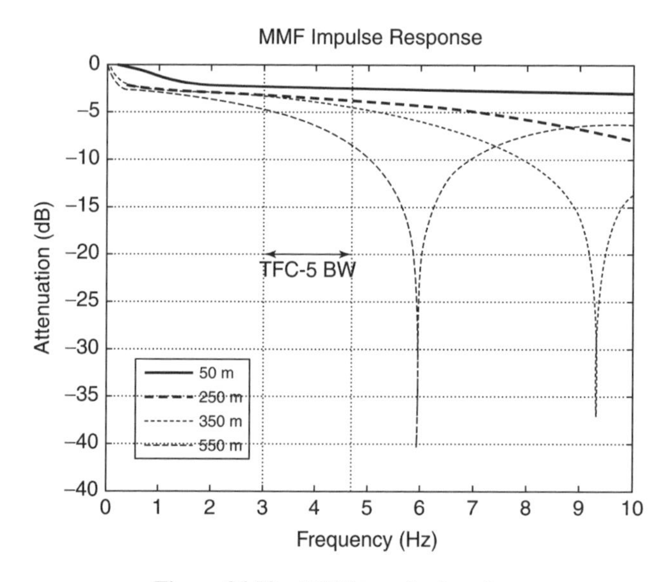

Figure 24.19 MMF transfer function.

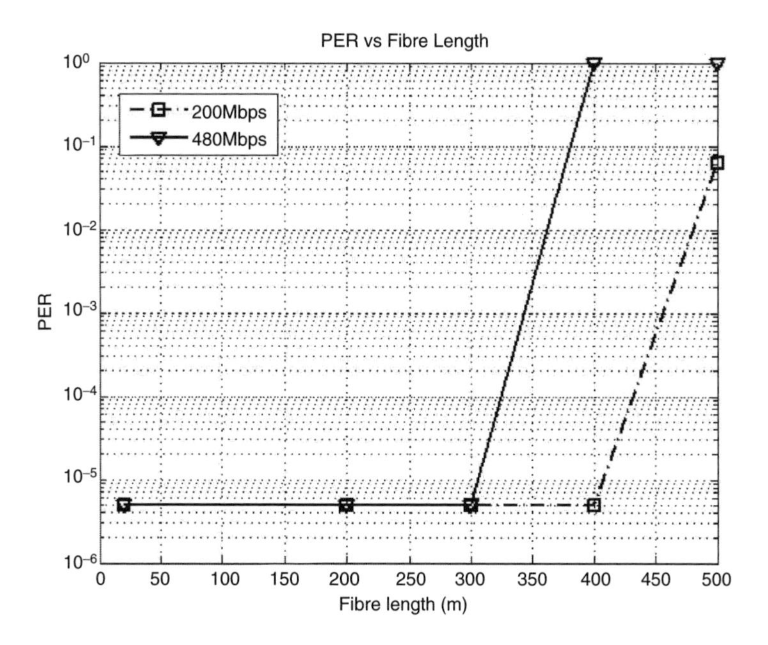

Figure 24.20 PER vs. link length.

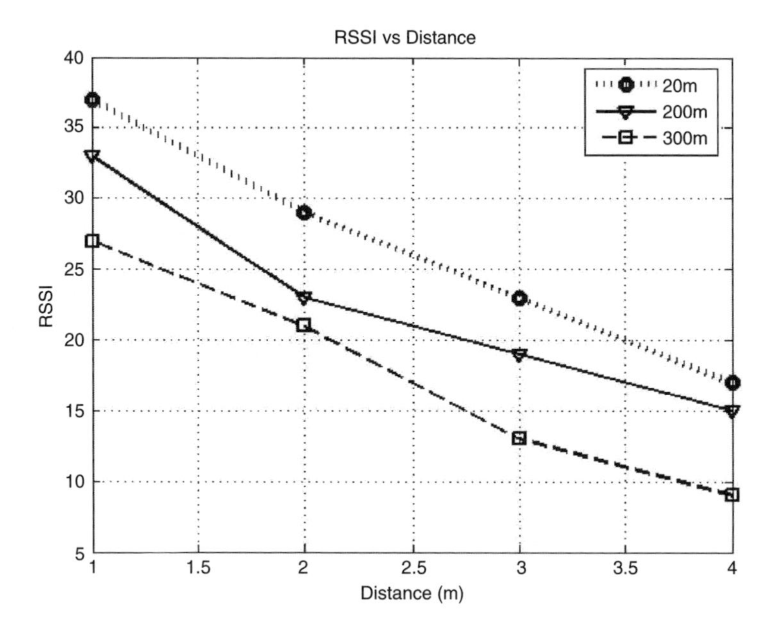

Figure 24.21 RSSI (480 Mb/s) vs. wireless link length.

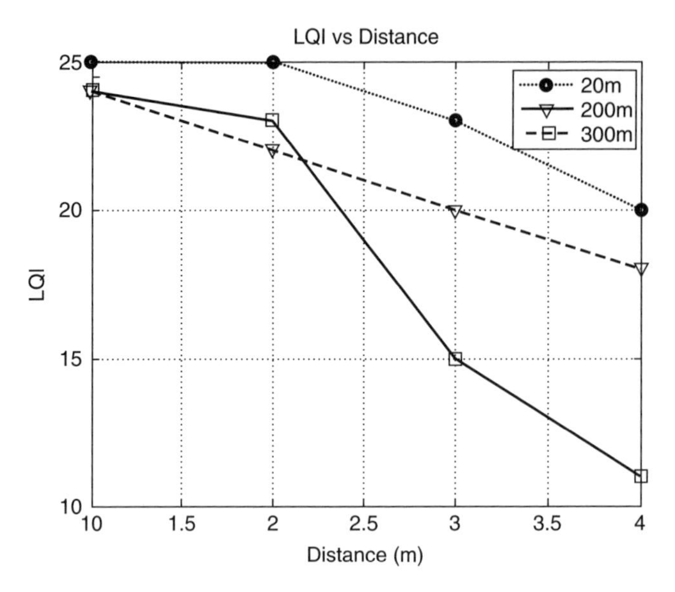

Figure 24.22 LQI (480 Mb/s) vs. wireless link length.

above the required receiver sensitivity. This high SSRI is not enough to guarantee the high SNR, because the signal can still be significantly corrupted by the channel distortions. The LQI shown in Figure 24.22 manifests this statement. The LQI is strongly degraded at the longer fibre lengths and leads to the high PER shown in Figures 24.17 and 24.18.

24.6.2.1 Conclusions

Optical range extension over 300 m MMF for a Wimedia-supported MB-OFDM UWB (480 Mb/s) radio system combined with a 1–4 m TFC-5 (Band #1, see Figure 24.1) wireless channel might use off-the-shelf VCSELs and commercial MB-OFDM transceivers. OFDM offers the prospect of low-cost and low-power wide-scale deployment with minimum management issues. The MB-OFDM UWB Tx/Rx mitigates the effects of link impairments whilst still providing full system condition reporting. As such, the techniques described above have a clear role to play in the current and future competitive environment for 3/4G and Bluetooth 3.0 WPAN implementations.

24.7 All-optical Generation of Ultra-wideband Impulse Radio

24.7.1 Introduction

UWB impulse radio (IR) modulation techniques are especially important because the UWB spectrum has been made available by the Federal Communication Commission (FCC) and it can be used with IRs developed to date [58]. Previously, most of the approaches to generating IR-UWB signals were based on electronic methods [59]. Recently, optically-based methods of the Gaussian IR-UWB monocycles and doublets [60] generation for low-cost high-data-rate

UWB wireless systems have been proposed [58, 59, 61–64]. Their advantages are the following: the decreasing of interference between electrical devices, low loss and light weight of opticals [58, 59]. Typically, an optically-based system generating Gaussian IR-UWB monocycles and doublets consists of an electro-optic phase modulator (EOM), a single-mode-fibre (SMF), Erbium doped fibre amplifier (EDFA), a Bragg grating (FBG) and a photodetector [59, 61, 62]. Generation of a Gaussian doublet pulse with a full width half maximum (FWHM) of about 40 ps has been demonstrated experimentally [61]. The shortfalls of this method are the necessity of the sophisticated circuit for the generation of short electric Gaussian pulses, the use of EOM, and the need for a long SMF (5 km) [62, 64].

We propose a theoretical analysis of a novel all-optical method of IR-UWB pulse generation in an integrated Mach–Zehnder interferometer (MZI) with quantum dot semiconductor optical amplifier (QD SOA) as an active element inserted into one arm of the integrated MZI, which results in an intensity-dependent optical signal interference at the output of MZI. The IR-UWB pulse generation process is based both on the cross phase modulation (XPM) and on the cross gain modulation (XGM) in QD SOA, characterized by an extremely high optical nonlinearity and high operation rate [65]. XGM gain and XPM phase shift are related due to the linewidth enhancement factor (LEF) α [66]. Unlike other all-optical methods, we need not optical fibres, FBG and EOM substantially reducing the cost and complexity of an IR-UWB generator. The theoretical analysis of the nonlinear MZI containing QD SOA based on the QD SOA dynamics and truncated equations for optical pulses propagation clearly shows that the proposed method permits generation of Gaussian IR-UWB doublet picosecond signals with an adjustable shape in the time and frequency domains. The successful operation of integrated SOA-MZI structures in an essentially nonlinear regime providing wavelength conversion (WC), all-optical logic gate and all-optical regeneration due to XGM has been demonstrated experimentally [67–69]. Hence, we believe that the proposed method of the IR-UWB pulse generation is feasible.

24.7.2 Operation Principles and Theoretical Approach

The block diagram of the proposed all-optical IR-UWB pulse generator consisting of a CW laser, a pulse laser and an MZI with a QD SOA in its upper arm is shown in Figure 24.23.

A CW signal of wavelength λ and optical power P_0 is split into two signals with equal optical power and fed into the two ports of the integrated MZI, one of which contains a QD SOA. The pulse laser produces a train of short Gaussian pulses, counter-propagating with respect to the input CW optical signal. The CW signal propagating through the upper arm of

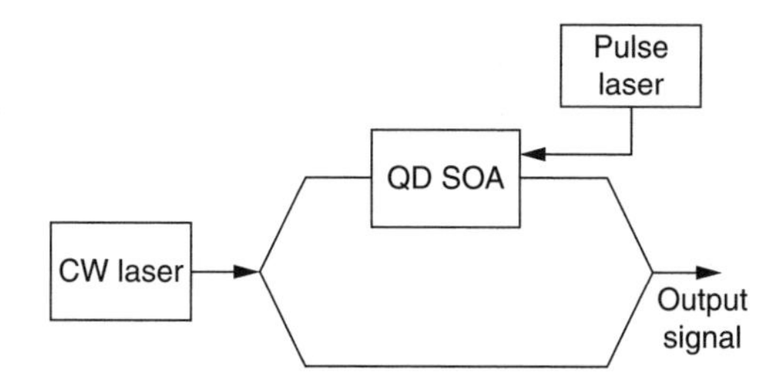

Figure 24.23 Block diagram of the proposed UWB-IR generator.

the MZI transforms into the Gaussian pulse at the output of the MZI due to XPM and XGM with the train of Gaussian pulses, whereas the optical signal in the linear lower arm of the MZI remains CW. Both these pulses interfere at the output of the MZI, and the output pulse shape is defined by the power-dependent phase difference $\Delta\phi(t) = \phi_1(t) - \phi_2(t)$, where $\phi_{1,2}(t)$ are the phase shifts in the upper and lower arms of the MZI, respectively. Evidently, the phase shift in the lower arm of the MZI is constant: $\phi_2 = \text{const}$. The MZI output optical power P_{out} is given by [70]:

$$P_{\text{out}} = \frac{P_0}{4}\left[G_1(t) + G_2(t) - 2\sqrt{G_1(t)G_2(t)}\cos\Delta\phi(t)\right] \tag{24.53}$$

where $G_{1,2}(t)$ are amplification factors of the upper and lower arms of the MZI. The upper arm amplification factor $\exp(g_{\text{sat}}L) \leq \exp(gL) \leq \exp(g_0 L)$, defined by the QD SOA gain g and its active region length L, is limited by the saturation gain g_{sat} and the maximum modal gain g_0, typical for the linear regime. The linear lower arm amplification factor is simply unity: $G_2 = 1$. The losses can be neglected due to a small length of integrated elements. Then the relation between the MZI phase shift and its amplification factor is given by $\Delta\phi(t) = -(\alpha/2)\ln G_1(t)$. QD SOA dynamics are discussed in detail in [65]. In particular, XGM process is accompanied by XPM [65]. The shape of the output pulse is determined by the time-dependence of $G_1(t)$, both directly and through $\Delta\phi(t)$ according to Eq. (24.53) [65], resulting in a Gaussian doublet under certain conditions. The system of rate equations for the electron occupation probabilities of the ground state (GS) and excited state (ES) in the conduction band of a QD and the electron concentration in the wetting layer (WL) has been thoroughly investigated [71] and it is not necessary to discuss it in detail in this paper. The optical signal propagation in a QD SOA is described by the following truncated equations for the slowly varying CW and pulse signal photon densities and phases $S_{CW,P} = P_{CW,P}/(\hbar\omega_{CW,P}(v_g)_{CW,P}A_{\text{eff}})$ and $\theta_{CW,P}$ [66]:

$$\frac{\partial S_{CW,P}(z,\tau)}{\partial z} = \left(g_{CW,P} - \alpha_{\text{int}}\right)S_{CW,P}(z,\tau) \tag{24.54}$$

$$\frac{\partial \theta_{CW,P}}{\partial z} = -\frac{\alpha}{2}g_{CW,P} \tag{24.55}$$

where $P_{CW,P}$ are the CW and pulse signal optical powers, respectively; A_{eff} is the QD SOA effective cross-section; $\omega_{CW,P}$, $(v_g)_{CW,P}$ are the CW and pulse signal group angular frequencies and velocities, respectively; $g_{CW,P}$ are the active medium (SOA) gains at the corresponding optical frequencies; α_{int} is the absorption coefficient of the SOA material. For the pulse propagation analysis, we replace the variables (z, t) with the retarded frame variables $\left(z, t = t \mp z/v_g\right)$. For optical pulses with duration $T > 10$ ps, the optical radiation of the pulse fills the entire active region of a QD SOA of length $L \leq 1$ mm and the propagation effects can be neglected [41]. Hence, in our case the photon densities

$$S_{CW,P}(z,\tau) = (S_{CW,P}(\tau))_{\text{in}}\exp\left[\int_0^z (g_{CW,P} - \alpha_{\text{int}})dz'\right]$$

can be averaged over the QD SOA length L, which yields:

$$S_{CW,P}(\tau) = \frac{1}{L}(S_{CW,P}(\tau))_{\text{in}} \int_0^L dz \exp\left[\int_0^z \left(g_{CW,P} - \alpha_{\text{int}}\right) dz'\right] \tag{24.56}$$

Solution of Equation (25.55) yields for the phases:

$$\theta_{CW,P}(\tau) = -\frac{\alpha}{2}\int_0^L g_{CW,P}dz$$

24.7.3 Simulation Results and Discussion

The study of the IR-UWB generator requires the simultaneous analysis of the QD SOA dynamics and Equations (24.53)–(25.55) . The system of dynamics and propagation equations has been solved numerically for the typical values of the QD SOA parameters [71]. In particular, the QD SOA active region length L and width W are, respectively, $L = 2$ mm, $W = 2$ μm; the confinement factor $\Gamma = 3 \times 10^{-2}$, the maximum gain $g_{\text{max}} = 11.5\,\text{cm}^{-1}$, $\alpha_{\text{int}} = 3\,\text{cm}^{-1}$, the fastest relaxation time for the transitions between ES and GS is $\tau_{21} = 0.16$ ps, while the largest relaxation time for the transition between ES and WL is about $\tau_{2w} \approx 1$ ns. The simulation results for the temporal variations of the optical signal power and phase in upper and lower arms of MZI and the output power at the different levels of the pulse laser power are shown in Figure 24.24. At high Gaussian pulse power levels QD SOA passes to the nonlinear saturation regime, where the amplification factor $G_1(t)$ and XPM phase shift $\Delta\phi(t)$ decrease to their

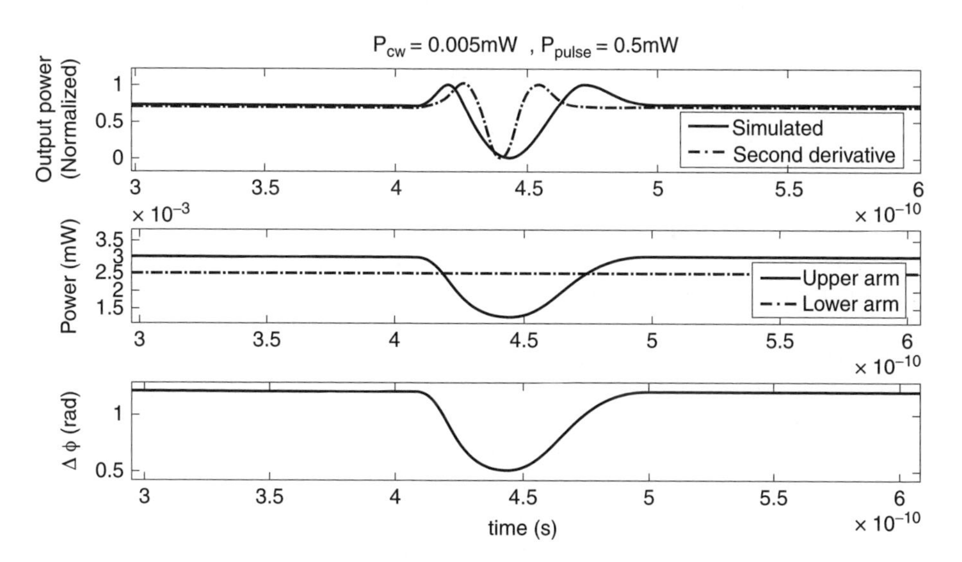

Figure 24.24 Gaussian doublet of the output power pulse (solid line) and the second derivative of the Gaussian pulse (dashed line) for the high pulse laser optical power of $P_p = 0.5$ mW.

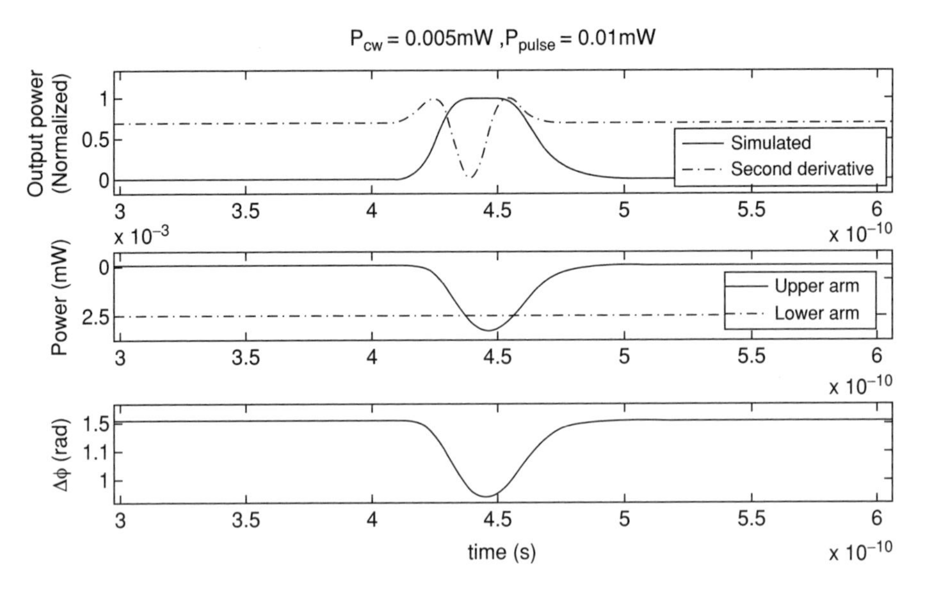

Figure 24.25 The low level of $P_p = 0.01$ mW – the Gaussian pulse remains unchanged.

minimum values. In this case, P_{out} also reaches its lowest level due to the maximum value of the oscillating term in Equation (24.53). In such a case the XPM process is dominant, and the Gaussian doublet occurs as seen in Figure 24.24. The local maxima of the doublet can be explained by the dominant role of the XGM process in the regions of small variation of $\Delta\phi(t)$. The variation of the oscillating term in Equation (24.53) is negligible. Generally, this term varies more slowly than $G_1(1)$ because of the logarithmic form of the argument $\Delta\phi(t)$, except in the QD SOA saturation regime, where $G_1(1)$ tends to its minimum value. In the opposite case of a weak Gaussian pulse, as shown in Figure 24.25, the QD-SOA operates in the linear regime, and the Gaussian pulse remains unchanged. Note that this situation corresponds to WC by monolithic integrated MZI experimentally realized in [67]. Different input power levels may result in different contributions of XGM and XPM processes and hence a different shape of the output pulses. The optical power in the linear lower arm of MZI – shown with the dashed line in Figures 24.24 and 24.25 – also does not vary.

The fast Fourier transforms (FFT) of the simulated IR-UWB signal (solid line) and second derivative of the standard Gaussian pulse (dashed line) are shown in Figure 24.26. The spectrum of the simulated IR-UWB signal manifests the filtering features of the proposed IR-UWB generator [60]. For the Gaussian pulse duration of about tens of picoseconds, the filtering behaviour of the proposed IR-UWB generator is caused by the fast transition relaxation time between GS and ES $\tau_{12} \approx 1$ ps in QD SOA limiting a rise time and a fall time of the pulse propagating through QD SOA. The operation rate of QD SOA is also strongly influenced by the SOA bias current and optical power [65, 71].

The theoretical analysis of a novel all-optical method of the IR-UWB signal generation based on XPM and XGM phenomena in QD SOA-based integrated MZI shows that, unlike other proposed all-optical methods, we need not optical fibres, FBG and EOM substantially reducing the cost and complexity of an IR-UWB generator. The IR-UWB signals generated

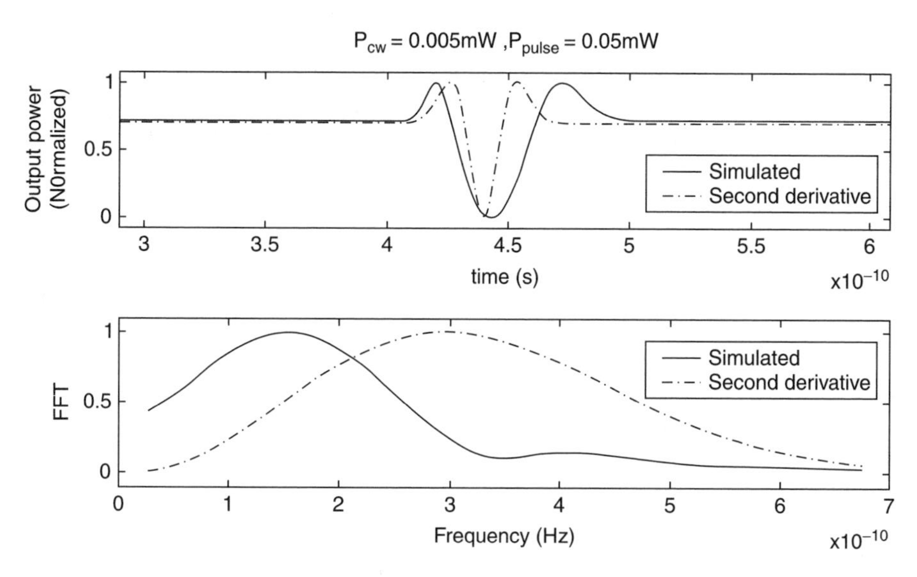

Figure 24.26 The fast Fourier transforms (FFT) of the simulated IR-UWB pulse (solid line) and second derivative of the standard Gaussian pulse (dashed line).

by the proposed QD SOA-based MZI structure have the form of the Gaussian doublet. The shape of the signal and its spectrum can be tailor-made for different applications by changing the QD SOA bias current and optical power.

24.8 Technology Trends to be Explored and Summary

The rapid growth of the next-generation mobile communication infrastructure and the hybrid radio-over-fibre systems for personal area networks creates new challenges both in signal processing technologies and in photonic components. In particular, the successful transmission of UWB signals requires the largest possible bandwidth up to 100 GHz and, at the same time, the mobility of wireless techniques. These requirements can be satisfied by using a UROOF technology, combining the advantages of optical signal transmission through low-loss opticals with the E/O and O/E signal conversion carried out by fast, efficient and low-cost modulators and detectors. As was argued above, microwave photonics, which has been rapidly developing in the last decade, is the most promising candidate for the realization of some seemingly contradictory conditions.

The processing of microwave signals at high frequencies up to 60 GHz, and even above that level, can be successfully realized in an optical domain. To that end, the high-level integration technologies of passive and active optical and electronic components, multichip modules, MMIC and so on are needed.

One of the existing obstacles to the integration of optical and microwave components in the same substrate in the framework of the common functional system is the incompatibility of the GaAs-based optical components technology and the Si components microelectronic technology.

As previously shown, while various photodetectors, such as PIN, metal–semiconductor–metal (MSM) and APD, can be used for high-speed applications, each of these devices has certain drawbacks which make them rather unsuitable for use in UWB-over-optical-fibre communications. Lateral PIN and MSM devices are simple to fabricate, but are characterized by low responsivity at wavelengths around 850 nm. APD devices, on the other hand, require relatively high bias voltages and have increased noise.

We believe that the most promising perspective for low-cost integrated opto-electronic(OE) system fabrication is a novel SiGe technology. The devices based on nanostructures, such as SiGe MWQ lasers, two-dimensional (2D) photonic crystal optical waveguides, high-speed opto-electronic receivers, SiGe/Si phototransistors and SiGe/Si quantum well waveguide EAMs, can be successfully integrated with microstrip and coplanar guiding systems for microwaves as well as with optical fibres. They can achieve exceptionally efficient performance, low cost and design optimization. Finally, MB-OFDM combined with multimode fibre (MMF) is a promising technology for UROOF that presents clear advantages over other alternatives such as IR-UWB and DS-UWB. Impressive theoretical and experimental results to counter both wireless multipath and modal dispersion based on Wimedia-defined MB-OFDM were reported in [50].

Other aspects of UROOF technologies that remain for further research are:

- Study of best-match UWB technology for UROOF converged system.
- Cost optimization of microwave photonics components and integrated silicon-photonic devices to enable low-cost residential applications.
- MAC enhancements and solving best multiple access for UROOF various multicell architectures.
- Coexistence of UROOF technologies over the combined wireless-fibre channels with legacy inherently digital optical communications.
- Use of optical multiple-input multiple-output (MIMO) combined with MB-OFDM over multimode fibre [72] to further extend the performance of UROOF in orders of magnitude.

References

1. Allen, B. (ed.) (2005) *Ultra Wideband: Technology and Future Perspectives*, WWRF White Paper V3.0.
2. Yang, L. and Giannakis, G.B. (2004) Ultra Wideband communications: an idea whose time has come, *IEEE Signal Processing Magazine*, **21** (6), 26–54.
3. ECMA-368 (2007) *High Rate Ultra Wideband PHY and MAC Standard.*
4. Siwiak, K. (2002) UWB Propagation notes to P802.15 SG3a, IEEE 802.15-02/328r0.
5. Cox, C.H. (2004) *Analog Optical Links Theory and Practice*, Cambridge University Press.
6. Cornoy, J.T., LoCicero, J.L. and Ucci, D.R. (1999) *Communication Techniques Using Monopulse Waveforms*, Vol. **2**, MILCOM'99, pp. 1185–91.
7. Oppermann, I., Hämäläinen, M. and Linatti, J. (Eds.) (2004) UWB Theory and Applications. Wiley & Sons, Ltd., Chichester, UK.
8. Fisher, R., McLaughlin, M. and Welborn, M. (2004) DS-UWB physical layer, IEEE P802.124.3a Working Group for Wireless Personal Area Networks IEEE P802.15-04/0137r3.
9. Kelly, J. (2004) *IEEE P802.124.3a Working Group for Wireless Personal Area Networks*. MB-OFDM Physical Layer Submission, IEEE 802.15-04/0122r4.
10. Saleh, A.M., Rustakoand, A.J. and Roman, R.S. (1987) Distributed antennas for indoor radio communications. *IEEE Trans. Commun.*, **COM-35**, np12.

11. Wake, D. (2002) Radio over systems for mobile applications, in *Radio over Technologies for Mobile Communications Networks* (eds H. Al-Raweshidy and S. Komaki), Artech, Boston, pp. 217–41.
12. Bong Kim, H. and Wolisz, A. (2005) *A Radio over Fiber Based Wireless Access Network Architecture for Rural Area*, 14th IST Mobile and Wireless Commun. Summit, Dresden.
13. Al-Raweshidy, H. and Komaki, S. (eds) (2002) *Radio over Technologies for Mobile Communications Networks*, Artech House.
14. Kim, S., Jang, H., Yonghoon, S. and Jeong, J. (2004) Performance evaluation for UWB signal transmission with different modulation schemes in multi-cell environment distributed using RoF technology. UWB Conf. 2004.
15. Varasi, M. (2003) Fast modulators, in *Microwave Photonics* (eds A. Vilcot, B. Cabon and J. Chazelas), Kluwer, Boston, pp. 57–73.
16. Kitayama, K.-I., Stohr, A., Kuri, T. *et al.* (2000) An approach to single component antenna base stations for broadband millimeter-wave -radio access systems. *IEEE Transactions on Microwave Theory and Techniques*, **48** (12), 2588–95.
17. WWRF (2001) *Book of Vision*.
18. Agrawal, G.P. (2002) *Fiber-Optic Communication Systems*. John Wiley & Sons, Inc., New York.
19. Cox, C.H. III (2003) Analog optical links: models, measures and limits of performance, in *Microwave Photonics* (eds A. Vilcot, B. Cabon and J. Chazelas), Kluwer, Boston, pp. 210–9.
20. Larsson, A., Carlson, C., Gustavsson, J. *et al.* (2004) Direct high frequency modulation of VCSELs and applications in optic RF and microwave links. *New J. of Physics*, **6**, 176–90.
21. Schow, C.L. *et al.* (2008) 300-Gb/s, 24-channel Full Duplex, 850 nm, CMOS based optical transceivers, OMK5, OFC/NFC San Diego.
22. Minot, C. (2003) Electroabsoprtion modulators and photo-oscillators for conversion of optics to millimeter waves, in *Microwave Photonics* (eds A. Vilcot, B. Cabon and J. Chazelas), Kluwer, Boston, pp. 73–81.
23. Chou, H.-F. and Bowers, J.E. (2007) High-speed OTDM and WDM networks using traveling-wave electroabsorption modulators. *IEEE Journal of Selected Topics in Quantum Electronics*, **13** (1), 58–69.
24. Takeuchi, H., Tsuzuki, K., Sato, K. *et al.* (1997) NRZ operation at 40 Gb/s of a compact module containing an MQW electroabsorption modulator integrated with a DFB laser. *IEEE Photonics Technology Letters*, **9** (5), 572–4.
25. Jager, D. (2003) High speed photodetection, in *Microwave Photonics* (eds A. Vilcot, B. Cabon and J. Chazelas), Kluwer, Boston, pp. 82–100.
26. Seeds, A.L. (2002) Microwave photonics. *IEEE Transactions on Microwave Theory and Techniques*, **50** (3), 877–87.
27. Malyshev, S. and Chizh, A. (2004) State of the art high-speed photodetectors for microwave photonics application. Microwaves, Radar and Wireless Communications, 2004 MICON 15 International Conference, Vol. **3**, Issue 17–19, pp. 765–75.
28. Boyer, B., Haidar, J., Vilcot, A. and Bouthinon, M. (1997) Tunable microwave load based on biased photoinduced plasma in silicon. *IEEE Trans. on MTT*, **45** (8), 1362–7.
29. Saddow, S. and Lee, C.H. (1995) Optical control of microwave-integrated circuits using high-speed GaAs and Si photoconductive switches. *IEEE Transactions on Microwave Theory and Techniques*, **43**(9), 2414–20.
30. El-Bataway, Y.M. and Deen, M.J. (2005) Analysis and circuit modeling of waveguide-separated absorption charge multiplication-avalanche photodetector (WG-SACM-APD). *IEEE Transactions on Electron Devices*, **52** (3), 335–44.
31. Maury, G., Hilt, A., Berceli, T. *et al.* (1997) Microwave-frequency conversion methods by optical interferometer and photodiode. IEEE Transactions on Microwave Theory and Techniques, Special Issue Microwave and Millimeter-Wave Photonics II, pp. 1481–5.
32. Esman, R.D. (2003) Microwave functions enabled by Photonics, in *Microwave Photonics* (eds A. Vilcot, B. Cabon and J. Chazelas), Kluwer, Boston, pp. 375–99.
33. Sun, C.K., Orazi, R.J. and Pappert, S.A. (1996) Efficient microwave frequency conversion using photonic link signal mixing. *IEEE Photonics Technology Letters*, **8** (1), 154–6.
34. Young, T., Conradi, J. and Tinga, W.R. (1996) BER characteristics of $\pi/4$ DQPSK microwave subcarrier signals on optical using Mach-Zehnder modulator nonlinear upconversion. *IEEE Photonics Technology Letters*, **8** (11), 1552–4.
35. Cabon, B., Le Guennec, Y. and Maury, G. (2005) Optical microwave mixing techniques for broadband and low cost radio-over-applications. Photonics North Conference, conference proceedings.

36. Le Guennec, Y., Maury, G., Yao, J. and Cabon, B. (2006) New optical microwave up-conversion solution in radio-over network for 60 GHz wireless applications. *IEEE J. of Lightwave Technology*, **24** (3), 996–8.
37. Westcott, R.J. (1967) Investigation of multiple FM/FDM carriers through a satellite TWT operating near to saturation. *Proc. IEEE*, **114**, 726–40.
38. Salgado, H.M. and O'Reilly, J.J. (1993) Performance assessment of FM broadcast subcarrier-multiplexed optical systems. *IEE Proc.-J*, **140**, 397–403.
39. Salgado, H.M. and O'Reilly, J.J. (1995) Performance assessment of subcarrier multiplexed optical systems, in *Analogue Optical Communications* (eds B. Wilson and I. Darwazeh), Peter Peregrinus Ltd.
40. Walker, S.D., Li, M., Boucouvalas, A.C. *et al.* (1990) Design techniques for subcarrier multiplexed broadcast optical networks. *Selected Areas in Communications, IEEE J.*, **8**, 1276–84.
41. Alameh, K. and Minasian, R.A. (1990) Optimum optical modulation index of laser transmitters in SCM systems. *Electron. Lett.*, **26**, 1273–5.
42. Saleh, A.A.M. (1989) Fundamental limit on number of channels in subcarrier multiplexed lightwave CATV system. *Electron. Lett.*, **25**, 776–7.
43. Phillips, M.R. and Darcie, T.E. (1991) Numerical simulation of clipping-induced distortion in analog lightwave systems. *IEEE Photonics Technol. Lett.*, **3**, 1153–5.
44. Mazo, J.E. (1992) Asymptotic distortion of clipped, dc-biased, gaussian noise. *IEEE Trans. on Commun.*, **40**, 1339–44.
45. Mendis, F. and Tan, B. (1990) Overmodulation in subcarrier multiplexed video FM broadband optical networks. *IEEE J. on Select. Areas in Commun.*, **8**, 1285–9.
46. Foerster, J. (2003) Channel modeling sub-committee report – Final, IEEEP802.15 Working group for Wireless Personal Area Network (WPAN), IEEE P802.15-02/490r1-SG3a.
47. Donlan, B.M., McKinstry, D.R. and Buehrer, R.M. (2006) The UWB indoor channel: Large and small scale modeling, *IEEE Trans. on Wireless Commun.*, **5** (10), 2863–73.
48. Salehi, B.E.A. and Teich, M.C. (1991) *Fundamentals of Photonics*, Wiley.
49. UROOF partners, deliverable D2.3, December 2007.
50. BenEzra, Y., Ran, M., Borochovtch, E. *et al.* (2008) *Wimendia-Defined, Ultra-Wideband Radio Transmission Over Optical*, OthD6.pdf NFC/OFC.
51. Smith, D.W., Borghesani, A., Moodie, D. *et al.* (2008) 480 Mbps ultra-wideband radio-over- transmission using a 1310/1550 nm reflective electro-absorption transducer and off-the-shelf components, PDP29, OFC/NFOEC, San Diego.
52. UROOF project D3.6a, D4.1 (http://www.ist-uroof.org/).
53. Pizzinat, A., Urvoas, P. and Charbonnier, B. (2007) *1.92 Gbit/s MB-OFDM Ultra Wide Band Radio Transmission over Low Bandwidth Multimode*, OThM6, OFC/NFOEC.
54. Lim, C.S., Yee, M.L. and Ong, L.C. (2006) Performance of Transmission of Ultra Wideband Signals Using Radio-over- System. *ITS Telecommunications Proceedings*, 250–3.
55. Ong, L.C., Yee, M.L. and Luo, B. (2006) Transmission of Ultra Wideband Signals through Radio-over- Systems. *Lasers and Electro-Optics Society, IEEE*, 522–3.
56. Federal Communications Commission (2007) http://www.fcc.gov/oet/info/rules/part15/part15-9-20-07.pdf.
57. Pepeljugonsky, P., Golowich, S.E., Ritger, A.J. *et al.* (2003) Modeling and Simulation of Next Generation Multimode Links. *J. Ligthw. Technol.*, **21** (5), 1242–55.
58. Lin, W.-P. and Chen, J.-Y. (2005) Implementation of a new ultra wide-band impulse system. *IEEE Photonics Technology Letters*, **17** (11), 2418–20.
59. Wang, Q. and Yao, J. (2006) UWB doublet generation using nonlinearly-biased electro-optic intensity modulator. *Electronic Letters*, **42** (22), 1304–6.
60. Ghavami, M., Michael, L.B. and Kohno, R. (2004) *Ultra Wideband Signals and Systems in Communication Engineering*. John Wiley & Sons, Ltd.
61. Zeng, F. and Yao, J. (2006) An approach to ultra wideband pulse generation and distribution over optical. *IEEE Photonics Technology Letters*, **18** (7), 823–5.
62. Zeng, F. and Yao, J. (2006) Ultra wideband impulse radio signal generation using a high-speed electrooptic phase modulator and a -Bragg-grating-based frequency discriminator. *IEEE Photonics Technology Letters*, **18** (19), 2062–4.
63. Le Guennec, Y. and Gary, R. (2007) Optical frequency conversion for millimeter-wave ultra-wideband-over-systems. *IEEE Photonics Technology Letters*, **19** (13), 996–8.

64. Zeng, F., Wang, O. and Yao, J.P. (2007) All-optical UWB impulse generation based on cross-phase modulation and frequency discrimination. *Electronic Letters*, **43** (2), 119–21.
65. Sugawara, M., Ebe, H., Hatori, N. *et al.* (2004) Theory of optical signal amplification and processing by quantum-dot semiconductor optical amplifiers. *Phys. Rev.*, **B69**, 235332-1-39.
66. Agrawal, G.P. and Olsson, N.A. (1989) Self-phase modulation and spectral broadening of optical pulses in semiconductor laser amplifiers. *IEEE Journal of Quantum Electronics*, **2** (11), 2297–2306.
67. Joergensen, C., Danielsen, S.L., Durhuus, T. *et al.* (1996) Wavelength conversion by optimized monolithic integrated Mach-Zehnder interferometer. *IEEE Photonics Technology Letters*, **8** (4), 21–3.
68. Chen, H., Zhu, G., Wang, Q. *et al.* (2002) All-optical logic XOR using differential scheme and Mach-Zender interferometer. *Electronic Letters*, **38** (21), 1271–3.
69. Kanellos, G.T., Petrantonakis, D., Tsiokos, D. *et al.* (2007) All-optical 3R burst-mode reception at 40 Gb/s using four integrated MZI switches. *Journal of Lightwave Technology*, **2** (1), 184–92.
70. Wang, Q., Zhu, G., Chen, H. *et al.* (2004) Study of all-optical XOR using Mach-Zehnder interferometer and differential scheme. *IEEE J. of Quant. Electr.*, **40** (6), 703–10.
71. Ben-Ezra, Y., Lembrikov, B.I. and Haridim, M. (2007) Specific features of XGM in QD-SOA. *IEEE Journal of Quantum Electronics*, **43** (8), 730–7.
72. Tarighat, A., Hsu, R.C.J., Shah, A. *et al.* (2007) Fundamentals and challenges of optical multiple-input multiple output multimodes. *IEEE Communication Magazine*, **45** (5), 57–63.

Contributions

Contributions to this section were received from the following persons:

Moshe Ran (Ed.) — Holon Institute of Technology (HIT), mran@hit.ac.il
Yossef Ben Ezra (Ed.) — Holon Institute of Technology (HIT), benezra@hit.ac.il
Boris I. Lembrikov (Ed.) — Holon Institute of Technology (HIT) borisle@hit.ac.il
Ronen Korman — Wisair, ronenk@wisair.com
Beatrice Cabon, Ghislaine Maury — Institut National Polytechnique de Grenoble cabon@enserg.fr, maury@enserg.fr
David E. Smith — Centre for Integrated photonics david.smith@ciphotonics.com
Roberto Llorente — Universidad Politécnica de Valencia rllorent@dcom.upv.es
Manuj Thakur — University of Essex mpthak@essex.ac.uk
Henrique Salgado — INESC Porto Research h.salgado@ieee.org

25

Visible Light Communications

Dominic O'Brien[1], Lubin Zeng[1], Hoa Le-Minh[1], Grahame Faulkner[1], Olivier Bouchet[2], Sebastien Randel[3], Joachim Walewski[3], Jose A. Rabadan Borges[4], Klaus-Dieter Langer[5], Jelena Grubor[5], Kyungwoo Lee[6] and Eun Tae Won[6]

[1] *University of Oxford, United Kingdom*
[2] *France Telecom, Cesson-Sévigné, France*
[3] *Siemens AG, Corporate Technology, Information and Communications, Germany*
[4] *ETSIT Universidad de Las Palmas de Gran Canaria, Spain*
[5] *Fraunhofer Institute for Telecommunications, Heinrich-Hertz-Institut, Germany*
[6] *Samsung Electronics Ltd, Korea*

25.1 Introduction

Visible Light Communication (VLC) has its origins in the rapid development of LEDs and sources for Solid-state Lighting (SSL). Much of the work is done in Japan, by members of the Visual Light Communications Consortium [1], although there is growing interest worldwide. There are predictions that efficient SSL could save six power stations in Japan and offer substantial savings in the UK [2]. SSL sources are also now extensively used for architectural lighting. Using multiple emitters of different colours in a single fixture allows overall colour (and intensity) to be controlled, and the long life of SSL devices makes them attractive for use where access to lamps is difficult and expensive. In automotive applications LEDs are extensively used for tail, brake and indicator lights, and are now being introduced for headlamps [3]. In this case LED sources should last the entire life of the car. Traffic signals also use LEDs for reasons of reliability and lifetime.

Another area of interest is short-range data links. The amount of memory in portable appliances is growing faster than the wireless data rate available to read and write information to it, so a very high-data-rate free-space link working over several centimetres could replace the need to plug appliances together, and offer a means to download DVDs and other media rapidly.

Short-Range Wireless Communications Rolf Kraemer and Marcos D. Katz

There are many challenges for VLC, both technical and regulatory, but the ability to add the function of communications to lighting systems is attractive, and may offer access to a new region of the electromagnetic spectrum for wireless communications.

This chapter contains an overview of the VLC channel, and the technical constraints and challenges faced in implementing VLC links. Potential applications and efforts to implement and standardize these techniques are then discussed.

25.2 VLC Link

25.2.1 Transmitter

The key component of the transmitter is a visible solid-state emitter. This can be either an LED or a semiconductor laser, depending on the application. LEDs are used where the source is used for illumination as well as communication, and both LEDs and lasers have been used for data links. In the case of lasers care has to be taken to meet eye safety regulations [4]. In the following section LEDs that are used for white-light illumination and communication are considered.

25.2.1.1 White-light LEDs

Two techniques are used to implement white-light LEDs. One is to combine light from red, green and blue (RGB) LEDs in the correct proportion; Figure 25.1(a) shows the spectrum of such a device. Typically these triplet devices consist of a single package with three emitters and combining optics, and they are often used in applications where variable colour emission is required.

These devices are attractive for VLC in that they offer the opportunity for transmitting different data on each LED (see for instance [5]). However, care must be taken to maintain colour balance by appropriate design and control of any modulation scheme that is developed.

The other technique is to use a single blue LED coated with, or sometimes embedded in, a layer of phosphor that emits yellow light. Some of the light from the blue LED is absorbed

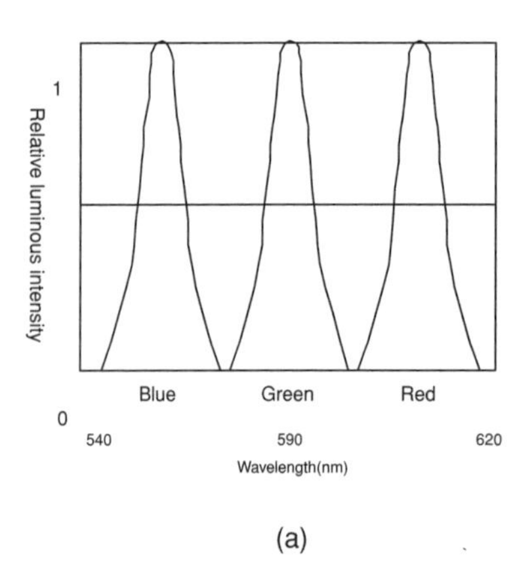

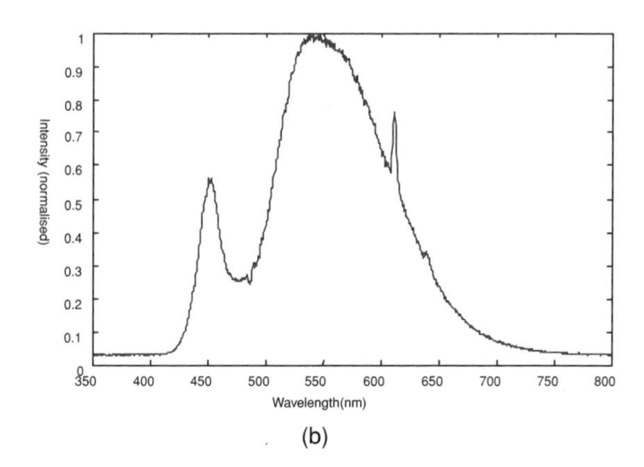

Figure 25.1 White-light LEDs: (a) Typical spectrum of an RGB triplet device; (b) Measured spectrum of a Single-chip LED (Luxeon Star).

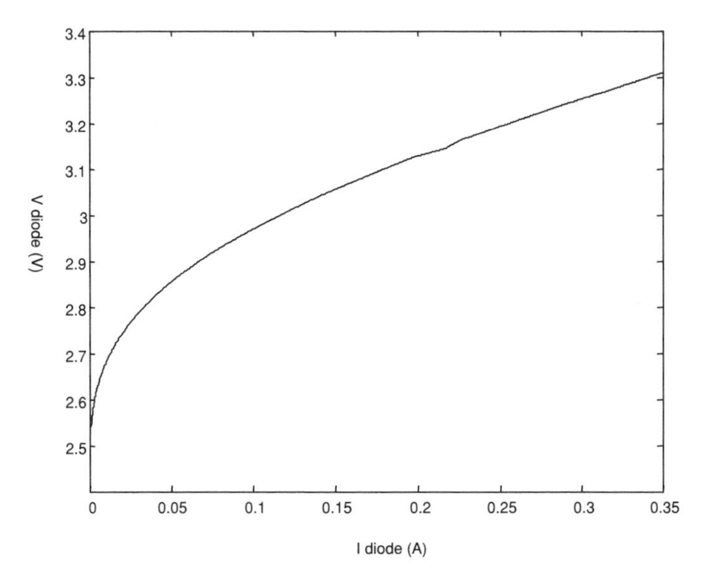

Figure 25.2 Measured V-I characteristics of LED (Luxeon Star) [6].

by the phosphor and this absorbed energy is emitted as yellow light. This mixes with the nonabsorbed blue component to create the required white colour. Figure 25.1(b) shows a typical spectrum from a Luxeon Star [6] device. In this case the blue LED emits at 440 nm and a coating of cerium-doped yttrium aluminium garnet phosphor is used to create the broad yellow spectrum.

At present, single-chip devices are favoured for purely illumination applications due to the lower complexity (and therefore cost). Widespread adoption of SSL is dependent on achieving higher energy efficiencies.

Figure 25.2 shows the Voltage Current (VI) characteristics of a Luxeon Star device. Typically an average current of several hundred milliamps is required to create the illumination levels found in office space (typically 200–500 lux is specified [7]). This can be provided by using a fixed bias current to the device, then adding a bipolar modulation current, which either turns the device off or increases the illumination output from its average level. Figure 25.3 shows a

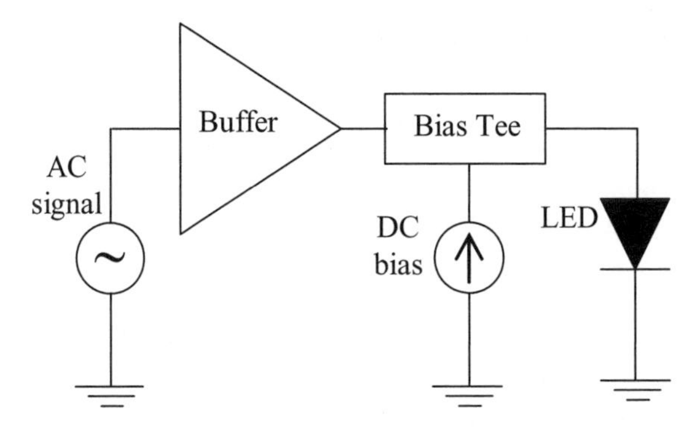

Figure 25.3 Typical driver circuit.

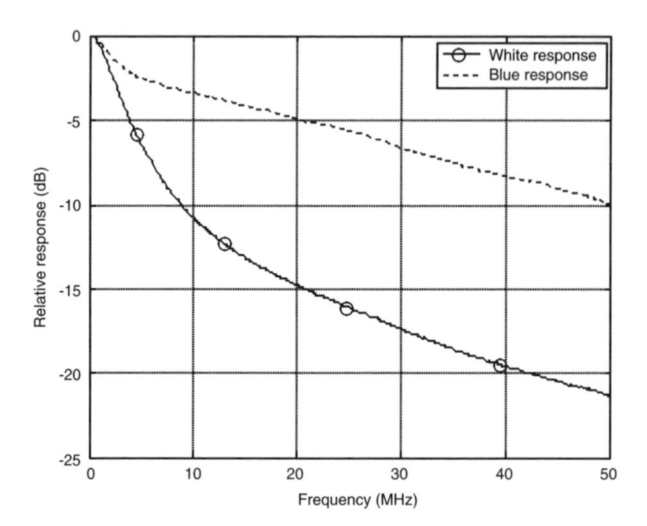

Figure 25.4 Measured small signal modulation bandwidth of Luxeon star device, showing unfiltered (white) and filtered (blue) responses. Responses are both normalised to their maximum.

typical circuit that can achieve this. Using separate bias and modulation current also allows the LED to be driven in its linear region, which is a requirement for complex modulation schemes such as Discrete Multitone Modulation (DMT).

The major challenge for VLC is the modulation bandwidth available from such large-area devices. The LED effective capacitance increases with area, and for both the RGB triplet devices and the blue emitter in the single-chip devices this leads to slow-rise times. In this case driving with a low impedance buffer can improve the time constant of the circuit, but this must be able to source the high currents required.

However, in the case of single-chip devices the phosphor ultimately limits the speed of overall optical response. Figure 25.4 gives the measured bandwidth of the blue and white optical response of a Luxeon Star device, showing that the bandwidth of the blue-only response is approximately 12 MHz, compared with approximately 3 MHz for the overall white response. Figure 25.5(a) gives the unfiltered response of the LED to a series of short electrical drive pulses from a circuit of the type shown in Figure 25.3, showing the long fall time due to the decay of the phosphor.

Several techniques to improve bandwidth have been investigated. The first is to filter the 'slow' phosphor response and use only the blue component for communications. Figure 25.5(b) shows the improvement in the response due to filtering, when compared with Figure 25.5(a). In [8] a 40 Mb/s data link that uses this technique is reported. Most optical communications systems use AC coupled receivers, so at high data rates the photocurrent caused by the slow phosphor component is largely removed due to this electrical filtering. However, this does not remove the shot noise caused by the yellow component, whereas blue filtering reduces this as the radiation does not reach the photodetector. Optical filtering also makes the receiver less susceptible to saturation due to the high level of DC photocurrent from the yellow light. The disadvantage of this technique is that the filter does not transmit all of the blue component,

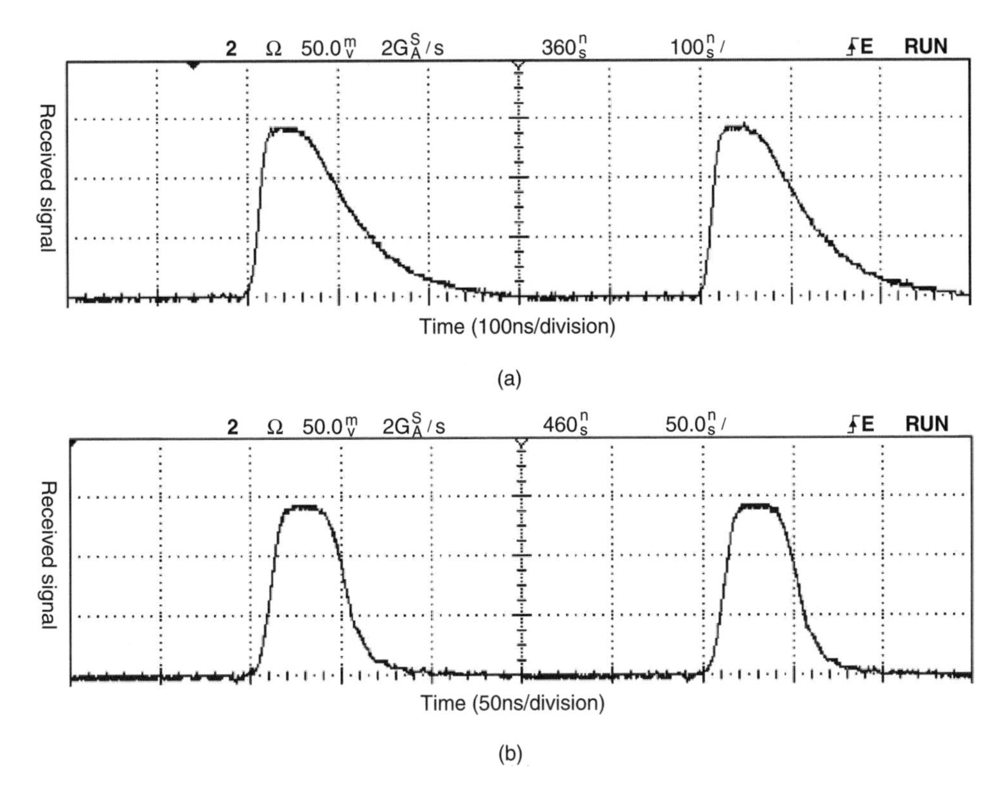

Figure 25.5 (a) Response from unfiltered LED; (b) Response from blue component of LED (note different timescales in plots).

which leads to reduced link margin. It also removes the low-frequency components that can be successfully transmitted by the light from the phosphor, thus reducing the communications efficiency of the source.

It is also possible to equalize the driving signal to the LED, and in [9] a link that uses an array of separately equalized LEDs is reported. Each LED is driven using a resonant technique that creates a maximum optical output at a specific frequency, and a combination of LEDs resonated at different frequencies allows an overall bandwidth of 25 MHz to be achieved, without blue filtering. This is used to create a 40 Mb/s data link operating over several metres.

This work indicates that whilst the bandwidth of the devices is low, optical filtering and more complex transmitter electronics should allow this to be increased substantially, whilst maintaining sufficient transmitted power for communications.

SSL sources typically require several hundred milliamps of current at several volts and these are supplied from commercial driver ICs and units [10, 11]. These typically use circuits which create electrical noise on the output LED drive current. For illumination this is not considered to be a problem as variation is too rapid to be perceived by the human eye. However, the effects of this noise in the data signal can be significant, in particular for complex modulation schemes that use some form of variable amplitude modulation (such as OFDM, DMT), so the noise is another impairment to be taken into account in the modulation design process.

Modulation schemes based on OOK (such as PPM or PWM) are therefore less likely to be less susceptible to a noisy channel.

In addition, dimming of these lights is normally achieved using Pulse-width Modulation, with frequencies of 10 KHz [10, 11]. Sources can then be off for a significant proportion of time, which means that data transmission must be coordinated with the dimming mechanism, or else other steps are required to ensure proper data transmission. (A scheme to do this is proposed in [12].)

The aim of the lighting community is to increase the efficiency of SSL sources. For LEDs, increasing carrier lifetime generally improves LED efficiency, but can also reduce modulation bandwidth [13]. This is one factor where the aims of VLC and illumination are in competition, and more work is required to understand the trade-offs that may exist, and to ensure developments take into account the needs of both communities.

25.2.1.2 Visible 'Coloured' Data Links

Several hundred million infrared data links corresponding to the Infrared Data Association (IrDA) specification [14] are installed in portable appliances, but there is growing interest in using visible light for such communications. In these cases the source might be an array of LEDs, such as a car tail-light, a signboard or a high-speed emitter for a point-and-shoot application.

For large-device-area display LEDs, high-speed operation is challenging. However, for point-and-shoot links where lower power is required, Red Resonant Cavity LEDs can be modulated at several hundred Mb/s [15], and more recently red Vertical Cavity Surface Emitting Lasers (VCSELs) have become available [15]. Several demonstrations of data links using red sources have been reported [16], and the availability of visible VCSELs should allow data rates to increase substantially.

25.2.2 The VLC Channel

25.2.2.1 In-room Systems

The VLC optical channel consists of a line-of-sight (LOS) or a number of LOS components corresponding to the paths from lighting units to the receiver, and, in addition, a diffuse component created by reflections of illumination from intermediate surfaces. The LOS component is straightforward to model. Figure 25.6 shows a typical geometry [17].

At data rates above 100 Mb/s the relative delay of the signal from different lighting units may become significant, and schemes such as OFDM may therefore be attractive. However, the available bandwidth from presently available devices makes data rates approaching 100 Mb/s difficult to implement, so that differential delay is not likely to be a problem in the near future.

There are several different approaches to modelling the diffuse component [8, 18–21]. In [20] the room is modelled as an integrating sphere. In this case the strength of the diffuse component needs to be calibrated against the LOS. Such an approach to modelling is described in more detail in [22]. In practice the diffuse component is small compared with the strong LOS and can therefore be ignored. Modelling in this paper shows that there is a very high

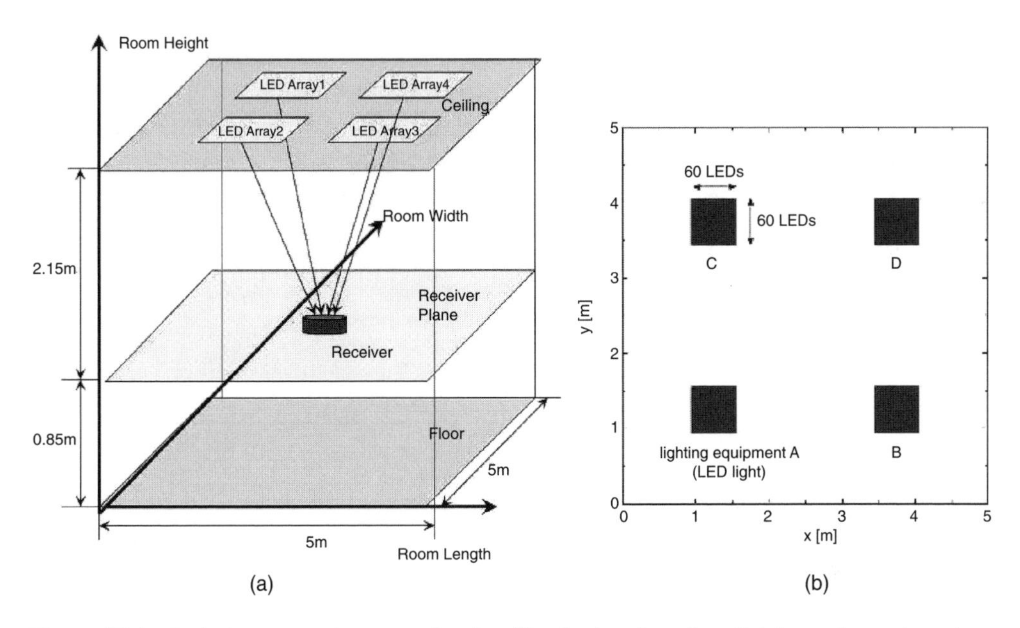

Figure 25.6 Typical room environment showing illumination from four lighting units, and receiver placed on 'communications plane' 1 m above floor.

link margin available. Figure 25.7 shows a typical signal-to-noise ratio (SNR) distribution, indicating there is a very high link margin for most modulation schemes. The particular parameters used in the simulation are show in Table 25.1.

In general the requirement of 200–500 lux for illumination purposes ensures that there will be very high levels of received signal, so that SNR is not an issue for the VLC channel.

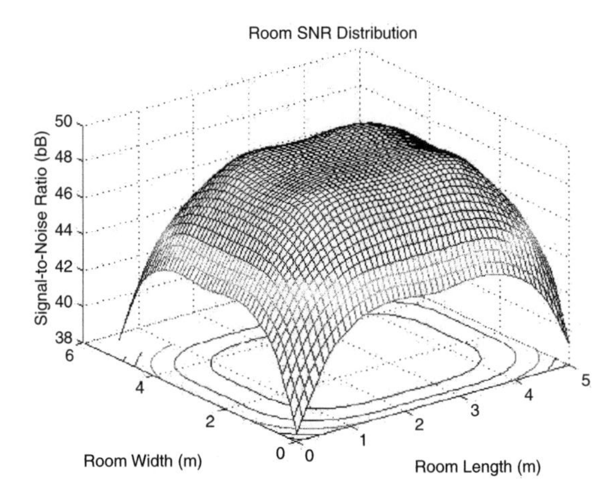

Figure 25.7 Typical received signal-to-noise ratio (SNR) in test environment.

Table 25.1 Simulation parameters

Parameter	Value	Parameter	Value
LED power	20 mW	Communications power	115.2 W
Lambertian order	1	Receiver Area	1 cm^2
No. of LEDs in 1 lighting unit	60 × 60 on 1 cm pitch	Input referred noise current	100 $pA/\sqrt{Hz}$
No. of lighting units	2 × 2 2.5 m apart (centre to centre)	Detector responsivity	0.4 A/W
Room	5 × 5 × 3 m high		
Receiver plane	0.85 m above ground		
Position of receiver	2.5 m, 2.5 m from centre		
Reflectivities	0.54 All surfaces		

25.2.2.2 Point-to-point Links

Point-to-point links are achieved through the user aligning the two terminals, so the field of view of the links must match the ability of the user to achieve this. Typical IrDA links have a field of view of 30°, as the wavelength used is outside the visible region of the spectrum. In the case of visible links this is assisted by the user, who can 'see' the data signal, which may allow narrower field of view and enhanced link margin. There may therefore be the opportunity of extended range and/or increased bit rate for these applications.

25.2.2.3 Ambient Light Interference

Ambient light interference affects all optical wireless channels to varying degrees. 'Natural' ambient light noise is from sunlight, and this causes extra photocurrent, which depends on the particular wavelength detected and the environment. This photocurrent is usually blocked by the AC coupling of the receiver. However, the white electrical spectrum of the shot noise that this light contributes is not blocked, and there is also the potential for photodetector saturation from the high light levels. In the VLC channel the daylight component of this noise is within the optical bandwidth of the transmission system, so optical filtering (which is traditionally used in IR OW systems) also reduces the desired signal. Modelling indicates there are only modest SNR gains to be made (several dB) by using an 'optimum' filter that maximises SNR by filtering typical ambient light due to daylight, indicating the effects of the interference are small.

There has been extensive characterisation and modelling of the noise from incandescent and fluorescent sources and their effects (see for instance [23–26]). Incandescent sources create shot noise due to the DC component, the electrical interference components at 100 Hz and harmonics of the 50 Hz mains frequency. However, these are usually small above 1 kHz and only the lower frequencies need to be considered. Fluorescent sources use ballasts that create interference photocurrent with significant spectral components up to 1 MHz, depending on the type of ballast. Mitigating these electrical harmonics is more complex, and requires careful electrical filter design. There is generally an optimum placement frequency for the high-pass filter response of the receiver. If this is too high, significant low-frequency signal energy is lost, causing baseline wander, but too low a frequency causes signal corruption due to the interference. An alternative is to use subcarrier modulation, as this moves signal energy away from the frequencies containing interference [27–31]. Spread spectrum techniques have also been used successfully [32–37].

The intensity of the source of interference illumination and the degree to which it is coupled into the receiver is another determining factor, and this of course is dependent on the particular environment. VLC has different interference constraints to traditional IR, as visible wavelengths that would normally be classified as interference are in the same spectral region as the signal. There has been little published work on the interaction between fluorescent, incandescent and natural ambient light with the VLC channel, and more investigation is required in this area.

25.2.3 Receiver

Figure 25.8 shows a block diagram of a typical receiver. Light is collected by a lens or nonimaging optical system [38] and concentrated onto a photodetector. The optical system increases the effective collection area for the desired field of view (FOV), subject to the constraints of the constant radiance theorem [39]. For VLC, silicon photodiodes are usually used as detectors. Both P-I-N diodes and avalanche photodetectors can be used, but at the high signal levels available P-I-N devices are usually sufficient.

There are many different types of silicon detector, with different wavelength responses, and with capacitance per unit area depending on the device structure. For an optical receiver the choice of detector and preamplifier is crucial if good performance is to be achieved. The bandwidth of the receiver is usually a function of the transit time of the detector, its capacitance and the input impedance of the preamplifier [40]. Increasing the area of the detector usually

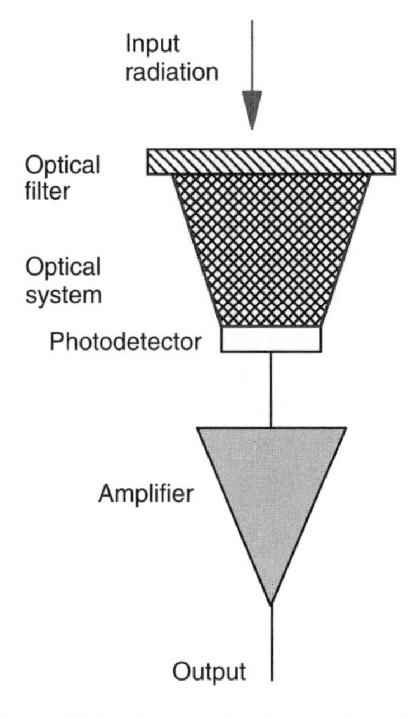

Figure 25.8 Schematic of optical receiver.

proportionally increases the power received, but often with decreased bandwidth. In the case of VLC the bandwidth requirements are modest, as channel bandwidth is usually limited at the transmitter, which allows a large detector area to be used.

Equalisation at the receiver has been investigated and this can provide significant increase in data rate. In [22] a simulated data rate in excess of 40 Mb/s was achieved using an unfiltered LED with a modulation bandwidth of 3 MHz. This is promising, but there is little work on VLC-specific receivers as yet and these issues require further examination.

25.2.4 Modulation

Various schemes have been investigated. NRZ-OOK has been used for several demonstrations [8, 21, 22], and this scheme has the advantages of simplicity and immunity to any LED nonlinearity. The high channel SNR makes complex modulation schemes attractive, and OFDM is investigated in [8, 21, 41–44]. The latter paper shows that 100 Mb/s is feasible within a 20 MHz channel bandwidth using a Discrete Multitone (DMT) approach, but that device nonlinearity reduces the expected performance. For low-bit-rate applications, PPM of various types [45, 46] has been considered, including schemes compatible with PWM dimming of the illuminating source [12]. As there are often many devices used in lighting fixtures, parallel transmission schemes are attractive. Schemes using an imaging receiver and transmission of patterns of LEDs have been proposed, both for traffic [47] and for other applications [48].

25.3 Potential Applications

VLC has applications where: (i) illumination, signalling or information display is the primary purpose of the source, and VLC provides extra capability; and (ii) using visible light for communications provides advantages over approaches using alternative wireless standards.

For Case (i) VLC is subsidiary to the main function, or at least must be complementary with it, so that any standards and developments must be closely integrated with the major purpose of the illumination. Traffic signals are one such major potential applications area for VLC. Most traffic signals use SSL sources for reasons of reliability, and several systems that use these to communicate with cars have been proposed and demonstrated [47, 49–54]. Car-to-car communication using head and tail lights has also been proposed [55, 56].

There have been a number of broadcasting applications using signboards. In [57] a board is used to broadcast audio signals, and in [5] a similar system is used to broadcast information. Optical ‘tag’ and ID systems have been proposed [58], and members of VLCC are also active in this area [59]. In this case the broadcasting is compatible with a source that is dimmed using PWM [58]. A modified PPM scheme that also implements PWM dimming is described in [12]. Using illumination in rooms for VLC offers local broadcasting with little modification of the lighting fixtures, and is therefore attractive to augment other wireless transmission techniques [17, 43, 60].

In Case (ii) visible point-and-shoot-type links may be competitive when compared with RF approaches, or else offer other advantages. These advantages could be perceived by the user (such as aesthetic) or technical (such as enabling user aiming and hence potentially improved link margins) [16]. For applications where security is a consideration, the ‘visible security’ of these links may also be advantageous. Retro-reflecting tag systems that use a visible source which allows the user to point the reader have also been fabricated [61], as well as visible point-and-shoot links for peripheral applications [22]. In these cases the link solely provides communications, so the standards can be optimised for this purpose.

25.4 Challenges

The technical challenges of implementing VLC are largely associated with the transmission bandwidth available from sources, and how VLC can operate in conjunction with illumination. In addition, provision of an uplink using VLC is challenging. Schemes that use retro-reflectors [62] and IR uplinks, as well as a cooperative scheme with RF standards [63], have been proposed. A European project addressing Gigabit Home Networks (OMEGA) [64] aims to investigate the cooperative use of both optical and RF wireless communications. Lighting techniques and standards are evolving, and VLC must also ensure that it can adapt, and provide transmission techniques that are compatible with future dimming and power supply techniques.

25.5 Standardisation

At present standardisation efforts are being undertaken by VLCC in Japan [59, 65], and an effort has just started within the IEEE [66]. Additionally, there is an effort to develop a white paper within the Wireless World Research Forum (WWRF) [67].

25.6 Conclusions

The future wireless landscape will consist of many different wireless communications standards, each suited to a particular purpose. VLC is well suited to broadcast where the signal level is naturally regulated by user control of illumination level. Full use of this will require close collaboration with the SSL industry and standards-setting bodies for illumination applications, and the traffic regulation authorities for ITS. In all cases the potential low-cost augmentation of capacity is an attractive prospect.

References

1. Visible Light Communications Consortium (2008) http://www.vlcc.net (accessed September 2008).
2. Market Transformation Programme (2006) *Light Emitting Diodes: Eco-design Innovation Roadmap*, http://www.mtprog.com/spm/files/download/byname/file/MTP_EIR_LEDs_Feb06.pdf (accessed September 2008).
3. Audi:Audi R8 promotional material (2008) http://www.audi.co.uk/audi/uk/en2/new_cars/r8/Design_Features.html (accessed September 2008).
4. IEC 60825-1 (2007) *Safety of Laser Products part 1*, British Standards Institution.
5. Park, S.-B., Jung, D.K., Shin, H.S. *et al.* (2007) *Information Broadcasting System based on Visible Light Signboard*. Proc. Wireless and Optical Communications, Montreal, Canada.
6. Lumileds (2006) *Power Light Source Luxeon Star*, Technical datasheet DS23, Philips.
7. Occupational-Safety-and-Health-Branch-Labour-Department (1999) *A Simple Guide to Health Risk Assessment - Office Environment Series OE 2/99*, http://www.oshc.org.hk/others/bookshelf/BB055E.pdf (accessed September 2008).
8. Grubor, J., Lee, S.C.J., Langer, K.-D. *et al.* (2007) *Wireless High-Speed Data Transmission with Phosphorescent White-Light LEDs. Post deadline session at European Conference on Optical Communications (PDS 3.6)*, pp. 1–2.

9. Le-Minh, H., O'Brien-Dc, Faulkner, G. *et al.* (2008) High-speed visible light communications using multiple-resonant equalization. *Photonics Technology Letters*, **20** (15), 1243–245.
10. Fairchild semiconductor (2006) FAN5608 LED driver data sheet.
11. National semiconductor (2006) Parallel white LED driver.
12. Lopez-Hernandez, F.J., Poves, E., Perez-Jimenez, R. and Rabadan, J. (2006) Low-cost diffuse wireless optical communication system based on white LED. Proc. 2006 IEEE Tenth International Symposium on Consumer Electronics. St. Petersburg, Russia, pp 1–4.
13. Schubert, E.F. (2006) *Light-Emitting Diodes*, Cambridge University Press, Cambridge, UK.
14. IrDA: http://www.irda.org.
15. Firecomms, http://www.firecomms.com.
16. O'Brien, D.C., Zeng, L., Bouchet, O. *et al.* (2006) *Visible Light Communications.* Wireless World Research Forum.
17. Komine, T. and Nakagawa, M. (2004) Fundamental analysis for visible-light communication system using LED lights. *IEEE Transactions on Consumer Electronics*, **50** (1), 100–7.
18. Carruthers, J.B. and Carroll, S.M. (2005) Statistical impulse response models for indoor optical wireless channels. *International Journal of Communication Systems*, **18** (3), 267–84.
19. Carruthers, J.B. and Kahn, J.M. (1997) Modeling of nondirected wireless infrared channels. *IEEE Trans. Commun.*, **45** (10), 1260–8.
20. Jungnickel, V., Pohl, V., Nonnig, S. and von Helmolt, C. (2002) A physical model of the wireless infrared communication channel. *IEEE Journal on Selected Areas in Communications*, **20** (3), 631–40.
21. Grubor, J., Gaete, J.-O., Waleski, J.S. *et al.* (2007) High-speed wireless indoor communication via visible light. *ITG Fachbericht*, 203–8.
22. Zeng, L., Le-Minh, H., O'Brien, D.C. *et al.* (2008) Improvement of Date Rate by using Equalization in an Indoor Visible Light Communication System. Accepted for presentation at IEEE International Conference on Circuits and Systems for Communications.
23. Ghassemlooy, Z., Hayes, A.R. and Wilson, B. (2003) Reducing the effects of intersymbol interference in diffuse DPIM optical wireless communications. *IEE Proceedings Optoelectronics*, **150** (5), 445–52.
24. Moreira, A.J.C., Valadas, R.T. and de-Oliveira-Duarte, A.M. (1997) Optical interference produced by artificial light. *Wireless Networks*, **3** (2), 131–40.
25. O'Brien, D.C., Street, A.M., Samaras, K. *et al.* (1997) *Smart Pixels for Optical Wireless Applications.* Proc. Proceedings of Spatial Light Modulators. TOPS, Incline Village, NV, USA, Vol. 14, pp. ix+284.
26. Tavares, A.M.R., Valadas, R.J.M.T. and de-Oliveira-Duarte, A.M. (1999) Performance of wireless infrared transmission systems considering both ambient light interference and intersymbol interference due to multipath dispersion. *Proceedings of the SPIE*, **3532**, 82–93.
27. Carruthers, J.B. and Kahn, J.M. (1996) Multiple-subcarrier modulation for nondirected wireless infrared communication. *IEEE Journal on Selected Areas in Communications*, **14** (3), 538–46.
28. Yu, B., Kahn, J.M. and You, R. (2002) Power-efficient multiple-subcarrier modulation scheme for optical wireless communications. *Proceedings of the SPIE*, **4873**, 41–53.
29. Ohtsuki, T. (2003) Multiple-subcarrier modulation in optical wireless communications. *IEEE Communications Magazine*, **41** (3), 74–9.
30. Grubor, J., Jungnickel, V. and Langer. K.D. (2006) *Rate-Adaptive Multiple Subcarrier-Based Transmission for Broadband Infrared Wireless Communication.* Proc. OFCNFOEC 2006. Optical Fiber Communication Conference and National Fiber Optic Engineers Conference. Anaheim, CA, USA.
31. Gonzalez, O., Perez-Jimenez, R., Rodriguez, S. *et al.* (2005) OFDM over indoor wireless optical channel. *IEE Proceedings Optoelectronics*, **152** (4), 199–204.
32. O'Farrell, T. and Kiatweerasakul, M. (2002) Wireless infrared transmission using direct sequence spread spectrum techniques. Proceedings of ISSSTA'98 International Symposium on Spread Spectrum Techniques and Applications. Proc., Sun City, South Africa, Vol. 3.
33. Vento, R., Rabadan, J., Perez-Jimenez, R. *et al.* (1999) Experimental characterization of a direct sequence spread spectrum system for in-house wireless infrared communications. *IEEE Transactions on Consumer Electronics*, **45** (4), 1038–45.
34. Wong, K.K., O'Farrell, T. and Kiatweerasakul, M. (2000) Infrared wireless communication using spread spectrum techniques. *IEE Proceedings in Optoelectronics*, **147** (4), 308–14.
35. Wong, K.K. and O'Farrell, T. (2003) Spread spectrum techniques for indoor wireless IR communications. *IEEE Wireless Communications*, **10** (2), 54–63.

36. Rabadan, J.A., Bacallado, M.A., Delgado, F. *et al.* (2004) Experimental characterization of a direct-sequence spread-spectrum optical wireless system based on pulse-conformation techniques for in-house communications. *IEEE Transactions on Consumer Electronics*, **50** (2), 484–90.
37. Delgado, F., Prez Jimenez, R., Rabadan, J. *et al.* (2002) Experimental characterization of a low-cost fast frequency-hopping spread-spectrum system for wireless in-house optical communications. *IEEE Transactions on Consumer Electronics*, **48** (1), 10–6.
38. Mohedano, R., Minano, J.C., Benitez, P. *et al.* (2000) Ultracompact nonimaging devices for optical wireless communications. *Optical Engineering*, **39** (10), 2740–7.
39. Welford, W.T. (1978) *The Optics of Nonimaging Concentrators : Light and Solar Energy*, Academic Press.
40. O'Brien, D.C. (2005) Improving the coverage and data rate in optical wireless communications. Proc. SPIE. Free space optical communications V, San Diego, pp. 58920X58921–58929.
41. Elgala, H., Mesleh, R., Haas, H. and Pricope, B. (2007) OFDM visible light wireless communication based on white LEDs. Proc. IEEE 65th Vehicular Technology Conference, Dublin, Ireland.
42. Afgani, M.Z., Haas, H., Elgala, H. and Knipp, D. (2006) Visible light communication using OFDM. *Proc. 2006 2nd International Conference on Testbeds and Research Infrastructures for the Development of Networks and Communities, Barcelona, Spain.*
43. Komine, T., Haruyama, S. and Nakagawa, M. (2006) Performance evaluation of narrowband OFDM on integrated system of power line communication and visible light wireless communication. Proc. International Symposium on Wireless Pervasive Computing, Phuket, Thailand.
44. Tanaka, Y., Komine, T., Haruyama, S. and Nakagawa, M. (2001) Indoor visible communication utilizing plural white LEDs as lighting. Proc. 12th IEEE International Symposium on Personal, Indoor and Mobile Radio Communications. PIMRC 2001. Proceedings. San Diego, CA, USA.
45. Sugiyama, H., Haruyama, S. and Nakagawa, M. (2006) Experimental investigation of modulation method for visible-light communications. *IEICE Transactions on Communications*, **E89-B** (12), 3393–400.
46. Yamanaka, D., Haruyama, S. and Nakagawa, M. (2006) Subcarrier modulation for visible-light communication using imaging sensor. Proceedings of the Third IASTED International Conference on Communications and Computer Networks, *Lima, Peru.*
47. Wook, H.B.C., Haruyama, S. and Nakagawa, M. (2006) Visible light communication with LED traffic lights using 2D image sensor. *IEICE Transactions on Fundamentals of Electronics, Communications and Computer Sciences*, **E89-A** (3), 654–9.
48. Okada, H., Masuda, K., Yamazato, T. and Katayama, M. (2005) Successive interference cancellation for hierarchical parallel optical wireless communication systems. Proc. 2005 Asia Pacific Conference on Communications, Perth, WA, Australia.
49. Akanegawa, M., Tanaka, Y. and Nakagawa, M. (2001) Basic study on traffic information system using LED traffic lights. *IEEE Transactions on Intelligent Transportation Systems*, **2** (4), 197–203.
50. Hara, T., Iwasaki, S., Yendo, T. *et al.* (2007) A New Receiving System of Visible Light Communication for ITS. Proc. Intelligent Vehicles Symposium, IEEE, pp. 474–9.
51. Iwasaki, S., Wada, M., Endo, T. *et al.* (2007) Basic Experiments on Paralle Wireless Optical Communication for ITS. Proc. Intelligent Vehicles Symposium, IEEE, pp. 321–6.
52. Kitano, S., Haruyama, S. and Nakagawa, M. (2003) LED road illumination communications system. Proc. IEEE 58th Vehicular Technology Conference. VTC 2003 Fall, Orlando, FL, USA.
53. Tomimoto, T. and Ogawa, H. (1997) Optical transmitter and receiver for intervehicle communication. *Oki Technical Review*, **63** (158), 7–10.
54. Wada, M., Yendo, T., Fujii, T. and Tanimoto, M. (2005) Road-to-vehicle communication using LED traffic light. Proc. IEEE Intelligent Vehicles Symposium Proceedings, Las Vegas, NV, USA. IEEE Intelligent Transportation Syst. Soc.
55. Saito, T., Haruyama, S. and Nakagawa, M. (2006) Inter-vehicle communication and ranging method using LED rear lights. Proceedings of the Fifth IASTED International Conference on Communication Systems and Networks. Palma de Mallorca, Spain.
56. EU project TST3-CT-2003-506316 (2006) Integrated communicatingsolid-stage light engine for use in automotive forward lighting andinformation exchange between vehicles and infrastructure. http://www.pb.izm.fraunhofer.de/p2sa/030_Projects/Optik/Pr_isle.html (accessed September 2008).
57. Pang, G., Ho, K.L., Kwan, T. and Yang, E. (1999) Visible light communication for audio systems. *IEEE Transactions on Consumer Electronics*, **45** (4), 1112–8.

58. Wong, D.W.K. and Chen, G. (2007) Illumination design of a white-light-emitting diode wireless transmission system. *Optical Engineering*, **46** (8), 1–6.
59. JEITA (2007) *CP-1221 Visible Light Communications System.*
60. O'Brien, D.C., Katz, M., Wang, P. *et al.* (2006) *Short-range optical wireless communications: Technologies for the Wireless Future: Wireless World Research Forum (WWRF)*, Vol. **2**, Wiley, pp. 277–96.
61. O'Brien, D.C., Sheard, S.J., Faulkner, G. and Edwards, D.J. (2004) Communications device retroreflecting incident modulated radiation, US patent 6,721,539.
62. Komine, T., Haruyama, S. and Nakagawa, M. (2003) Bidirectional visible-light communication using corner cube modulator. Proc. Wireless and Optical Communication (WOC). Banff, Canada. IASTED.
63. O'Brien, D.C. (2007) Cooperation and cognition in optical wireless communications. in *Cognitive Wireless Networks: Concepts, Methodologies and Visions - Inspiring the Age of Enlightenment of Wireless Communications* (ed. M.K.A.F. Fitzek), Springer.
64. EU FP7 project (2008) *Home Gigabit Access (OMEGA)*, http://www.ict-omega.eu (accessed September 2008).
65. JEITA (2007) *CP-1222 Visible Light ID System.*
66. IEEE 802.15 (2008) *WPAN Visual Light Communication Interest Group (IGvlc)*, http://www.IEEE802.org/15/pub/IGvlc.html.
67. Wireless World Research Forum (2008) http://www.wireless-world-research.org/?id=92 (accessed September 2008).

Index